Contractor's Pricing Guide:
Residential Repair & Remodeling Costs with RSMeans data

Thomas Lane, Senior Editor

G⊘RDIAN®

2019

Chief Data Officer
Noam Reininger

Engineering Director
Bob Mewis (*1, 3, 4, 5, 11, 12*)

Contributing Editors
Brian Adams (*21, 22*)
Christopher Babbitt
Sam Babbitt
Michelle Curran
Matthew Doheny (*8, 9, 10*)
John Gomes (*13, 41*)
Derrick Hale, PE (*2, 31, 32, 33, 34, 35, 44, 46*)
Michael Henry

Joseph Kelble (*14, 23, 25*)
Charles Kibbee
Gerard Lafond, PE
Thomas Lane (*6, 7*)
Jake MacDonald
Elisa Mello
Michael Ouillette (*26, 27, 28, 48*)
Gabe Sirota
Matthew Sorrentino
Kevin Souza
David Yazbek

Product Manager
Andrea Sillah

Production Manager
Debbie Panarelli

Production
Jonathan Forgit
Mary Lou Geary
Sharon Larsen
Sheryl Rose

Data Quality Manager
Joseph Ingargiola

Technical Support
Kedar Gaikwad
Todd Klapprodt
John Liu

Cover Design
Blaire Collins

Data Analytics
Tim Duggan
Todd Glowac
Matthew Kelliher-Gibson

Numbers in italics are the divisional responsibilities for each editor. Please contact the designated editor directly with any questions.

GORDIAN®

Contractor's Pricing Guide:
Residential Repair & Remodeling Costs
with RSMeans data

RSMeans data from Gordian
Construction Publishers & Consultants
1099 Hingham Street, Suite 201
Rockland, MA 02370
United States of America
1.800.448.8182
RSMeans.com

Printed in the United States of America
ISSN 1542-9156
ISBN 978-1-946872-69-2

0349A $65.99 per copy (in United States)
Price is subject to change without prior notice.

Table of Contents

Foreword

The Value of RSMeans data from Gordian

Since 1942, RSMeans data has been the industry-standard materials, labor, and equipment cost information database for contractors, facility owners and managers, architects, engineers, and anyone else that requires the latest localized construction cost information. More than 75 years later, the objective remains the same: to provide facility and construction professionals with the most current and comprehensive construction cost database possible.

With the constant influx of new construction methods and materials, in addition to ever-changing labor and material costs, last year's cost data is not reliable for today's designs, estimates, or budgets. Gordian's cost engineers apply real-world construction experience to identify and quantify new building products and methodologies, adjust productivity rates, and adjust costs to local market conditions across the nation. This adds up to more than 22,000 hours in cost research annually. This unparalleled construction cost expertise is why so many facility and construction professionals rely on RSMeans data year over year.

About Gordian

Gordian originated in the spirit of innovation and a strong commitment to helping clients reach and exceed their construction goals. In 1982, Gordian's chairman and founder, Harry H. Mellon, created Job Order Contracting while serving as chief engineer at the Supreme Headquarters Allied Powers Europe. Job Order Contracting is a unique indefinite delivery/indefinite quantity (IDIQ) process, which enables facility owners to complete a substantial number of repair, maintenance, and construction projects with a single, competitively awarded contract. Realizing facility and infrastructure owners across various industries could greatly benefit from the time and cost saving advantages of this innovative construction procurement solution, he established Gordian in 1990.

Continuing the commitment to providing the most relevant and accurate facility and construction data, software, and expertise in the industry, Gordian enhanced the fortitude of its data with the acquisition of RSMeans in 2014. And in an effort to expand its facility management capabilities, Gordian acquired Sightlines, the leading provider of facilities benchmarking data and analysis, in 2015.

Our Offerings

Gordian is the leader in facility and construction cost data, software, and expertise for all phases of the building life cycle. From planning to design, procurement, construction, and operations, Gordian's solutions help clients maximize efficiency, optimize cost savings, and increase building quality with its highly specialized data engineers, software, and unique proprietary data sets.

Our Commitment

At Gordian, we do more than talk about the quality of our data and the usefulness of its application. We stand behind all of our RSMeans data—from historical cost indexes to construction materials and techniques—to craft current costs and predict future trends. If you have any questions about our products or services, please call us toll-free at 800.448.8182 or visit our website at gordian.com.

How the Cost data Is Built: An Overview

A Powerful Construction Tool

You have in your hands one of the most powerful construction tools available today. A successful project is built on the foundation of an accurate and dependable estimate. This data will enable you to construct such an estimate.

For the casual user the data is designed to be:

- quickly and easily understood so you can get right to your estimate.
- filled with valuable information so you can understand the necessary factors that go into the cost estimate.

For the regular user, the data is designed to be:

- a handy reference that can be quickly referred to for key costs.
- a comprehensive, fully reliable source of current construction costs, so you'll be prepared to estimate any project.
- a source for project costs, product selections, and alternate materials and methods.

To meet all of these requirements we have organized the data into the following clearly defined sections.

Unit Price Section

The cost data has been divided into 23 sections representing the order of construction. Within each section is a listing of components and variations of those components that are applicable to that section. Costs are shown for the various actions that must be done to that component.

Reference Section

This section includes information on Location Factors and a listing of abbreviations.

Location Factors

You can adjust total project costs to over 900 locations throughout the U.S. and Canada by using the data in this section.

Abbreviations

A listing of the abbreviations used throughout, along with the terms they represent, is included.

Index

A comprehensive listing of all terms and subjects will help you quickly find what you need when you are not sure where it falls in the order of construction.

The Scope of this Data

This data is designed to be as comprehensive and easy to use as possible. To that end we have made certain assumptions and limited its scope in two key ways:

1. We have established material prices based on a national average.
2. We have computed labor costs based on residential wage rates.

Project Size

This data is intended for use by those involved primarily in residential repair and remodeling construction costing between $10,000 and $100,000.

With reasonable exercise of judgment, the figures can be used for any building work. For other types of projects, such as new home construction or commercial buildings, consult the appropriate RSMeans data set for more information.

How to Use the Cost Data: The Details

What's Behind the Numbers? The Development of Cost Data

RSMeans data engineers continually monitor developments in the construction industry in order to ensure reliable, thorough, and up-to-date cost information. While overall construction costs may vary relative to general economic conditions, price fluctuations within the industry are dependent upon many factors. Individual price variations may, in fact, be opposite to overall economic trends. Therefore, costs are constantly tracked and complete updates are performed yearly. Also, new items are frequently added in response to changes in materials and methods.

Costs in U.S. Dollars

All costs represent U.S. national averages and are given in U.S. dollars. The City Cost Index (CCI) with RSMeans data or Location Factors can be used to adjust costs to a particular location. The CCI or Location Factors for Canada can be used to adjust U.S. national averages to local costs in Canadian dollars. No exchange rate conversion is necessary because it has already been factored in.

Material Costs

RSMeans data engineers contact manufacturers, dealers, distributors, and contractors all across the U.S. and Canada to determine national average material costs. If you have access to current material costs for your specific location, you may wish to make adjustments to reflect differences from the national average. Included within material costs are fasteners for a normal installation. RSMeans data engineers use manufacturers' recommendations, written specifications, and/or standard construction practices for the sizing and spacing of fasteners. Adjustments to material costs may be required for your specific application or location. The manufacturer's warranty is assumed. Extended warranties are not included in the material costs. **Material costs do not include sales tax.**

Labor Costs

Labor costs are based upon a mathematical average of trade-specific wages in 30 major U.S. cities. The type of wage (union, open shop, or residential) is identified on the inside back cover of printed publications or selected by the estimator when using the electronic products. Markups for the wages can also be found on the inside back cover of printed publications and/or under the labor references found in the electronic products.

- If wage rates in your area vary from those used, or if rate increases are expected within a given year, labor costs should be adjusted accordingly.

Labor costs reflect productivity based on actual working conditions. In addition to actual installation, these figures include time spent during a normal weekday on tasks, such as material receiving and handling, mobilization at the site, site movement, breaks, and cleanup.

Productivity data is developed over an extended period so as not to be influenced by abnormal variations and reflects a typical average.

Equipment Costs

Equipment costs include not only rental but also operating costs for equipment under normal use. The operating costs include parts and labor for routine servicing, such as the repair and replacement of pumps, filters, and worn lines. Normal operating expendables, such as fuel, lubricants, tires, and electricity (where applicable), are also included. Extraordinary operating expendables with highly variable wear patterns, such as diamond bits and blades, are excluded. These costs are included under materials. Equipment rental rates are obtained from industry sources throughout North America—contractors, suppliers, dealers, manufacturers, and distributors.

Rental rates can also be treated as reimbursement costs for contractor-owned equipment. Owned equipment costs include depreciation, loan payments, interest, taxes, insurance, storage, and major repairs.

Equipment costs do not include operators' wages.

Factors Affecting Costs

Costs can vary depending upon a number of variables. Here's a listing of some factors that affect costs and points to consider.

Quality—The prices for materials and the workmanship upon which productivity is based represent sound construction work. They are also in line with industry standards and manufacturer specifications and are frequently used by federal, state, and local governments.

Overtime—We have made no allowance for overtime. If you anticipate premium time or work beyond normal working hours, be sure to make an appropriate adjustment to your labor costs.

Productivity—The productivity, daily output, and labor-hour figures for each line item are based on an eight-hour work day in daylight hours in moderate temperatures and up to a 14' working height unless otherwise indicated. For work that extends beyond normal work hours or is performed under adverse conditions, productivity may decrease.

Size of Project—The size, scope of work, and type of construction project will have a significant impact on cost. Economies of scale can reduce costs for large projects. Unit costs can often run higher for small projects.

Location—Material prices are for metropolitan areas. However, in dense urban areas, traffic and site storage limitations may increase costs. Beyond a 20-mile radius of metropolitan areas, extra trucking or transportation charges may also increase the material costs slightly. On the other hand, lower wage rates may be in effect. Be sure to consider both of these factors when preparing an estimate, particularly if the job site is located in a central city or remote rural location. In addition, highly specialized subcontract items may require travel and per-diem expenses for mechanics.

Other Factors—

- season of year
- contractor management
- weather conditions
- local union restrictions
- building code requirements
- availability of:
  - adequate energy
  - skilled labor
  - building materials
- owner's special requirements/restrictions
- safety requirements
- environmental considerations
- access

Unpredictable Factors—General business conditions influence "in-place" costs of all items. Substitute materials and construction methods may have to be employed. These may affect the installed cost and/or life cycle costs. Such factors may be difficult to evaluate and cannot necessarily be predicted on the basis of the job's location in a particular section of the country. Thus, where these factors apply, you may find significant but unavoidable cost variations for which you will have to apply a measure of judgment to your estimate.

Final Checklist

Estimating can be a straightforward process provided you remember the basics. Here's a checklist of some of the steps you should remember to complete before finalizing your estimate.

Did you remember to:

- factor in the City Cost Index or Location Factors for your locale?
- take into consideration which items have been marked up and by how much?
- mark up the entire estimate sufficiently for your purposes?

- read the background information on techniques and technical matters that could impact your project time span and cost?
- include all components of your project in the final estimate?
- double check your figures for accuracy?
- call RSMeans data engineers if you have any questions about your estimate or the data you've used? Remember, Gordian stands behind all of our products, including our extensive RSMeans data solutions. If you have any questions about your estimate, about the costs you've used from our data, or even about the technical aspects of the job that may affect your estimate, feel free to call the Gordian RSMeans editors at 1.800.448.8182.

Included with Purchase

RSMeans Data Online
Free Trial

rsmeans.com/2019freetrial

Unit Price Section

Table of Contents

Table of Contents (cont.)

RSMeans data: Unit Prices—How They Work

The prices in this publication include all overhead and profit mark-ups for the installing contractor. If a general contractor is involved, an additional 10% should be added to the cost shown in the "Total" column.

The costs for material, labor, and any equipment are shown here. Material costs include delivery and a 10% mark-up. No sales taxes are included. Labor costs include all mark-ups as shown on the inside back cover of this data set. Equipment costs also include a 10% mark-up.

The cost in the Total column should be multiplied by the quantity of each item to determine the total cost for that item. The sum of all items' total costs becomes the project's total cost.

The data set is arranged into 23 categories that follow the order of construction for residential projects. The category is shown as white letters on a black background at the top of each page.

A brief specification of the cost item is shown here.

Graphics representing some of the cost items are found at the beginning of each category.

The cost data is arranged in a tree structure format. Major cost sections are left justified and shown as bold headings within a grey background. Cost items and sub-cost items are found under the bold headings. The complete description should read as "bold heading, cost item, sub-cost item."

For example, the identified item's description is "Beam/Girder, Solid Wood Beam, 2" x 8"."

Each action performed to the cost item is conveniently listed to the right of the description. See the explanation below for the scope of work included as part of the action.

The unit of measure upon which all the costs are based is shown here. An abbreviations list is included within the Reference Section found at the back of this data set.

Rough Frame / Structure

Rough Framing

Beam / Girder		Unit	Material	Labor	Equip.	Total	Specification
Solid Wood Beam							
2" x 8"							
	Demolish	L.F.		.98		.98	Cost includes material and labor to install 2" x 8" beam.
	Install	L.F.	1.07	1.45		2.52	
	Demolish and Install	L.F.	1.07	2.43		3.50	
	Clean	L.F.	.06	.32		.38	
	Paint	L.F.	.15	2.11		2.26	
	Minimum Charge	Job		236		236	
2" x 10"							
	Demolish	L.F.		1.21		1.21	Cost includes material and labor to install 2" x 10" beam.
	Install	L.F.	1.60	1.57		3.17	
	Demolish and Install	L.F.	1.60	2.78		4.38	
	Clean	L.F.	.06	.39		.45	
	Paint	L.F.	.20	1.30		1.50	
	Minimum Charge	Job		236		236	
2" x 12"							
	Demolish	L.F.		1.47		1.47	Cost includes material and labor to install 2" x 12" beam.
	Install	L.F.	2.01	1.71		3.72	
	Demolish and Install	L.F.	2.01	3.18		5.19	
	Clean	L.F.	.07	.45		.52	
	Paint	L.F.	.23	3.13		3.36	
	Minimum Charge	Job		236		236	
4" x 8"							
	Demolish	L.F.		3.01		3.01	Cost includes material and labor to install 4" x 8" beam.
	Install	L.F.	4.72	2.17	.52	7.41	
	Demolish and Install	L.F.	4.72	5.18	.52	10.42	
	Clean	L.F.	.06	.39		.45	
	Paint	L.F.	.20	1.30		1.50	
	Minimum Charge	Job		236		236	
4" x 10"							
	Demolish	L.F.		3.76		3.76	Cost includes material and labor to install 4" x 10" beam.
	Install	L.F.	5.55	2.29	.55	8.39	
	Demolish and Install	L.F.	5.55	6.05	.55	12.15	
	Clean	L.F.	.07	.45		.52	
	Paint	L.F.	.22	1.52		1.74	
	Minimum Charge	Job		236		236	
4" x 12"							
	Demolish	L.F.		4.51		4.51	Cost includes material and labor to install 4" x 12" beam.
	Install	L.F.	6.10	2.41	.58	9.09	
	Demolish and Install	L.F.	6.10	6.92	.58	13.60	
	Clean	L.F.	.08	.52		.60	
	Paint	L.F.	.23	3.13		3.36	
	Minimum Charge	Job		236		236	

Action	Scope
Demolish	Includes removing of the item and hauling the debris to a truck, dumpster, or storage area.
Install	Includes the installation of the item with normal placement and spacing of fasteners.
Demolish and Install	Includes the demolish action and the install action.
Reinstall	Includes the reinstallation of a removed item.
Clean	Includes cleaning of the item.
Paint	Includes up to two coats of finish on the item (unless otherwise indicated).
Minimum Charge	Includes the minimum labor or the minimum labor and equipment cost. Use this charge for small quantities of material.

Job Costs

	Scaffolding	Temporary Electrical Service	Air Compressor

Construction Fees		Unit	Material	Labor	Equip.	Total	Specification
Asbestos Test	Minimum Charge	Job		275		275	Minimum charge for testing.
Architectural	Minimum Charge	Day	2075			2075	Minimum labor charge for architect.
Engineering	Minimum Charge	Day	1900			1900	Minimum labor charge for field engineer.

Construction Permits		Unit	Material	Labor	Equip.	Total	Specification
Job (less than 5K)	Minimum Charge	Job	59			59	Permit fees, job less than 5K.
Job (5K to 10K)	Minimum Charge	Job	120			120	Permit fees, job more than 5K, but less than 10K.
Job (10K to 25K)	Minimum Charge	Job	238			238	Permit fees, job more than 10K, but less than 25K.
Job (25K to 50K)	Minimum Charge	Job	455			455	Permit fees, job more than 25K, but less than 50K.
Job (50K to 100K)	Minimum Charge	Job	835			835	Permit fees, job more than 50K, but less than 100K.
Job (100K to 250K)	Minimum Charge	Job	1700			1700	Permit fees, job more than 100K, but less than 250K.
Job (250K to 500K)	Minimum Charge	Job	3275			3275	Permit fees, job more than 250K, but less than 500K.
Job (500K to 1M)	Minimum Charge	Job	5825			5825	Permit fees, job more than 500K, but less than 1 million.

For customer support on your Contractor's Pricing Guide: Residential Repair & Remodeling, call 800.448.8182.

1

Job Costs

Trade Labor		Unit	Material	Labor	Equip.	Total	Specification
Carpenter							
Daily (8 hours)							
	Install	Day		470		470	Daily labor rate for one carpenter.
	Minimum Charge	Job		236		236	
Weekly (40 hours)							
	Install	Week		2350		2350	Weekly labor rate for one carpenter.
	Minimum Charge	Job		236		236	
Drywaller							
Daily (8 hours)							
	Install	Day		470		470	Daily labor rate for one carpenter.
	Minimum Charge	Job		236		236	
Weekly (40 hours)							
	Install	Week		2350		2350	Weekly labor rate for one carpenter.
	Minimum Charge	Job		236		236	
Roofer							
Daily (8 hours)							
	Install	Day		445		445	Daily labor rate for one skilled roofer.
	Minimum Charge	Job		221		221	
Weekly (40 hours)							
	Install	Week		2225		2225	Weekly labor rate for one skilled
	Minimum Charge	Job		221		221	roofer.
Painter							
Daily (8 hours)							
	Install	Day		390		390	Daily labor rate for one painter.
	Minimum Charge	Job		196		196	
Weekly (40 hours)							
	Install	Week		1950		1950	Weekly labor rate for one painter.
	Minimum Charge	Job		196		196	
Mason							
Daily (8 hours)							
	Install	Day		470		470	Daily labor rate for one skilled mason.
	Minimum Charge	Job		235		235	
Weekly (40 hours)							
	Install	Week		2350		2350	Weekly labor rate for one skilled
	Minimum Charge	Job		235		235	mason.
Electrician							
Daily (8 hours)							
	Install	Day		540		540	Daily labor rate for an electrician.
	Minimum Charge	Job		270		270	
Weekly (40 hours)							
	Install	Week		2700		2700	Weekly labor rate for an electrician.
	Minimum Charge	Job		270		270	
Plumber							
Daily (8 hours)							
	Install	Day		530		530	Daily labor rate for a plumber.
	Minimum Charge	Job		265		265	
Weekly (40 hours)							
	Install	Week		2650		2650	Weekly labor rate for a plumber.
	Minimum Charge	Job		265		265	
Common Laborer							
Daily (8 hours)							
	Install	Day		365		365	Daily labor rate for a common
	Minimum Charge	Job		182		182	laborer.

For customer support on your Contractor's Pricing Guide: Residential Repair & Remodeling, call 800.448.8182.

Job Costs

Trade Labor

Trade Labor		Unit	Material	Labor	Equip.	Total	Specification
Weekly (40 hours)							
	Install	Week		1825		1825	Weekly labor rate for a common
	Minimum Charge	Job		182		182	laborer.

Temporary Utilities

Temporary Utilities		Unit	Material	Labor	Equip.	Total	Specification
Power Pole							
Job Site Electricity							
	Install	Month	59			59	Includes average cost of electricity per
	Minimum Charge	Job		270		270	month.
Water Service							
Job Site Water							
	Install	Month	76			76	Includes average costs for water for
	Minimum Charge	Month	76			76	one month.
Telephone							
	Install	Month	95.50			95.50	Includes average cost for local and
	Minimum Charge	Month	95.50			95.50	long distance telephone service for
							one month.
Chemical Toilet							
	Install	Month			320	320	Monthly rental of portable toilet.
	Minimum Charge	Day			15.90	15.90	
Office Trailer							
Place and Remove							
	Install	Job	390			390	Includes cost to a contractor to deliver,
	Minimum Charge	Job		216		216	place and level an office trailer and
							remove at end of job.
Storage Container							
Place and Remove							
	Install	Job	220			220	Includes cost to a contractor to deliver
	Minimum Charge	Job		216		216	and place a storage container and
							remove at end of job.
Fencing							
	Install	L.F.	4.20			4.20	Includes material costs for temporary
	Minimum Charge	Job		182		182	fencing.
Railing							
	Install	L.F.	19.95	12.70	.60	33.25	Includes material and labor to install
	Minimum Charge	Job		182		182	primed steel pipe.

Equipment Rental

Equipment Rental		Unit	Material	Labor	Equip.	Total	Specification
Flatbed Truck							
Daily Rental							
	Install	Day			215	215	Daily rental of flatbed truck.
	Minimum Charge	Job			215	215	
Weekly Rental							
	Install	Week			1075	1075	Weekly rental of flatbed truck.
	Minimum Charge	Job			215	215	
Monthly Rental							
	Install	Month			4300	4300	Monthly rental of flatbed truck.
	Minimum Charge	Job			215	215	

For customer support on your Contractor's Pricing Guide: Residential Repair & Remodeling, call 800.448.8182.

Equipment Rental		Unit	Material	Labor	Equip.	Total	Specification
Dump Truck							
Daily Rental							
	Install	Day			375	375	Daily rental of dump truck.
	Minimum Charge	Day			375	375	
Weekly Rental							
	Install	Week			1875	1875	Weekly rental of dump truck.
	Minimum Charge	Day			375	375	
Monthly Rental							
	Install	Month			7475	7475	Monthly rental of dump truck.
	Minimum Charge	Day			375	375	
Forklift							
Daily Rental							
	Install	Day			305	305	Daily rental of forklift.
	Minimum Charge	Day			305	305	
Weekly Rental							
	Install	Week			1525	1525	Weekly rental of forklift.
	Minimum Charge	Day			305	305	
Monthly Rental							
	Install	Month			6075	6075	Monthly rental of forklift.
	Minimum Charge	Day			305	305	
Bobcat							
Daily Rental							
	Install	Day			192	192	Daily rental of wheeled, skid steer
	Minimum Charge	Day			192	192	loader.
Weekly Rental							
	Install	Week			960	960	Weekly rental of wheeled, skid steer
	Minimum Charge	Day			192	192	loader.
Monthly Rental							
	Install	Month			3850	3850	Monthly rental of wheeled, skid steer
	Minimum Charge	Day			192	192	loader.
Air Compressor							
Daily Rental							
	Install	Day			117	117	Daily rental of air compressor.
	Minimum Charge	Day			117	117	
Weekly Rental							
	Install	Week			585	585	Weekly rental of air compressor.
	Minimum Charge	Day			117	117	
Monthly Rental							
	Install	Month			2350	2350	Monthly rental of air compressor.
	Minimum Charge	Day			117	117	
Jackhammer							
Daily Rental							
	Install	Day			11.85	11.85	Daily rental of air tool, jackhammer.
	Minimum Charge	Day			11.85	11.85	
Weekly Rental							
	Install	Week			59	59	Weekly rental of air tool, jackhammer.
	Minimum Charge	Day			11.85	11.85	
Monthly Rental							
	Install	Month			237	237	Monthly rental of air tool,
	Minimum Charge	Day			11.85	11.85	jackhammer.
Concrete Mixer							
Daily Rental							
	Install	Day			149	149	Daily rental of concrete mixer.
	Minimum Charge	Day			149	149	

Equipment Rental		Unit	Material	Labor	Equip.	Total	Specification
Weekly Rental							
	Install	Week			745	745	Weekly rental of concrete mixer.
	Minimum Charge	Day			149	149	
Monthly Rental							
	Install	Month			2975	2975	Monthly rental of concrete mixer.
	Minimum Charge	Day			149	149	
Concrete Bucket							
Daily Rental							
	Install	Day			24	24	Daily rental of concrete bucket.
	Minimum Charge	Day			24	24	
Weekly Rental							
	Install	Week			120	120	Weekly rental of concrete bucket.
	Minimum Charge	Day			24	24	
Monthly Rental							
	Install	Month			480	480	Monthly rental of concrete bucket.
	Minimum Charge	Day			24	24	
Concrete Pump							
Daily Rental							
	Install	Day			930	930	Daily rental of concrete pump.
	Minimum Charge	Day			930	930	
Weekly Rental							
	Install	Week			4650	4650	Weekly rental of concrete pump.
	Minimum Charge	Day			930	930	
Monthly Rental							
	Install	Month			18600	18600	Monthly rental of concrete pump.
	Minimum Charge	Day			930	930	
Generator							
Daily Rental							
	Install	Day			29.50	29.50	Daily rental of generator.
	Minimum Charge	Day			29.50	29.50	
Weekly Rental							
	Install	Week			147	147	Weekly rental of generator.
	Minimum Charge	Day			29.50	29.50	
Monthly Rental							
	Install	Month			585	585	Monthly rental of generator.
	Minimum Charge	Day			29.50	29.50	
Welder							
Daily Rental							
	Install	Day			44	44	Daily rental of welder.
	Minimum Charge	Day			44	44	
Weekly Rental							
	Install	Week			221	221	Weekly rental of welder.
	Minimum Charge	Day			44	44	
Monthly Rental							
	Install	Month			885	885	Monthly rental of welder.
	Minimum Charge	Day			44	44	
Sandblaster							
Daily Rental							
	Install	Day			23	23	Daily rental of sandblaster.
	Minimum Charge	Day			23	23	
Weekly Rental							
	Install	Week			114	114	Weekly rental of sandblaster.
	Minimum Charge	Day			23	23	

For customer support on your Contractor's Pricing Guide: Residential Repair & Remodeling, call 800.448.8182.

Job Costs

Equipment Rental

Equipment Rental		Unit	Material	Labor	Equip.	Total	Specification
Monthly Rental							
	Install	Month			455	455	Monthly rental of sandblaster.
	Minimum Charge	Day			23	23	
Space Heater							
Daily Rental							
	Install	Day			18.15	18.15	Daily rental of space heater.
	Minimum Charge	Day			22.50	22.50	
Weekly Rental							
	Install	Day			91	91	Weekly rental of space heater.
	Minimum Charge	Day			22.50	22.50	
Monthly Rental							
	Install	Day			365	365	Monthly rental of space heater.
	Minimum Charge	Day			22.50	22.50	

Move / Reset Contents

Move / Reset Contents		Unit	Material	Labor	Equip.	Total	Specification
Small Room							
	Install	Room		44		44	Includes cost to remove and reset the
	Minimum Charge	Job		182		182	contents of a small room.
Average Room							
	Install	Room		53.50		53.50	Includes cost to remove and reset the
	Minimum Charge	Job		182		182	contents of an average room.
Large Room							
	Install	Room		73.50		73.50	Includes cost to remove and reset the
	Minimum Charge	Job		182		182	contents of a large room.
Extra Large Room							
	Install	Room		117		117	Includes cost to remove and reset the
	Minimum Charge	Job		182		182	contents of an extra large room.
Store Contents							
Weekly							
	Install	Week	297			297	Storage fee, per week.
Monthly							
	Install	Month	890			890	Storage fee, per month.

Cover and Protect

Cover and Protect		Unit	Material	Labor	Equip.	Total	Specification
Cover / Protect Walls							
	Install	S.F. Wall		.14		.14	Includes labor costs to cover and
	Minimum Charge	Job		182		182	protect walls.
Cover / Protect Floors							
	Install	S.F. Flr.		.14		.14	Includes labor costs to cover and
	Minimum Charge	Job		182		182	protect floors.

Temporary Bracing

Temporary Bracing		Unit	Material	Labor	Equip.	Total	Specification
Shim Support Piers							
	Install	Ea.	4.62	140		144.62	Includes labor and material to install
	Minimum Charge	Job		420		420	shims support for piers.

Job Costs

Jack and Relevel

		Unit	Material	Labor	Equip.	Total	Specification
Single Story							
	Install	S.F.	1.83	1.67		3.50	Includes labor and equipment to jack
	Minimum Charge	Job		420		420	and relevel single-story building.
Two-story							
	Install	S.F.	1.83	2.79		4.62	Includes labor and equipment to jack
	Minimum Charge	Job		420		420	and relevel two-story building.

Dumpster Rental

		Unit	Material	Labor	Equip.	Total	Specification
Small Load							
	Install	Week			620	620	Dumpster, weekly rental, 2 dumps, 5
	Minimum Charge	Week			620	620	yard container.
Medium Load							
	Install	Week			705	705	Dumpster, weekly rental, 2 dumps, 10
	Minimum Charge	Week			705	705	yard container.
Large Load							
	Install	Week			1075	1075	Dumpster, weekly rental, 2 dumps, 30
	Minimum Charge	Week			1075	1075	yard container.

Debris Hauling

		Unit	Material	Labor	Equip.	Total	Specification
Per Ton							
	Install	Ton		115	46	161	Includes hauling of debris by trailer or
	Minimum Charge	Ton		860	345	1205	dump truck.
Per Cubic Yard							
	Install	C.Y.		38.50	15.40	53.90	Includes hauling of debris by trailer or
	Minimum Charge	Job		860	345	1205	dump truck.
Per Pick-up Truck Load							
	Install	Ea.		202		202	Includes hauling of debris by pick-up
	Minimum Charge	Ea.		202		202	truck.

Dump Fees

		Unit	Material	Labor	Equip.	Total	Specification
Per Cubic Yard							
	Install	C.Y.	54			54	Dump charges, building materials.
	Minimum Charge	C.Y.	54			54	
Per Ton							
	Install	Ton	139			139	Dump charges, building materials.
	Minimum Charge	Ton	139			139	

Scaffolding

		Unit	Material	Labor	Equip.	Total	Specification
4' to 6' High							
	Install	Month	67			67	Includes monthly rental of scaffolding,
	Minimum Charge	Month	67			67	steel tubular, 30" wide, 7' long, 5' high.
7' to 11' High							
	Install	Month	78.50			78.50	Includes monthly rental of scaffolding,
	Minimum Charge	Month	78.50			78.50	steel tubular, 30" wide, 7' long, 10' high.

For customer support on your Contractor's Pricing Guide: Residential Repair & Remodeling, call 800.448.8182.

Job Costs

Scaffolding		Unit	Material	Labor	Equip.	Total	Specification
12' to 16' High							
	Install	Month	91			91	Includes monthly rental of scaffolding, steel tubular, 30" wide, 7' long, 15' high.
	Minimum Charge	Month	91			91	
17' to 21' High							
	Install	Month	117			117	Includes monthly rental of scaffolding, steel tubular, 30" wide, 7' long, 20' high.
	Minimum Charge	Month	117			117	
22' to 26' High							
	Install	Month	128			128	Includes monthly rental of scaffolding, steel tubular, 30" wide, 7' long, 25' high.
	Minimum Charge	Month	128			128	
27' to 30' High							
	Install	Month	140			140	Includes monthly rental of scaffolding, steel tubular, 30" wide, 7' long, 30' high.
	Minimum Charge	Month	140			140	

Construction Clean-up		Unit	Material	Labor	Equip.	Total	Specification
Initial							
	Clean	S.F.		.45		.45	Includes labor for general clean-up.
	Minimum Charge	Job		182		182	
Progressive							
	Clean	S.F.		.23		.23	Includes labor for progressive site/job clean-up on a weekly basis.
	Minimum Charge	Job		182		182	

For customer support on your Contractor's Pricing Guide: Residential Repair & Remodeling, call 800.448.8182.

Building Demolition		Unit	Material	Labor	Equip.	Total	Specification
Wood-framed Building							
1 Story Demolition							
	Demolish	S.F.		1.74	1.80	3.54	Includes demolition of 1 floor only
	Minimum Charge	Job		2750	2850	5600	light wood-framed building, including hauling of debris. Dump fees not included.
2nd Story Demolition							
	Demolish	S.F.		1.77	1.84	3.61	Includes demolition of 2nd floor only
	Minimum Charge	Job		2750	2850	5600	light wood-framed building, including hauling of debris. Dump fees not included.
3rd Story Demolition							
	Demolish	S.F.		2.30	2.39	4.69	Includes demolition of 3rd floor only
	Minimum Charge	Job		2750	2850	5600	light wood-framed building, including hauling of debris. Dump fees not included.
Masonry Building							
	Demolish	S.F.		2.47	2.57	5.04	Includes demolition of one story
	Minimum Charge	Job		2750	2850	5600	masonry building, including hauling of debris. Dump fees not included.
Concrete Building							
Unreinforced							
	Demolish	S.F.		5	1.11	6.11	Includes demolition of a one story
	Minimum Charge	Job		2750	2850	5600	non-reinforced concrete building, including hauling of debris. Dump fees not included.
Reinforced							
	Demolish	S.F.		5.75	1.27	7.02	Includes demolition of a one story
	Minimum Charge	Job		2750	2850	5600	reinforced concrete building, including hauling of debris. Dump fees not included.
Steel Building							
	Demolish	S.F.		3.30	3.43	6.73	Includes demolition of a metal
	Minimum Charge	Job		2750	2850	5600	building, including hauling of debris. Dump fees not included. No salvage value assumed for scrap metal.

9

For customer support on your Contractor's Pricing Guide: Residential Repair & Remodeling, call 800.448.8182.

Job Preparation

General Clean-up

General Clean-up	Unit	Material	Labor	Equip.	Total	Specification
Remove Debris						
Clean	S.F.		1.15	.21	1.36	Includes removal of debris caused by the loss including contents items and loose debris.
Minimum Charge	Job		575	107	682	
Muck-out						
Clean	S.F.		.71		.71	Includes labor to remove 2″ of muck and mud after a flood loss.
Minimum Charge	Job		182		182	
Deodorize / Disinfect						
Clean	S.F. Flr.		.24		.24	Includes deodorizing building with application of mildicide and disinfectant to floors and walls - will vary depending on the extent of deodorization required.
Minimum Charge	Job		182		182	
Clean Walls						
Heavy						
Clean	S.F.		.16		.16	Includes labor and material for heavy cleaning (multiple applications) with detergent and solvent.
Minimum Charge	Job		182		182	
Clean Ceiling						
Light						
Clean	S.F.		.26		.26	Includes labor and material to clean light to moderate smoke and soot.
Minimum Charge	Job		182		182	
Heavy						
Clean	S.F.		.39		.39	Includes labor and material to clean heavy smoke and soot.
Minimum Charge	Job		182		182	

Flood Clean-up

Flood Clean-up	Unit	Material	Labor	Equip.	Total	Specification
Emergency Service Call						
After Hours / Weekend						
Minimum Charge	Job		242		242	Includes minimum charges for deflooding clean-up after hours or on weekends.
Grey Water / Sewage						
Minimum Charge	Job		182		182	Includes minimum charges for deflooding clean-up.
Water Extraction						
Clean (non-grey)						
Clean	S.F.	.04	.73		.77	Includes labor to remove clean uncontaminated water from the loss.
Minimum Charge	Job		182		182	
Grey (non-solids)						
Clean	S.F.	.08	1.45		1.53	Includes labor to remove non-solid grey water from the loss.
Minimum Charge	Job		182		182	
Sewage / Solids						
Clean	S.F. Flr.	.13	1.82		1.95	Includes labor to remove water from the loss that may contain sewage or solids.
Minimum Charge	Job		182		182	

For customer support on your Contractor's Pricing Guide: Residential Repair & Remodeling, call 800.448.8182.

Job Preparation

Flood Clean-up		Unit	Material	Labor	Equip.	Total	Specification
Mildicide Walls							
Topographical	Clean	S.F. Wall		.33		.33	Clean-up of deflooding including
	Minimum Charge	Job		182		182	topographical mildicide application.
Injection Treatments	Clean	S.F. Flr.	48	5.60		53.60	Clean-up of deflooding including
	Minimum Charge	Job		182		182	mildicide injection.
Carpet Cleaning							
Uncontaminated	Clean	S.F. Flr.		.23		.23	Includes labor to clean carpet.
	Minimum Charge	Job		182		182	
Contaminated	Clean	S.F. Flr.		.36		.36	Includes labor to clean contaminated
	Minimum Charge	Job		182		182	carpet.
Stairway	Clean	Ea.	.10	4.28		4.38	Includes labor to clean contaminated
	Minimum Charge	Job		182		182	carpet on a stairway.
Carpet Treatment							
Disinfect / Deodorize	Clean	S.F. Flr.		.23		.23	Includes disinfecting and deodorizing
	Minimum Charge	Job		182		182	of carpet.
Disinfect For Sewage	Clean	S.F. Flr.		.59		.59	Includes labor to disinfect and
	Minimum Charge	Job		182		182	deodorize carpet exposed to sewage.
Mildicide	Clean	S.F. Flr.	.02	.23		.25	Includes labor and materials to apply
	Minimum Charge	Job		182		182	mildicide to carpet.
Thermal Fog Area	Clean	S.F. Flr.	.01	.56		.57	Includes labor and materials to
	Minimum Charge	Job		182		182	perform thermal fogging during
							deflooding clean-up.
Wet Fog / ULV Area	Clean	S.F. Flr.	.03	.32		.35	Includes labor and materials to
	Minimum Charge	Job		182		182	perform wet fogging / ULV during
							deflooding clean-up.
Steam Clean Fixtures	Clean	Ea.	.01	36.50		36.51	Includes labor and materials to steam
	Minimum Charge	Job		182		182	clean fixtures during deflooding
							clean-up.
Pressure Wash Sewage	Clean	S.F. Flr.		3.31		3.31	Includes labor to pressure wash
	Minimum Charge	Job		182		182	sewage contamination.
Air Mover	Minimum Charge	Day			45.50	45.50	Daily rental of air mover.
Dehumidifier							
Medium Size	Minimum Charge	Day			59.50	59.50	Daily rental of dehumidifier, medium.
Large Size	Minimum Charge	Day			112	112	Daily rental of dehumidifier, large.

11

Job Preparation

Flood Clean-up		Unit	Material	Labor	Equip.	Total	Specification
Air Purifier							
HEPA Filter-Bacteria							HEPA filter for air purifier.
	Minimum Charge	Ea.	335			335	

Asbestos Removal		Unit	Material	Labor	Equip.	Total	Specification
Asbestos Analysis							
	Minimum Charge	Job		275		275	Minimum charge for testing.
Asbestos Pipe Insulation							
1/2″ to 3/4″ Diameter							
	Demolish	L.F.	.30	7.10		7.40	Minimum charge to remove asbestos. Includes full Tyvek suits for workers changed four times per eight-hour shift with respirators, air monitoring and supervision by qualified professionals.
	Minimum Charge	Job		4400		4400	
1″ to 3″ Diameter							
	Demolish	L.F.	.91	22		22.91	Minimum charge to remove asbestos. Includes full Tyvek suits for workers changed four times per eight-hour shift with respirators, air monitoring and supervision by qualified professionals.
	Minimum Charge	Job		4400		4400	
Asbestos-based Siding							
	Demolish	S.F.	2.04	49		51.04	Minimum charge to remove asbestos. Includes full Tyvek suits for workers changed four times per eight-hour shift with respirators, air monitoring and supervision by qualified professionals.
	Minimum Charge	Job		4400		4400	
Asbestos-based Plaster							
	Demolish	S.F.	1.56	37.50		39.06	Minimum charge to remove asbestos. Includes full Tyvek suits for workers changed four times per eight-hour shift with respirators, air monitoring and supervision by qualified professionals.
	Minimum Charge	Job		4400		4400	
Encapsulate Asbestos Ceiling							
	Install	S.F.	.46	.21		.67	Includes encapsulation with penetrating sealant sprayed onto ceiling and walls with airless sprayer.
	Minimum Charge	Job		4400		4400	
Scrape Acoustical Ceiling							
	Demolish	S.F.	.06	1.24		1.30	Minimum charge to remove asbestos. Includes full Tyvek suits for workers changed four times per eight-hour shift with respirators, air monitoring and supervision by qualified professionals.
	Minimum Charge	Job		4400		4400	

For customer support on your Contractor's Pricing Guide: Residential Repair & Remodeling, call 800.448.8182.

Job Preparation

Asbestos Removal

Asbestos Removal		Unit	Material	Labor	Equip.	Total	Specification
Seal Ceiling Asbestos							
	Install	S.F.	2.09	4.14		6.23	Includes attaching plastic cover to
	Minimum Charge	Job		4400		4400	ceiling area and sealing where asbestos exists, taping all seams, caulk or foam insulation in cracks and negative air system.
Seal Wall Asbestos							
	Install	S.F.	2.09	3.68		5.77	Includes attaching plastic cover to wall
	Minimum Charge	Job		4400		4400	area and sealing where asbestos exists, taping all seams, caulk or foam insulation in cracks and negative air system.
Seal Floor Asbestos							
	Install	S.F.	3.34	7.35		10.69	Includes attaching plastic cover to
	Minimum Charge	Job		4400		4400	floor area and sealing where asbestos exists, taping all seams, caulk or foam insulation in cracks and negative air system.
Negative Air Vent System							
	Install	Day			47	47	Daily rental of negative air vent
	Minimum Charge	Week			234	234	system.
HEPA Vacuum Cleaner							
	Install	Day			18.80	18.80	Daily rental of HEPA vacuum, 16
	Minimum Charge	Week			94	94	gallon.
Airless Sprayer							
	Install	Day			73	73	Daily rental of airless sprayer.
	Minimum Charge	Week			365	365	
Decontamination Unit							
	Install	Day			138	138	Daily rental of decontamination unit.
	Minimum Charge	Week			690	690	
Light Stand							
	Install	Day			18.65	18.65	Daily rental of floodlight w/tripod.
	Minimum Charge	Week			93	93	

Room Demolition

Room Demolition		Unit	Material	Labor	Equip.	Total	Specification
Strip Typical Room							
	Demolish	S.F.		3.64		3.64	Includes 4 labor hours for general
	Minimum Charge	Job		202		202	demolition.
Strip Typical Bathroom							
	Demolish	S.F.		4.85		4.85	Includes 4 labor hours for general
	Minimum Charge	Job		202		202	demolition.
Strip Typical Kitchen							
	Demolish	S.F.		4.66		4.66	Includes 4 labor hours for general
	Minimum Charge	Job		202		202	demolition.
Strip Typical Laundry							
	Demolish	S.F.		4.16		4.16	Includes 4 labor hours for general
	Minimum Charge	Job		202		202	demolition.

Job Preparation

Selective Demolition		Unit	Material	Labor	Equip.	Total	Specification
Concrete Slab							
4" Unreinforced							
	Demolish	S.F.		.25	.54	.79	Includes labor and equipment to
	Minimum Charge	Job		555	1225	1780	remove unreinforced concrete slab and haul debris to truck or dumpster.
6" Unreinforced							
	Demolish	S.F.		.28	.61	.89	Includes labor and equipment to
	Minimum Charge	Job		555	1225	1780	remove unreinforced concrete slab and haul debris to truck or dumpster.
8" Unreinforced							
	Demolish	S.F.		.32	.70	1.02	Includes labor and equipment to
	Minimum Charge	Job		555	1225	1780	remove unreinforced concrete slab and haul debris to truck or dumpster.
4" Reinforced							
	Demolish	S.F.		.25	.54	.79	Includes labor and equipment to
	Minimum Charge	Job		555	1225	1780	remove reinforced concrete slab and haul debris to truck or dumpster.
6" Reinforced							
	Demolish	S.F.		.28	.61	.89	Includes labor and equipment to
	Minimum Charge	Job		555	1225	1780	remove reinforced concrete slab and haul debris to truck or dumpster.
8" Reinforced							
	Demolish	S.F.		.32	.70	1.02	Includes labor and equipment to
	Minimum Charge	Job		555	1225	1780	remove reinforced concrete slab and haul debris to truck or dumpster.
Brick Wall							
4" Thick							
	Demolish	S.F.		3.35	.47	3.82	Includes labor and equipment to
	Minimum Charge	Job		1025	144	1169	remove brick wall and haul debris to a central location for disposal.
8" Thick							
	Demolish	S.F.		4.61	.65	5.26	Includes labor and equipment to
	Minimum Charge	Job		1025	144	1169	remove brick wall and haul debris to a central location for disposal.
12" Thick							
	Demolish	S.F.		6.15	.86	7.01	Includes labor and equipment to
	Minimum Charge	Job		1025	144	1169	remove brick wall and haul debris to a central location for disposal.
Concrete Masonry Wall							
4" Thick							
	Demolish	S.F.		3.73	.52	4.25	Includes minimum labor and
	Minimum Charge	Job		1025	144	1169	equipment to remove 4" block wall and haul debris to a central location for disposal.
6" Thick							
	Demolish	S.F.		4.61	.65	5.26	Includes minimum labor and
	Minimum Charge	Job		1025	144	1169	equipment to remove 6" block wall and haul debris to a central location for disposal.

Job Preparation

Selective Demolition

Selective Demolition		Unit	Material	Labor	Equip.	Total	Specification
8" Thick							
	Demolish	S.F.		5.70	.80	6.50	Includes minimum labor and equipment to remove 8" block wall and haul debris to a central location for disposal.
	Minimum Charge	Job		1025	144	1169	
12" Thick							
	Demolish	S.F.		8.20	1.15	9.35	Includes minimum labor and equipment to remove 12" block wall and haul debris to a central location for disposal.
	Minimum Charge	Job		1025	144	1169	
Tree Removal							
Small, 4" - 8"							
	Demolish	Ea.		130	103	233	Minimum charge, common labor trade.
	Minimum Charge	Job		182		182	
Medium, 9" - 12"							
	Demolish	Ea.		195	154	349	Minimum charge, common labor trade.
	Minimum Charge	Job		182		182	
Large, 13" - 18"							
	Demolish	Ea.		234	185	419	Minimum charge, tree removal.
	Minimum Charge	Job		600	213	813	
Very Large, 19" - 24"							
	Demolish	Ea.		292	232	524	Minimum charge, tree removal.
	Minimum Charge	Job		600	213	813	
Stump Removal							
	Demolish	Ea.		85	140	225	Includes minimum charges for a site demolition crew and backhoe / loader.
	Minimum Charge	Job		330	98	428	

15

Backhoe/Loader - Wheel Type   Tractor Loader - Wheel Type   Backhoe - Crawler Type

Excavation		Unit	Material	Labor	Equip.	Total	Specification
Footings							
Hand Trenching							
	Install	L.F.		2.27		2.27	Includes hand excavation for footing
	Minimum Charge	Job		91		91	or trench up to 10" deep and 12" wide.
Machine Trenching							
	Install	L.F.		.86	.35	1.21	Includes machine excavation for
	Minimum Charge	Job		430	176	606	footing or trench up to 10" deep and 12" wide.
Trenching w / Backfill							
Hand							
	Install	L.F.		3.64		3.64	Includes hand excavation, backfill and
	Minimum Charge	Job		91		91	compaction.
Machine							
	Install	L.F.		4.28	10.10	14.38	Includes machine excavation, backfill
	Minimum Charge	Job		182	35	217	and compaction.

Backfill		Unit	Material	Labor	Equip.	Total	Specification
Hand (no compaction)							
	Install	L.C.Y.		30.50		30.50	Includes backfill by hand in average
	Minimum Charge	Job		182		182	soil without compaction.
Hand (w / compaction)							
	Install	E.C.Y.		43		43	Includes backfill by hand in average
	Minimum Charge	Job		182		182	soil with compaction.
Machine (no compaction)							
	Install	L.C.Y.		4.28	1.76	6.04	Includes backfill, by 55 HP wheel
	Minimum Charge	Job		430	176	606	loader, of trenches from loose material piled adjacent without compaction.
Machine (w / compaction)							
	Install	L.C.Y.		11.40	4.70	16.10	Includes backfill, by 55 HP wheel
	Minimum Charge	Job		430	176	606	loader, of trenches from loose material piled adjacent with compaction by vibrating tampers.

17

For customer support on your Contractor's Pricing Guide: Residential Repair & Remodeling, call 800.448.8182.

Concrete Footing and Formwork

Wood Pile

Piles	Unit	Material	Labor	Equip.	Total	Specification
Treated Wood, 12″ Butt						
To 30′						
Demolish	L.F.		5.75	1.70	7.45	Includes material, labor and
Install	V.L.F.	17.75	5.35	3.44	26.54	equipment to install up to 30′ long
Demolish and Install	V.L.F.	17.75	11.10	5.14	33.99	treated wood piling.
Paint	L.F.	.56	2.24		2.80	
Minimum Charge	Job		3350	2150	5500	
30′ to 40′						
Demolish	L.F.		5.75	1.70	7.45	Includes material, labor and
Install	V.L.F.	18.10	4.78	3.07	25.95	equipment to install 30′ to 40′ long
Demolish and Install	V.L.F.	18.10	10.53	4.77	33.40	treated wood piling.
Paint	L.F.	.56	2.24		2.80	
Minimum Charge	Job		3350	2150	5500	
40′ to 50′						
Demolish	L.F.		5.75	1.70	7.45	Includes material, labor and
Install	V.L.F.	19.15	4.64	2.99	26.78	equipment to install 40′ to 50′ long
Demolish and Install	V.L.F.	19.15	10.39	4.69	34.23	treated wood piling.
Paint	L.F.	.56	2.24		2.80	
Minimum Charge	Job		3350	2150	5500	
50′ to 60′						
Demolish	L.F.		5.75	1.70	7.45	Includes material, labor and
Install	V.L.F.	21.50	4.18	2.69	28.37	equipment to install 50′ to 60′ long
Demolish and Install	V.L.F.	21.50	9.93	4.39	35.82	treated wood piling.
Paint	L.F.	.56	2.24		2.80	
Minimum Charge	Job		3350	2150	5500	
60′ to 80′						
Demolish	L.F.		5.75	1.70	7.45	Includes material, labor and
Install	V.L.F.	27.50	3.98	3.19	34.67	equipment to install 60′ to 80′ long
Demolish and Install	V.L.F.	27.50	9.73	4.89	42.12	treated wood piling.
Paint	L.F.	.56	2.24		2.80	
Minimum Charge	Job		3350	2150	5500	
Untreated Wood, 12″ Butt						
To 30′						
Demolish	L.F.		5.75	1.70	7.45	Includes material, labor and
Install	V.L.F.	14.05	5.35	3.44	22.84	equipment to install up to 30′ long
Demolish and Install	V.L.F.	14.05	11.10	5.14	30.29	untreated wood piling.
Paint	L.F.	.56	2.24		2.80	
Minimum Charge	Job		3350	2150	5500	

Piles	Unit	Material	Labor	Equip.	Total	Specification
30' to 40'						
Demolish	L.F.		5.75	1.70	7.45	Includes material, labor and
Install	V.L.F.	14.80	4.78	3.07	22.65	equipment to install 30' to 40' long
Demolish and Install	V.L.F.	14.80	10.53	4.77	30.10	untreated wood piling.
Paint	L.F.	.56	2.24		2.80	
Minimum Charge	Job		3350	2150	5500	
40' to 50'						
Demolish	L.F.		5.75	1.70	7.45	Includes material, labor and
Install	V.L.F.	15.55	4.64	2.99	23.18	equipment to install 40' to 50' long
Demolish and Install	V.L.F.	15.55	10.39	4.69	30.63	untreated wood piling.
Paint	L.F.	.56	2.24		2.80	
Minimum Charge	Job		3350	2150	5500	
50' to 60'						
Demolish	L.F.		5.75	1.70	7.45	Includes material, labor and
Install	V.L.F.	18.70	4.18	2.69	25.57	equipment to install 50' to 60' long
Demolish and Install	V.L.F.	18.70	9.93	4.39	33.02	untreated wood piling.
Paint	L.F.	.56	2.24		2.80	
Minimum Charge	Job		3350	2150	5500	
60' to 80'						
Demolish	L.F.		5.75	1.70	7.45	Includes material, labor and
Install	V.L.F.	21.50	3.98	2.56	28.04	equipment to install 60' to 80' long
Demolish and Install	V.L.F.	21.50	9.73	4.26	35.49	untreated wood piling.
Paint	L.F.	.56	2.24		2.80	
Minimum Charge	Job		3350	2150	5500	
Precast Concrete						
10" Square						
Demolish	L.F.		5.90	1.74	7.64	Includes material, labor and
Install	V.L.F.	16.30	4.78	3.07	24.15	equipment to install a precast,
Demolish and Install	V.L.F.	16.30	10.68	4.81	31.79	prestressed, 40' long, 10" square
Paint	L.F.	.56	2.24		2.80	concrete pile.
Minimum Charge	Job		3350	2150	5500	
12" Square						
Demolish	L.F.		6.30	1.86	8.16	Includes material, labor and
Install	V.L.F.	24	4.92	3.16	32.08	equipment to install a precast,
Demolish and Install	V.L.F.	24	11.22	5.02	40.24	prestressed, 40' long, 12" square
Paint	L.F.	.56	2.24		2.80	concrete pile.
Minimum Charge	Job		3350	2150	5500	
14" Square						
Demolish	L.F.		6.50	1.91	8.41	Includes material, labor and
Install	V.L.F.	28	5.55	3.58	37.13	equipment to install a precast,
Demolish and Install	V.L.F.	28	12.05	5.49	45.54	prestressed, 40' long, 14" square
Paint	L.F.	.56	2.24		2.80	concrete pile.
Minimum Charge	Job		3350	2150	5500	
16" Square						
Demolish	L.F.		6.80	2.01	8.81	Includes material, labor and
Install	V.L.F.	34.50	5.95	3.84	44.29	equipment to install a precast,
Demolish and Install	V.L.F.	34.50	12.75	5.85	53.10	prestressed, 40' long, 16" square
Paint	L.F.	.56	2.24		2.80	concrete pile.
Minimum Charge	Job		3350	2150	5500	
18" Square						
Demolish	L.F.		7.40	2.17	9.57	Includes material, labor and
Install	V.L.F.	48	6.45	5.15	59.60	equipment to install a precast,
Demolish and Install	V.L.F.	48	13.85	7.32	69.17	prestressed, 40' long, 18" square
Paint	L.F.	.56	2.24		2.80	concrete pile.
Minimum Charge	Job		3350	2150	5500	

Piles	Unit	Material	Labor	Equip.	Total	Specification
Cross Bracing						
Treated 2" x 6"						
Demolish	L.F.		1.19		1.19	Cost includes material and labor to
Install	L.F.	.84	6.75		7.59	install 2" x 6" lumber for cross-bracing
Demolish and Install	L.F.	.84	7.94		8.78	of foundation piling or posts.
Reinstall	L.F.		5.40		5.40	
Clean	S.F.		.45		.45	
Paint	L.F.	.34	1.50		1.84	
Minimum Charge	Job		835		835	
Treated 2" x 8"						
Demolish	L.F.		1.19		1.19	Cost includes material and labor to
Install	L.F.	1.23	7		8.23	install 2" x 8" lumber for cross-bracing
Demolish and Install	L.F.	1.23	8.19		9.42	of foundation piling or posts.
Reinstall	L.F.		5.58		5.58	
Clean	S.F.		.45		.45	
Paint	L.F.	.45	1.56		2.01	
Minimum Charge	Job		835		835	
Treated 2" x 10"						
Demolish	L.F.		1.22		1.22	Cost includes material and labor to
Install	L.F.	1.54	7.20		8.74	install 2" x 10" lumber for
Demolish and Install	L.F.	1.54	8.42		9.96	cross-bracing of foundation piling or
Reinstall	L.F.		5.77		5.77	posts.
Clean	S.F.		.45		.45	
Paint	L.F.	.55	1.60		2.15	
Minimum Charge	Job		835		835	
Treated 2" x 12"						
Demolish	L.F.		1.47		1.47	Cost includes material and labor to
Install	L.F.	2.29	7.50		9.79	install 2" x 12" lumber for
Demolish and Install	L.F.	2.29	8.97		11.26	cross-bracing of foundation piling or
Reinstall	L.F.		5.98		5.98	posts.
Clean	S.F.		.45		.45	
Paint	L.F.	.64	1.63		2.27	
Minimum Charge	Job		835		835	
Metal Pipe						
6" Concrete Filled						
Demolish	L.F.		8.20	2.42	10.62	Includes material, labor and
Install	V.L.F.	23.50	6.95	4.48	34.93	equipment to install concrete filled
Demolish and Install	V.L.F.	23.50	15.15	6.90	45.55	steel pipe pile.
Paint	L.F.	.36	2.24		2.60	
Minimum Charge	Job		3350	2150	5500	
6" Unfilled						
Demolish	L.F.		8.20	2.42	10.62	Includes material, labor and
Install	V.L.F.	18.75	6.35	4.10	29.20	equipment to install unfilled steel pipe
Demolish and Install	V.L.F.	18.75	14.55	6.52	39.82	pile.
Paint	L.F.	.36	2.24		2.60	
Minimum Charge	Job		3350	2150	5500	
8" Concrete Filled						
Demolish	L.F.		8.20	2.42	10.62	Includes material, labor and
Install	V.L.F.	26.50	7.25	4.67	38.42	equipment to install concrete filled
Demolish and Install	V.L.F.	26.50	15.45	7.09	49.04	steel pipe pile.
Paint	L.F.	.36	2.24		2.60	
Minimum Charge	Job		3350	2150	5500	
8" Unfilled						
Demolish	L.F.		8.20	2.42	10.62	Includes material, labor and
Install	V.L.F.	23	6.70	4.30	34	equipment to install unfilled steel pipe
Demolish and Install	V.L.F.	23	14.90	6.72	44.62	pile.
Paint	L.F.	.36	2.24		2.60	
Minimum Charge	Job		3350	2150	5500	

Piles		Unit	Material	Labor	Equip.	Total	Specification
10″ Concrete Filled							
	Demolish	L.F.		8.45	2.49	10.94	Includes material, labor and
	Install	V.L.F.	32	7.45	4.78	44.23	equipment to install concrete filled
	Demolish and Install	V.L.F.	32	15.90	7.27	55.17	steel pipe pile.
	Paint	L.F.	.36	2.24		2.60	
	Minimum Charge	Job		3350	2150	5500	
10″ Unfilled							
	Demolish	L.F.		8.45	2.49	10.94	Includes material, labor and
	Install	V.L.F.	27	6.70	4.30	38	equipment to install unfilled steel pipe
	Demolish and Install	V.L.F.	27	15.15	6.79	48.94	pile.
	Paint	L.F.	.36	2.24		2.60	
	Minimum Charge	Job		3350	2150	5500	
12″ Concrete Filled							
	Demolish	L.F.		8.45	2.49	10.94	Includes material, labor and
	Install	V.L.F.	39	8.05	5.20	52.25	equipment to install concrete filled
	Demolish and Install	V.L.F.	39	16.50	7.69	63.19	steel pipe pile.
	Paint	L.F.	.36	2.24		2.60	
	Minimum Charge	Job		3350	2150	5500	
12″ Unfilled							
	Demolish	L.F.		8.45	2.49	10.94	Includes material, labor and
	Install	V.L.F.	32.50	7.05	4.53	44.08	equipment to install unfilled steel pipe
	Demolish and Install	V.L.F.	32.50	15.50	7.02	55.02	pile.
	Paint	L.F.	.36	2.24		2.60	
	Minimum Charge	Job		3350	2150	5500	

Steel H Section

		Unit	Material	Labor	Equip.	Total	Specification
H Section 8 x 8 x 36#							
	Demolish	L.F.		5.70	1.69	7.39	Includes material, labor and
	Install	V.L.F.	18.30	5.20	3.36	26.86	equipment to install steel H section
	Demolish and Install	V.L.F.	18.30	10.90	5.05	34.25	pile.
	Paint	S.F.	.11	.33		.44	
	Minimum Charge	Job		3350	2150	5500	
H Section 10 x 10 x 57#							
	Demolish	L.F.		6.25	1.85	8.10	Includes material, labor and
	Install	V.L.F.	29	5.50	3.52	38.02	equipment to install steel H section
	Demolish and Install	V.L.F.	29	11.75	5.37	46.12	pile.
	Paint	S.F.	.11	.33		.44	
	Minimum Charge	Job		3350	2150	5500	
H Section 12 x 12 x 74#							
	Demolish	L.F.		6.50	1.91	8.41	Includes material, labor and
	Install	V.L.F.	39	5.65	4.54	49.19	equipment to install steel H section
	Demolish and Install	V.L.F.	39	12.15	6.45	57.60	pile.
	Paint	S.F.	.11	.33		.44	
	Minimum Charge	Job		3350	2150	5500	
H Section 14 x 14 x 89#							
	Demolish	L.F.		6.80	2.01	8.81	Includes material, labor and
	Install	V.L.F.	47	6.20	4.96	58.16	equipment to install steel H section
	Demolish and Install	V.L.F.	47	13	6.97	66.97	pile.
	Paint	S.F.	.11	.33		.44	
	Minimum Charge	Job		3350	2150	5500	

Foundation

Foundation Post

	Unit	Material	Labor	Equip.	Total	Specification
Wood Foundation Post						
4" x 4"						
Demolish	L.F.		.93	.27	1.20	Cost includes material and labor to
Install	L.F.	1.41	6.05		7.46	install 4" x 4" treated wood post by
Demolish and Install	L.F.	1.41	6.98	.27	8.66	hand in normal soil conditions.
Reinstall	L.F.		4.85		4.85	
Minimum Charge	Job		835		835	
4" x 6"						
Demolish	L.F.		.93	.27	1.20	Cost includes material and labor to
Install	L.F.	2.18	6.05		8.23	install 4" x 6" treated wood post by
Demolish and Install	L.F.	2.18	6.98	.27	9.43	hand in normal soil conditions.
Reinstall	L.F.		4.85		4.85	
Minimum Charge	Job		835		835	
6" x 6"						
Demolish	L.F.		.93	.27	1.20	Cost includes material and labor to
Install	L.F.	3.25	6.05		9.30	install 6" x 6" treated wood post by
Demolish and Install	L.F.	3.25	6.98	.27	10.50	hand in normal soil conditions.
Reinstall	L.F.		4.85		4.85	
Minimum Charge	Job		835		835	
Water Jet						
Demolish	Ea.		108		108	Includes material, labor and
Install	Ea.	75.50	830		905.50	equipment to water jet a 6" x 6" wood
Demolish and Install	Ea.	75.50	938		1013.50	post under an existing building. Cost
Reinstall	Ea.		831.09		831.09	may vary depending upon location
Minimum Charge	Job		725		725	and access.
Shim						
Install	Ea.	1.78	83.50		85.28	Includes labor and material to trim or
Minimum Charge	Job		835		835	readjust existing foundation posts
						under an existing building. Access of
						2-4 foot assumed.
Cross-bracing						
Treated 2" x 6"						
Demolish	L.F.		1.19		1.19	Cost includes material and labor to
Install	L.F.	.84	6.75		7.59	install 2" x 6" lumber for cross-bracing
Demolish and Install	L.F.	.84	7.94		8.78	of foundation piling or posts.
Reinstall	L.F.		5.40		5.40	
Clean	S.F.		.45		.45	
Paint	L.F.	.34	1.50		1.84	
Minimum Charge	Job		835		835	
Treated 2" x 8"						
Demolish	L.F.		1.19		1.19	Cost includes material and labor to
Install	L.F.	1.23	7		8.23	install 2" x 8" lumber for cross-bracing
Demolish and Install	L.F.	1.23	8.19		9.42	of foundation piling or posts.
Reinstall	L.F.		5.58		5.58	
Clean	S.F.		.45		.45	
Paint	L.F.	.45	1.56		2.01	
Minimum Charge	Job		835		835	
Treated 2" x 10"						
Demolish	L.F.		1.22		1.22	Cost includes material and labor to
Install	L.F.	1.54	7.20		8.74	install 2" x 10" lumber for
Demolish and Install	L.F.	1.54	8.42		9.96	cross-bracing of foundation piling or
Reinstall	L.F.		5.77		5.77	posts.
Clean	S.F.		.45		.45	
Paint	L.F.	.55	1.60		2.15	
Minimum Charge	Job		835		835	

23

Foundation

Foundation Post

	Unit	Material	Labor	Equip.	Total	Specification
Treated 2″ x 12″						
Demolish	L.F.		1.47		1.47	Cost includes material and labor to
Install	L.F.	2.29	7.50		9.79	install 2″ x 12″ lumber for
Demolish and Install	L.F.	2.29	8.97		11.26	cross-bracing of foundation piling or
Reinstall	L.F.		5.98		5.98	posts.
Clean	S.F.		.45		.45	
Paint	L.F.	.64	1.63		2.27	
Minimum Charge	Job		835		835	

Concrete Footing

	Unit	Material	Labor	Equip.	Total	Specification
Continuous						
12″ w x 6″ d						
Demolish	L.F.		.61	1.36	1.97	Includes material and labor to install
Install	L.F.	5	3.92	.02	8.94	continuous reinforced concrete footing
Demolish and Install	L.F.	5	4.53	1.38	10.91	poured by chute cast against earth
Minimum Charge	Job		940		940	including reinforcing, forming and finishing.
12″ w x 12″ d						
Demolish	L.F.		.74	1.63	2.37	Includes material and labor to install
Install	L.F.	10.95	7.85	.04	18.84	continuous reinforced concrete footing
Demolish and Install	L.F.	10.95	8.59	1.67	21.21	poured by chute cast against earth
Minimum Charge	Job		940		940	including reinforcing, forming and finishing.
18″ w x 10″ d						
Demolish	L.F.		.74	1.63	2.37	Includes material and labor to install
Install	L.F.	10.25	9.65	.04	19.94	continuous reinforced concrete footing
Demolish and Install	L.F.	10.25	10.39	1.67	22.31	poured by chute cast against earth
Minimum Charge	Job		940		940	including reinforcing, forming and finishing.
24″ w x 24″ d						
Demolish	L.F.		1.11	2.44	3.55	Includes material and labor to install
Install	L.F.	31	31.50	.15	62.65	continuous reinforced concrete footing
Demolish and Install	L.F.	31	32.61	2.59	66.20	poured by chute cast against earth
Minimum Charge	Job		940		940	including reinforcing, forming and finishing.
Stem Wall						
Single Story						
Demolish	L.F.		16.25	13	29.25	Includes material and labor to install
Install	L.F.	25.50	86	68.50	180	stem wall 24″ above and 18″ below
Demolish and Install	L.F.	25.50	102.25	81.50	209.25	grade for a single story structure.
Minimum Charge	Job		940		940	Includes 12″ wide and 8″ deep footing.
Two Story						
Demolish	L.F.		16.25	13	29.25	Includes material and labor to install
Install	L.F.	31.50	123	99	253.50	stem wall 24″ above and 18″ below
Demolish and Install	L.F.	31.50	139.25	112	282.75	grade typical for a two story structure,
Minimum Charge	Job		940		940	includes 18″ wide and 10″ deep footing.

For customer support on your Contractor's Pricing Guide: Residential Repair & Remodeling, call 800.448.8182.

Concrete Slab		Unit	Material	Labor	Equip.	Total	Specification
Reinforced							
4"							
	Demolish	S.F.		.25	.54	.79	Cost includes material and labor to
	Install	S.F.	2.63	.47		3.10	install 4" reinforced concrete
	Demolish and Install	S.F.	2.63	.72	.54	3.89	slab-on-grade poured by chute
	Clean	S.F.	.10	.40		.50	including forms, vapor barrier, wire
	Paint	S.F.	.19	.86		1.05	mesh, 3000 PSI concrete, float finish,
	Minimum Charge	Job		231		231	and curing.
6"							
	Demolish	S.F.		.28	.61	.89	Cost includes material and labor to
	Install	S.F.	3.60	.71	.01	4.32	install 6" reinforced concrete
	Demolish and Install	S.F.	3.60	.99	.62	5.21	slab-on-grade poured by chute
	Clean	S.F.	.10	.40		.50	including forms, vapor barrier, wire
	Paint	S.F.	.19	.86		1.05	mesh, 3000 PSI concrete, float finish,
	Minimum Charge	Job		231		231	and curing.
8"							
	Demolish	S.F.		.32	.70	1.02	Cost includes material and labor to
	Install	S.F.	4.66	.93	.01	5.60	install 8" reinforced concrete
	Demolish and Install	S.F.	4.66	1.25	.71	6.62	slab-on-grade poured by chute
	Clean	S.F.	.10	.40		.50	including forms, vapor barrier, wire
	Paint	S.F.	.19	.86		1.05	mesh, 3000 PSI concrete, float finish,
	Minimum Charge	Job		231		231	and curing.
Unreinforced							
4"							
	Demolish	S.F.		.25	.54	.79	Cost includes material and labor to
	Install	S.F.	1.78	1.13	.01	2.92	install 4" concrete slab-on-grade
	Demolish and Install	S.F.	1.78	1.38	.55	3.71	poured by chute, including forms,
	Clean	S.F.	.10	.40		.50	vapor barrier, 3000 PSI concrete, float
	Paint	S.F.	.19	.86		1.05	finish, and curing.
	Minimum Charge	Job		231		231	
6"							
	Demolish	S.F.		.28	.61	.89	Cost includes material and labor to
	Install	S.F.	2.61	1.16	.01	3.78	install 6" concrete slab-on-grade
	Demolish and Install	S.F.	2.61	1.44	.62	4.67	poured by chute, including forms,
	Clean	S.F.	.10	.40		.50	vapor barrier, 3000 PSI concrete, float
	Paint	S.F.	.19	.86		1.05	finish, and curing.
	Minimum Charge	Job		231		231	
8"							
	Demolish	S.F.		.32	.70	1.02	Cost includes material and labor to
	Install	S.F.	3.56	1.22	.01	4.79	install 8" concrete slab-on-grade
	Demolish and Install	S.F.	3.56	1.54	.71	5.81	poured by chute, including forms,
	Clean	S.F.	.10	.40		.50	vapor barrier, 3000 PSI concrete, float
	Paint	S.F.	.19	.86		1.05	finish, and curing.
	Minimum Charge	Job		231		231	
Lightweight							
	Demolish	S.F.		1.08	.24	1.32	Includes material and labor to install
	Install	S.F.	1.54	3.43		4.97	lightweight concrete, 3" - 4" thick.
	Demolish and Install	S.F.	1.54	4.51	.24	6.29	
	Clean	S.F.	.10	.40		.50	
	Paint	S.F.	.19	.86		1.05	
	Minimum Charge	Job		231		231	

25

Concrete Slab	Unit	Material	Labor	Equip.	Total	Specification
Gunite						Includes material and labor to install
Demolish	S.F.		1.33	.39	1.72	gunite on flat plane per inch of
Install	S.F.	.44	1.45	.21	2.10	thickness using 1" as base price. No
Demolish and Install	S.F.	.44	2.78	.60	3.82	forms or reinforcing are included.
Clean	S.F.	.10	.40		.50	
Paint	S.F.	.19	.86		1.05	
Minimum Charge	Job		1275		1275	
Epoxy Inject						
Install	L.F.	6.35	7.70		14.05	Includes labor and material to inject
Minimum Charge	Job		231		231	epoxy into cracks.
Slabjacking						
Install	S.F.	.56	1.68	.12	2.36	Includes labor, material and
Minimum Charge	Job		231		231	equipment to install slabjacking under
						an existing building.
Scrape and Paint						
Paint	S.F.	.23	.98		1.21	Includes labor and materials to scrape
Minimum Charge	Job		196		196	and paint a concrete floor.
Vapor Barrier						
Install	S.F.	.03	.13		.16	Cost includes material and labor to
Minimum Charge	Job		236		236	install 4 mil polyethylene vapor barrier
						10' wide sheets with 6" overlaps.

Rough Framing

Beam / Girder	Unit	Material	Labor	Equip.	Total	Specification
Solid Wood Beam						
2″ x 8″						
Demolish	L.F.		.98		.98	Cost includes material and labor to
Install	L.F.	1.07	1.45		2.52	install 2″ x 8″ beam.
Demolish and Install	L.F.	1.07	2.43		3.50	
Clean	L.F.	.06	.32		.38	
Paint	L.F.	.15	2.11		2.26	
Minimum Charge	Job		236		236	
2″ x 10″						
Demolish	L.F.		1.21		1.21	Cost includes material and labor to
Install	L.F.	1.60	1.57		3.17	install 2″ x 10″ beam.
Demolish and Install	L.F.	1.60	2.78		4.38	
Clean	L.F.	.06	.39		.45	
Paint	L.F.	.20	1.30		1.50	
Minimum Charge	Job		236		236	
2″ x 12″						
Demolish	L.F.		1.47		1.47	Cost includes material and labor to
Install	L.F.	2.01	1.71		3.72	install 2″ x 12″ beam.
Demolish and Install	L.F.	2.01	3.18		5.19	
Clean	L.F.	.07	.45		.52	
Paint	L.F.	.23	3.13		3.36	
Minimum Charge	Job		236		236	
4″ x 8″						
Demolish	L.F.		3.01		3.01	Cost includes material and labor to
Install	L.F.	4.72	2.17	.52	7.41	install 4″ x 8″ beam.
Demolish and Install	L.F.	4.72	5.18	.52	10.42	
Clean	L.F.	.06	.39		.45	
Paint	L.F.	.20	1.30		1.50	
Minimum Charge	Job		236		236	
4″ x 10″						
Demolish	L.F.		3.76		3.76	Cost includes material and labor to
Install	L.F.	5.55	2.29	.55	8.39	install 4″ x 10″ beam.
Demolish and Install	L.F.	5.55	6.05	.55	12.15	
Clean	L.F.	.07	.45		.52	
Paint	L.F.	.22	1.52		1.74	
Minimum Charge	Job		236		236	
4″ x 12″						
Demolish	L.F.		4.51		4.51	Cost includes material and labor to
Install	L.F.	6.10	2.41	.58	9.09	install 4″ x 12″ beam.
Demolish and Install	L.F.	6.10	6.92	.58	13.60	
Clean	L.F.	.08	.52		.60	
Paint	L.F.	.23	3.13		3.36	
Minimum Charge	Job		236		236	

27

Rough Frame / Structure

Beam / Girder		Unit	Material	Labor	Equip.	Total	Specification
6" x 8"							
	Demolish	L.F.		4.51		4.51	Cost includes material and labor to
	Install	L.F.	8.85	4.14	.99	13.98	install 6" x 8" beam.
	Demolish and Install	L.F.	8.85	8.65	.99	18.49	
	Clean	L.F.	.07	.45		.52	
	Paint	L.F.	.22	1.52		1.74	
	Minimum Charge	Job		236		236	
6" x 10"							
	Demolish	L.F.		5.65		5.65	Cost includes material and labor to
	Install	L.F.	4.79	1.88		6.67	install 6" x 10" triple beam.
	Demolish and Install	L.F.	4.79	7.53		12.32	
	Clean	L.F.	.08	.52		.60	
	Paint	L.F.	.25	1.74		1.99	
	Minimum Charge	Job		236		236	
6" x 12"							
	Demolish	L.F.		6.70		6.70	Cost includes material and labor to
	Install	L.F.	6.05	1.98		8.03	install 6" x 12" triple beam.
	Demolish and Install	L.F.	6.05	8.68		14.73	
	Clean	L.F.	.09	.58		.67	
	Paint	L.F.	.29	1.96		2.25	
	Minimum Charge	Job		236		236	
Steel I-type							
W8 x 31							
	Demolish	L.F.		10.40	2.84	13.24	Includes material and labor to install
	Install	L.F.	49.50	3.04	1.73	54.27	W shaped steel beam / girder.
	Demolish and Install	L.F.	49.50	13.44	4.57	67.51	
	Minimum Charge	Job		550		550	
W8 x 48							
	Demolish	L.F.		10.40	2.84	13.24	Includes material and labor to install
	Install	L.F.	77	3.18	1.81	81.99	W shaped steel beam / girder.
	Demolish and Install	L.F.	77	13.58	4.65	95.23	
	Minimum Charge	Job		550		550	
W8 x 67							
	Demolish	L.F.		10.40	2.84	13.24	Includes material and labor to install
	Install	L.F.	107	3.34	1.90	112.24	W shaped steel beam / girder.
	Demolish and Install	L.F.	107	13.74	4.74	125.48	
	Minimum Charge	Job		550		550	
W10 x 45							
	Demolish	L.F.		10.40	2.84	13.24	Includes material and labor to install
	Install	L.F.	72	3.18	1.81	76.99	W shaped steel beam / girder.
	Demolish and Install	L.F.	72	13.58	4.65	90.23	
	Minimum Charge	Job		550		550	
W10 x 68							
	Demolish	L.F.		10.40	2.84	13.24	Includes material and labor to install
	Install	L.F.	109	3.34	1.90	114.24	W shaped steel beam / girder.
	Demolish and Install	L.F.	109	13.74	4.74	127.48	
	Minimum Charge	Job		550		550	
W12 x 50							
	Demolish	L.F.		10.40	2.84	13.24	Includes material and labor to install
	Install	L.F.	80	3.18	1.81	84.99	W shaped steel beam / girder.
	Demolish and Install	L.F.	80	13.58	4.65	98.23	
	Minimum Charge	Job		550		550	
W12 x 87							
	Demolish	L.F.		15.30	4.18	19.48	Includes material and labor to install
	Install	L.F.	139	3.34	1.90	144.24	W shaped steel beam / girder.
	Demolish and Install	L.F.	139	18.64	6.08	163.72	
	Minimum Charge	Job		550		550	

For customer support on your Contractor's Pricing Guide: Residential Repair & Remodeling, call 800.448.8182.

Rough Frame / Structure

Beam / Girder

	Unit	Material	Labor	Equip.	Total	Specification
W12 x 120						
Demolish	L.F.		15.30	4.18	19.48	Includes material and labor to install
Install	L.F.	191	3.42	1.95	196.37	W shaped steel beam / girder.
Demolish and Install	L.F.	191	18.72	6.13	215.85	
Minimum Charge	Job		550		550	
W14 x 74						
Demolish	L.F.		15.30	4.18	19.48	Includes material and labor to install
Install	L.F.	119	3.34	1.90	124.24	W shaped steel beam / girder.
Demolish and Install	L.F.	119	18.64	6.08	143.72	
Minimum Charge	Job		550		550	
W14 x 120						
Demolish	L.F.		15.30	4.18	19.48	Includes material and labor to install
Install	L.F.	191	3.42	1.95	196.37	W shaped steel beam / girder.
Demolish and Install	L.F.	191	18.72	6.13	215.85	
Minimum Charge	Job		550		550	

Column

	Unit	Material	Labor	Equip.	Total	Specification
Concrete						
Small Diameter						
Demolish	L.F.		17.10	2.39	19.49	Includes material and labor to install
Install	L.F.	211	15.60	2.56	229.16	small diameter concrete column.
Demolish and Install	L.F.	211	32.70	4.95	248.65	
Minimum Charge	Job		780	128	908	
Large Diameter						
Demolish	L.F.		51	7.20	58.20	Includes material and labor to install
Install	L.F.	415	18.35	3.01	436.36	large diameter concrete column.
Demolish and Install	L.F.	415	69.35	10.21	494.56	
Minimum Charge	Job		780	128	908	

Floor Framing System

	Unit	Material	Labor	Equip.	Total	Specification
12" O.C.						
2" x 6" Joists						
Demolish	S.F.		.83		.83	Cost includes material and labor to
Install	S.F.	.92	.75		1.67	install 2" x 6" joists, 12" O.C.
Demolish and Install	S.F.	.92	1.58		2.50	including box or band joist. Blocking
Reinstall	S.F.		.60		.60	or bridging not included.
Minimum Charge	Job		236		236	
2" x 8" Joists						
Demolish	S.F.		.86		.86	Cost includes material and labor to
Install	S.F.	1.25	.86		2.11	install 2" x 8" joists, 12" O.C.
Demolish and Install	S.F.	1.25	1.72		2.97	including box or band joist. Blocking
Reinstall	S.F.		.69		.69	or bridging not included.
Minimum Charge	Job		236		236	
2" x 10" Joists						
Demolish	S.F.		.89		.89	Cost includes material and labor to
Install	S.F.	1.87	1.05		2.92	install 2" x 10" joists, 12" O.C.
Demolish and Install	S.F.	1.87	1.94		3.81	including box or band joist. Blocking
Reinstall	S.F.		.84		.84	or bridging not included.
Minimum Charge	Job		236		236	

For customer support on your Contractor's Pricing Guide: Residential Repair & Remodeling, call 800.448.8182.

Floor Framing System		Unit	Material	Labor	Equip.	Total	Specification
2″ x 12″ Joists							
	Demolish	S.F.		.92		.92	Cost includes material and labor to
	Install	S.F.	2.35	1.08		3.43	install 2″ x 12″ joists, 12″ O.C.
	Demolish and Install	S.F.	2.35	2		4.35	including box or band joist. Blocking
	Reinstall	S.F.		.86		.86	or bridging not included.
	Minimum Charge	Job		236		236	
Block / Bridge							
	Demolish	Ea.		1.26		1.26	Includes material and labor to install
	Install	Ea.	2.13	2.65		4.78	set of cross bridging or per block of
	Demolish and Install	Ea.	2.13	3.91		6.04	solid bridging for 2″ x 10″ joists cut to
	Minimum Charge	Job		236		236	size on site.
16″ O.C.							
2″ x 6″ Joists							
	Demolish	S.F.		.62		.62	Cost includes material and labor to
	Install	S.F.	.72	.56		1.28	install 2″ x 6″ joists including box or
	Demolish and Install	S.F.	.72	1.18		1.90	band joist installed 16″ O.C. Does not
	Reinstall	S.F.		.45		.45	include beams, blocking or bridging.
	Minimum Charge	Job		236		236	
2″ x 8″ Joists							
	Demolish	S.F.		.65		.65	Cost includes material and labor to
	Install	S.F.	.96	.64		1.60	install 2″ x 8″ joists including box or
	Demolish and Install	S.F.	.96	1.29		2.25	band joist installed 16″ O.C. Does not
	Reinstall	S.F.		.51		.51	include beams, blocking or bridging.
	Minimum Charge	Job		236		236	
2″ x 10″ Joists							
	Demolish	S.F.		.66		.66	Cost includes material and labor to
	Install	S.F.	1.43	.79		2.22	install 2″ x 10″ joists including box or
	Demolish and Install	S.F.	1.43	1.45		2.88	band joist 16″ O.C. Does not include
	Reinstall	S.F.		.63		.63	beams, blocking or bridging.
	Minimum Charge	Job		236		236	
2″ x 12″ Joists							
	Demolish	S.F.		.69		.69	Cost includes material and labor to
	Install	S.F.	1.80	.81		2.61	install 2″ x 12″ joists including box or
	Demolish and Install	S.F.	1.80	1.50		3.30	band joist installed 16″ O.C. Does not
	Reinstall	S.F.		.65		.65	include beams, blocking or bridging.
	Minimum Charge	Job		236		236	
Block / Bridge							
	Demolish	Ea.		1.26		1.26	Includes material and labor to install
	Install	Ea.	2.13	2.65		4.78	set of cross bridging or per block of
	Demolish and Install	Ea.	2.13	3.91		6.04	solid bridging for 2″ x 10″ joists cut to
	Minimum Charge	Job		236		236	size on site.
24″ O.C.							
2″ x 6″ Joists							
	Demolish	S.F.		.42		.42	Cost includes material and labor to
	Install	S.F.	.51	.38		.89	install 2″ x 6″ joists including box or
	Demolish and Install	S.F.	.51	.80		1.31	band joist installed 24″ O.C. Does not
	Reinstall	S.F.		.30		.30	include beams, blocking or bridging.
	Minimum Charge	Job		236		236	
2″ x 8″ Joists							
	Demolish	S.F.		.43		.43	Cost includes material and labor to
	Install	S.F.	.68	.43		1.11	install 2″ x 8″ joists including box or
	Demolish and Install	S.F.	.68	.86		1.54	band joist installed 24″ O.C. Does not
	Reinstall	S.F.		.34		.34	include beams, blocking or bridging.
	Minimum Charge	Job		236		236	

Floor Framing System		Unit	Material	Labor	Equip.	Total	Specification
2″ x 10″ Joists							
	Demolish	S.F.		.44		.44	Cost includes material and labor to
	Install	S.F.	1.02	.52		1.54	install 2″ x 10″ joists including box or
	Demolish and Install	S.F.	1.02	.96		1.98	band joist installed 24″ O.C. Does not
	Reinstall	S.F.		.42		.42	include beams, blocking or bridging.
	Minimum Charge	Job		236		236	
2″ x 12″ Joists							
	Demolish	S.F.		.46		.46	Cost includes material and labor to
	Install	S.F.	1.29	.54		1.83	install 2″ x 12″ joists including box or
	Demolish and Install	S.F.	1.29	1		2.29	band joist installed 24″ O.C. Does not
	Reinstall	S.F.		.43		.43	include beams, blocking or bridging.
	Minimum Charge	Job		236		236	
Block / Bridge							
	Demolish	Ea.		1.26		1.26	Includes material and labor to install
	Install	Ea.	2.13	2.65		4.78	set of cross bridging or per block of
	Demolish and Install	Ea.	2.13	3.91		6.04	solid bridging for 2″ x 10″ joists cut to
	Minimum Charge	Job		236		236	size on site.
Engineered Lumber, Joist							
9-1/2″							
	Demolish	S.F.		.66		.66	Includes material and labor to install
	Install	S.F.	1.75	.47		2.22	engineered lumber truss / joists per
	Demolish and Install	S.F.	1.75	1.13		2.88	S.F. of floor area based on joists 16″
	Reinstall	S.F.		.38		.38	O.C. Beams, supports and bridging
	Minimum Charge	Job		236		236	are not included.
11-7/8″							
	Demolish	S.F.		.69		.69	Includes material and labor to install
	Install	S.F.	2	.48		2.48	engineered lumber truss / joists per
	Demolish and Install	S.F.	2	1.17		3.17	S.F. of floor area based on joists 16″
	Reinstall	S.F.		.39		.39	O.C. Beams, supports and bridging
	Minimum Charge	Job		236		236	are not included.
14″							
	Demolish	S.F.		.71		.71	Includes material and labor to install
	Install	S.F.	2.56	.52		3.08	engineered lumber truss / joists per
	Demolish and Install	S.F.	2.56	1.23		3.79	S.F. of floor area based on joists 16″
	Reinstall	S.F.		.41		.41	O.C. Beams, supports and bridging
	Minimum Charge	Job		236		236	are not included.
16″							
	Demolish	S.F.		.74		.74	Includes material and labor to install
	Install	S.F.	4.08	.54		4.62	engineered lumber truss / joists per
	Demolish and Install	S.F.	4.08	1.28		5.36	S.F. of floor area based on joists 16″
	Reinstall	S.F.		.43		.43	O.C. Beams, supports and bridging
	Minimum Charge	Job		236		236	are not included.
Block / Bridge							
	Demolish	Pr.		1.26		1.26	Includes material and labor to install
	Install	Pr.	1.93	3.49		5.42	set of steel, one-nail type cross
	Demolish and Install	Pr.	1.93	4.75		6.68	bridging for trusses placed 16″ O.C.
	Minimum Charge	Job		236		236	
Block / Bridge							
2″ x 6″							
	Demolish	Ea.		1.26		1.26	Includes material and labor to install
	Install	Ea.	1.06	2.12		3.18	set of cross bridging or per block of
	Demolish and Install	Ea.	1.06	3.38		4.44	solid bridging for 2″ x 6″ joists cut to
	Minimum Charge	Job		236		236	size on site.

31

For customer support on your Contractor's Pricing Guide: Residential Repair & Remodeling, call 800.448.8182.

Floor Framing System		Unit	Material	Labor	Equip.	Total	Specification
2″ x 8″							
	Demolish	Ea.		1.26		1.26	Includes material and labor to install
	Install	Ea.	1.43	2.36		3.79	set of cross bridging or per block of
	Demolish and Install	Ea.	1.43	3.62		5.05	solid bridging for 2″ x 8″ joists cut to
	Minimum Charge	Job		236		236	size on site.
2″ x 10″							
	Demolish	Ea.		1.26		1.26	Includes material and labor to install
	Install	Ea.	2.13	2.65		4.78	set of cross bridging or per block of
	Demolish and Install	Ea.	2.13	3.91		6.04	solid bridging for 2″ x 10″ joists cut to
	Minimum Charge	Job		236		236	size on site.
2″ x 12″							
	Demolish	Ea.		1.26		1.26	Includes material and labor to install
	Install	Ea.	2.70	3.12		5.82	set of cross bridging or per block of
	Demolish and Install	Ea.	2.70	4.38		7.08	solid bridging for 2″ x 12″ joists cut to
	Minimum Charge	Job		236		236	size on site.
Ledger Strips							
1″ x 2″							
	Demolish	L.F.		.34		.34	Cost includes material and labor to
	Install	L.F.	.30	1.37		1.67	install 1″ x 2″ ledger strip nailed to the
	Demolish and Install	L.F.	.30	1.71		2.01	face of studs, beams or joist.
	Minimum Charge	Job		236		236	
1″ x 3″							
	Demolish	L.F.		.34		.34	Includes material and labor to install
	Install	L.F.	.58	1.57		2.15	up to 1″ x 4″ ledger strip nailed to the
	Demolish and Install	L.F.	.58	1.91		2.49	face of studs, beams or joist.
	Minimum Charge	Job		236		236	
1″ x 4″							
	Demolish	L.F.		.34		.34	Includes material and labor to install
	Install	L.F.	.58	1.57		2.15	up to 1″ x 4″ ledger strip nailed to the
	Demolish and Install	L.F.	.58	1.91		2.49	face of studs, beams or joist.
	Minimum Charge	Job		236		236	
2″ x 2″							
	Demolish	L.F.		.37		.37	Cost includes material and labor to
	Install	L.F.	.46	1.43		1.89	install 2″ x 2″ ledger strip nailed to the
	Demolish and Install	L.F.	.46	1.80		2.26	face of studs, beams or joist.
	Minimum Charge	Job		236		236	
2″ x 4″							
	Demolish	L.F.		.40		.40	Cost includes material and labor to
	Install	L.F.	.47	1.88		2.35	install 2″ x 4″ ledger strip nailed to the
	Demolish and Install	L.F.	.47	2.28		2.75	face of studs, beams or joist.
	Reinstall	L.F.		1.51		1.51	
	Minimum Charge	Job		236		236	
Ledger Boards							
2″ x 6″							
	Demolish	L.F.		.40		.40	Cost includes material and labor to
	Install	L.F.	5.60	2.95		8.55	install 2″ x 6″ ledger board fastened to
	Demolish and Install	L.F.	5.60	3.35		8.95	a wall, joists or studs.
	Reinstall	L.F.		2.36		2.36	
	Minimum Charge	Job		236		236	
4″ x 6″							
	Demolish	L.F.		.67		.67	Cost includes material and labor to
	Install	L.F.	9.80	3.37		13.17	install 4″ x 6″ ledger board fastened to
	Demolish and Install	L.F.	9.80	4.04		13.84	a wall, joists or studs.
	Reinstall	L.F.		2.69		2.69	
	Minimum Charge	Job		236		236	

Rough Frame / Structure

Floor Framing System

Floor Framing System	Unit	Material	Labor	Equip.	Total	Specification
4" x 8"						
Demolish	L.F.		.90		.90	Cost includes material and labor to
Install	L.F.	11.05	3.93		14.98	install 4" x 8" ledger board fastened to
Demolish and Install	L.F.	11.05	4.83		15.88	a wall, joists or studs.
Reinstall	L.F.		3.14		3.14	
Minimum Charge	Job		236		236	
Sill Plate Per L.F.						
2" x 4"						
Demolish	L.F.		.35		.35	Cost includes material and labor to
Install	L.F.	.66	1.71		2.37	install 2" x 4" pressure treated lumber,
Demolish and Install	L.F.	.66	2.06		2.72	drilled and installed on foundation
Reinstall	L.F.		1.37		1.37	bolts 48" O.C. Bolts, nuts and washers
Minimum Charge	Job		236		236	are not included.
2" x 6"						
Demolish	L.F.		.52		.52	Cost includes material and labor to
Install	L.F.	.83	1.88		2.71	install 2" x 6" pressure treated lumber,
Demolish and Install	L.F.	.83	2.40		3.23	drilled and installed on foundation
Reinstall	L.F.		1.51		1.51	bolts 48" O.C. Bolts, nuts and washers
Minimum Charge	Job		236		236	not included.
2" x 8"						
Demolish	L.F.		.69		.69	Cost includes material and labor to
Install	L.F.	1.22	2.09		3.31	install 2" x 8" pressure treated lumber,
Demolish and Install	L.F.	1.22	2.78		4	drilled and installed on foundation
Reinstall	L.F.		1.68		1.68	bolts 48" O.C. Bolts, nuts and washers
Minimum Charge	Job		236		236	not included.
Earthquake Strapping						
Install	Ea.	3	2.95		5.95	Includes labor and material to replace
Minimum Charge	Job		236		236	an earthquake strap.

Subflooring

Subflooring	Unit	Material	Labor	Equip.	Total	Specification
Plywood						
1/2"						
Demolish	S.F.		1.05		1.05	Cost includes material and labor to
Install	S.F. Flr.	.53	.51		1.04	install 1/2" plywood subfloor, CD
Demolish and Install	S.F. Flr.	.53	1.56		2.09	standard interior grade, plugged and
Minimum Charge	Job		236		236	touch sanded.
5/8"						
Demolish	S.F.		1.06		1.06	Cost includes material and labor to
Install	S.F. Flr.	.65	.56		1.21	install 5/8" plywood subfloor, CD
Demolish and Install	S.F. Flr.	.65	1.62		2.27	standard interior grade, plugged and
Reinstall	S.F. Flr.		.45		.45	touch sanded.
Minimum Charge	Job		236		236	
3/4"						
Demolish	S.F.		1.08		1.08	Cost includes material and labor to
Install	S.F. Flr.	.79	.61		1.40	install 3/4" plywood subfloor, CD
Demolish and Install	S.F. Flr.	.79	1.69		2.48	standard interior grade, plugged and
Reinstall	S.F. Flr.		.49		.49	touch sanded.
Minimum Charge	Job		236		236	

33

Rough Frame / Structure

Subflooring

Subflooring		Unit	Material	Labor	Equip.	Total	Specification
Particle Board							
1/2"							
	Demolish	S.F.		.53		.53	Cost includes material and labor to
	Install	S.F. Flr.	.46	.63		1.09	install 1/2" particle board subfloor.
	Demolish and Install	S.F. Flr.	.46	1.16		1.62	
	Minimum Charge	Job		236		236	
5/8"							
	Demolish	S.F.		.53		.53	Cost includes material and labor to
	Install	S.F. Flr.	.59	.69		1.28	install 5/8" particle board subfloor.
	Demolish and Install	S.F. Flr.	.59	1.22		1.81	
	Minimum Charge	Job		236		236	
3/4"							
	Demolish	S.F.		.54		.54	Cost includes material and labor to
	Install	S.F. Flr.	.74	.75		1.49	install 3/4" particle board subfloor.
	Demolish and Install	S.F. Flr.	.74	1.29		2.03	
	Minimum Charge	Job		236		236	
Plank Board							
	Demolish	S.F.		1.61		1.61	Cost includes material and labor to
	Install	S.F. Flr.	1.88	1.05		2.93	install 1" x 6" standard grade plank
	Demolish and Install	S.F. Flr.	1.88	2.66		4.54	flooring.
	Reinstall	S.F. Flr.		1.05		1.05	
	Minimum Charge	Job		236		236	
Felt Underlay							
	Demolish	S.F.		.09		.09	Includes material and labor to install
	Install	S.F.	.06	.13		.19	#15 felt building paper.
	Demolish and Install	S.F.	.06	.22		.28	
	Minimum Charge	Job		236		236	
Prep (for flooring)							
	Install	S.F.		.42		.42	Preparation of subflooring for
	Minimum Charge	Job		222		222	installation of new finished flooring. Cost reflects average time to prep area.

Wall Framing System

Wall Framing System		Unit	Material	Labor	Equip.	Total	Specification
2" x 4"							
8' High							
	Demolish	L.F.		5.40		5.40	Cost includes material and labor to
	Install	L.F.	4.88	7.85		12.73	install 2" x 4" x 8' high wall system,
	Demolish and Install	L.F.	4.88	13.25		18.13	including studs, treated bottom plate,
	Reinstall	L.F.		6.28		6.28	double top plate and one row of fire
	Minimum Charge	Job		236		236	blocking.
9' High							
	Demolish	L.F.		6.05		6.05	Cost includes material and labor to
	Install	L.F.	5.25	7.85		13.10	install 2" x 4" x 9' wall system
	Demolish and Install	L.F.	5.25	13.90		19.15	including studs, treated bottom plate,
	Reinstall	L.F.		6.28		6.28	double top plate and one row of fire
	Minimum Charge	Job		236		236	blocking.
10' High							
	Demolish	L.F.		6.75		6.75	Cost includes material and labor to
	Install	L.F.	5.60	7.85		13.45	install 2" x 4" x 10' wall system
	Demolish and Install	L.F.	5.60	14.60		20.20	including studs, treated bottom plate,
	Reinstall	L.F.		6.28		6.28	double top plate and one row of fire
	Minimum Charge	Job		236		236	blocking.

Rough Frame / Structure

Wall Framing System	Unit	Material	Labor	Equip.	Total	Specification
12' High						
Demolish	L.F.		8.10		8.10	Cost includes material and labor to
Install	L.F.	6.25	9.80		16.05	install 2" x 4" x 12' wall system
Demolish and Install	L.F.	6.25	17.90		24.15	including studs, treated bottom plate,
Reinstall	L.F.		7.85		7.85	double top plate and one row of fire
Minimum Charge	Job		236		236	blocking.
2" x 6"						
8' High						
Demolish	L.F.		5.05		5.05	Cost includes material and labor to
Install	L.F.	8	8.75		16.75	install 2" x 6" x 8' wall system
Demolish and Install	L.F.	8	13.80		21.80	including studs, treated bottom plate,
Reinstall	L.F.		6.98		6.98	double top plate and one row of fire
Minimum Charge	Job		236		236	blocking.
9' High						
Demolish	L.F.		5.70		5.70	Cost includes material and labor to
Install	L.F.	8.60	8.75		17.35	install 2" x 6" x 9' wall system
Demolish and Install	L.F.	8.60	14.45		23.05	including studs, treated bottom plate,
Reinstall	L.F.		6.98		6.98	double top plate and one row of fire
Minimum Charge	Job		236		236	blocking.
10' High						
Demolish	L.F.		6.20		6.20	Cost includes material and labor to
Install	L.F.	9.20	8.75		17.95	install 2" x 6" x 10' wall system
Demolish and Install	L.F.	9.20	14.95		24.15	including studs, treated bottom plate,
Reinstall	L.F.		6.98		6.98	double top plate and one row of fire
Minimum Charge	Job		236		236	blocking.
12' High						
Demolish	L.F.		7.35		7.35	Cost includes material and labor to
Install	L.F.	10.35	10.70		21.05	install 2" x 6" x 12' wall system
Demolish and Install	L.F.	10.35	18.05		28.40	including studs, treated bottom plate,
Reinstall	L.F.		8.57		8.57	double top plate and one row of fire
Minimum Charge	Job		236		236	blocking.
Fireblock						
2" x 4", 16" O.C. System						
Demolish	L.F.		.67		.67	Cost includes material and labor to
Install	L.F.	.47	1.57		2.04	install 2" x 4" fireblocks in wood frame
Demolish and Install	L.F.	.47	2.24		2.71	walls per L.F. of wall to be blocked.
Reinstall	L.F.		1.26		1.26	
Minimum Charge	Job		236		236	
2" x 4", 24" O.C. System						
Demolish	L.F.		.67		.67	Cost includes material and labor to
Install	L.F.	.47	1.57		2.04	install 2" x 4" fireblocks in wood frame
Demolish and Install	L.F.	.47	2.24		2.71	walls per L.F. of wall to be blocked.
Reinstall	L.F.		1.26		1.26	
Minimum Charge	Job		236		236	
2" x 6", 16" O.C. System						
Demolish	L.F.		1.01		1.01	Cost includes material and labor to
Install	L.F.	.79	1.57		2.36	install 2" x 6" fireblocks in wood frame
Demolish and Install	L.F.	.79	2.58		3.37	walls per L.F. of wall to be blocked.
Reinstall	L.F.		1.26		1.26	
Minimum Charge	Job		236		236	
2" x 6", 24" O.C. System						
Demolish	L.F.		1.01		1.01	Cost includes material and labor to
Install	L.F.	.79	1.57		2.36	install 2" x 6" fireblocks in wood frame
Demolish and Install	L.F.	.79	2.58		3.37	walls per L.F. of wall to be blocked.
Reinstall	L.F.		1.26		1.26	
Minimum Charge	Job		236		236	

35

Wall Framing System		Unit	Material	Labor	Equip.	Total	Specification
Bracing							
1" x 3"							
	Demolish	L.F.		.38		.38	Cost includes material and labor to
	Install	L.F.	.48	2.95		3.43	install 1" x 3" let-in bracing.
	Demolish and Install	L.F.	.48	3.33		3.81	
	Minimum Charge	Job		236		236	
1" x 4"							
	Demolish	L.F.		.38		.38	Cost includes material and labor to
	Install	L.F.	.58	2.95		3.53	install 1" x 4" let-in bracing.
	Demolish and Install	L.F.	.58	3.33		3.91	
	Minimum Charge	Job		236		236	
1" x 6"							
	Demolish	L.F.		.38		.38	Cost includes material and labor to
	Install	L.F.	.87	3.14		4.01	install 1" x 6" let-in bracing.
	Demolish and Install	L.F.	.87	3.52		4.39	
	Minimum Charge	Job		236		236	
2" x 3"							
	Demolish	L.F.		.50		.50	Cost includes material and labor to
	Install	L.F.	.46	3.14		3.60	install 2" x 3" let-in bracing.
	Demolish and Install	L.F.	.46	3.64		4.10	
	Reinstall	L.F.		2.51		2.51	
	Minimum Charge	Job		236		236	
2" x 4"							
	Demolish	L.F.		.50		.50	Cost includes material and labor to
	Install	L.F.	.47	3.14		3.61	install 2" x 4" let-in bracing.
	Demolish and Install	L.F.	.47	3.64		4.11	
	Reinstall	L.F.		2.51		2.51	
	Minimum Charge	Job		236		236	
2" x 6"							
	Demolish	L.F.		.50		.50	Cost includes material and labor to
	Install	L.F.	.79	3.37		4.16	install 2" x 6" let-in bracing.
	Demolish and Install	L.F.	.79	3.87		4.66	
	Reinstall	L.F.		2.69		2.69	
	Minimum Charge	Job		236		236	
2" x 8"							
	Demolish	L.F.		.50		.50	Cost includes material and labor to
	Install	L.F.	1.07	3.37		4.44	install 2" x 8" let-in bracing.
	Demolish and Install	L.F.	1.07	3.87		4.94	
	Reinstall	L.F.		2.69		2.69	
	Minimum Charge	Job		236		236	
Earthquake Strapping							
	Install	Ea.	3	2.95		5.95	Includes labor and material to replace
	Minimum Charge	Job		236		236	an earthquake strap.
Hurricane Clips							
	Install	Ea.	1.12	3.25		4.37	Includes labor and material to install a
	Minimum Charge	Job		236		236	hurricane clip.
Stud							
2" x 4"							
	Demolish	L.F.		.40		.40	Cost includes material and labor to
	Install	L.F.	.46	.86		1.32	install 2" x 4" wall stud per L.F. of stud.
	Demolish and Install	L.F.	.46	1.26		1.72	
	Reinstall	L.F.		.69		.69	
	Minimum Charge	Job		236		236	

Wall Framing System		Unit	Material	Labor	Equip.	Total	Specification
2" x 6"							
	Demolish	L.F.		.50		.50	Cost includes material and labor to
	Install	L.F.	.79	.94		1.73	install 2" x 6" stud per L.F. of stud.
	Demolish and Install	L.F.	.79	1.44		2.23	
	Reinstall	L.F.		.75		.75	
	Minimum Charge	Job		236		236	
Plates							
2" x 4"							
	Demolish	L.F.		.37		.37	Cost includes material and labor to
	Install	L.F.	.47	1.18		1.65	install 2" x 4" plate per L.F. of plate.
	Demolish and Install	L.F.	.47	1.55		2.02	
	Reinstall	L.F.		.94		.94	
	Minimum Charge	Job		236		236	
2" x 6"							
	Demolish	L.F.		.38		.38	Cost includes material and labor to
	Install	L.F.	.79	1.26		2.05	install 2" x 6" plate per L.F. of plate.
	Demolish and Install	L.F.	.79	1.64		2.43	
	Reinstall	L.F.		1.01		1.01	
	Minimum Charge	Job		236		236	
Headers							
2" x 6"							
	Demolish	L.F.		.36		.36	Includes material and labor to install
	Install	L.F.	.79	2.62		3.41	header over wall openings and
	Demolish and Install	L.F.	.79	2.98		3.77	around floor, ceiling and roof
	Clean	L.F.	.01	.46		.47	openings or flush beam.
	Paint	L.F.	.11	1.56		1.67	
	Minimum Charge	Job		236		236	
2" x 8"							
	Demolish	L.F.		.38		.38	Includes material and labor to install
	Install	L.F.	1.07	2.77		3.84	header over wall openings and
	Demolish and Install	L.F.	1.07	3.15		4.22	around floor, ceiling and roof
	Clean	L.F.	.01	.46		.47	openings or flush beam.
	Paint	L.F.	.15	2.11		2.26	
	Minimum Charge	Job		236		236	
2" x 10"							
	Demolish	L.F.		.38		.38	Includes material and labor to install
	Install	L.F.	1.60	2.95		4.55	header over wall openings and
	Demolish and Install	L.F.	1.60	3.33		4.93	around floor, ceiling and roof
	Clean	L.F.	.01	.46		.47	openings or flush beam.
	Paint	L.F.	.19	2.61		2.80	
	Minimum Charge	Job		236		236	
2" x 12"							
	Demolish	L.F.		.40		.40	Includes material and labor to install
	Install	L.F.	2.01	3.14		5.15	header over wall openings and
	Demolish and Install	L.F.	2.01	3.54		5.55	around floor, ceiling and roof
	Clean	L.F.	.01	.47		.48	openings or flush beam.
	Paint	L.F.	.23	3.13		3.36	
	Minimum Charge	Job		236		236	
4" x 8"							
	Demolish	L.F.		.72		.72	Includes material and labor to install
	Install	L.F.	4.72	3.62		8.34	header over wall openings and
	Demolish and Install	L.F.	4.72	4.34		9.06	around floor, ceiling and roof
	Clean	L.F.	.01	.46		.47	openings or flush beam.
	Paint	L.F.	.15	2.11		2.26	
	Minimum Charge	Job		236		236	

37

Wall Framing System

Wall Framing System	Unit	Material	Labor	Equip.	Total	Specification
4" x 10"						
Demolish	L.F.		.77		.77	Includes material and labor to install
Install	L.F.	5.55	3.93		9.48	header over wall openings and
Demolish and Install	L.F.	5.55	4.70		10.25	around floor, ceiling and roof
Clean	L.F.	.01	.46		.47	openings or flush beam.
Paint	L.F.	.19	2.61		2.80	
Minimum Charge	Job		236		236	
4" x 12"						
Demolish	L.F.		.81		.81	Includes material and labor to install
Install	L.F.	6.10	4.96		11.06	header over wall openings and
Demolish and Install	L.F.	6.10	5.77		11.87	around floor, ceiling and roof
Clean	L.F.	.01	.47		.48	openings or flush beam.
Paint	L.F.	.23	3.13		3.36	
Minimum Charge	Job		236		236	
6" x 8"						
Demolish	L.F.		.72		.72	Includes material and labor to install
Install	L.F.	8.85	2.62		11.47	header over wall openings and
Demolish and Install	L.F.	8.85	3.34		12.19	around floor, ceiling and roof
Clean	L.F.	.01	.46		.47	openings or flush beam.
Paint	L.F.	.15	2.11		2.26	
Minimum Charge	Job		236		236	
6" x 10"						
Demolish	L.F.		.77		.77	Includes material and labor to install
Install	L.F.	7.70	5.70		13.40	header over wall openings and
Demolish and Install	L.F.	7.70	6.47		14.17	around floor, ceiling and roof
Clean	L.F.	.01	.46		.47	openings or flush beam.
Paint	L.F.	.19	2.61		2.80	
Minimum Charge	Job		236		236	
6" x 12"						
Demolish	L.F.		.81		.81	Includes material and labor to install
Install	L.F.	9.75	6.75		16.50	header over wall openings and
Demolish and Install	L.F.	9.75	7.56		17.31	around floor, ceiling and roof
Clean	L.F.	.01	.47		.48	openings or flush beam.
Paint	L.F.	.23	1.09		1.32	
Minimum Charge	Job		236		236	

Rough-in Opening

Rough-in Opening	Unit	Material	Labor	Equip.	Total	Specification
Door w / 2" x 4" Lumber						
3' Wide						
Demolish	Ea.		12.65		12.65	Includes material and labor to install
Install	Ea.	20	14.75		34.75	header, double studs each side,
Demolish and Install	Ea.	20	27.40		47.40	cripples, blocking and nails, up to 3'
Reinstall	Ea.		11.78		11.78	opening in 2" x 4" stud wall 8' high.
Minimum Charge	Job		236		236	
4' Wide						
Demolish	Ea.		12.65		12.65	Includes material and labor to install
Install	Ea.	21.50	14.75		36.25	header, double studs each side,
Demolish and Install	Ea.	21.50	27.40		48.90	cripples, blocking and nails, up to 4'
Reinstall	Ea.		11.78		11.78	opening in 2" x 4" stud wall 8' high.
Minimum Charge	Job		236		236	
5' Wide						
Demolish	Ea.		12.65		12.65	Includes material and labor to install
Install	Ea.	26.50	14.75		41.25	header, double studs each side,
Demolish and Install	Ea.	26.50	27.40		53.90	cripples, blocking and nails, up to 5'
Reinstall	Ea.		11.78		11.78	opening in 2" x 4" stud wall 8' high.
Minimum Charge	Job		236		236	

Rough Frame / Structure

Rough-in Opening		Unit	Material	Labor	Equip.	Total	Specification
6' Wide							
	Demolish	Ea.		12.65		12.65	Includes material and labor to install
	Install	Ea.	28	14.75		42.75	header, double studs each side,
	Demolish and Install	Ea.	28	27.40		55.40	cripples, blocking and nails, up to 6'
	Reinstall	Ea.		11.78		11.78	opening in 2" x 4" stud wall 8' high.
	Minimum Charge	Job		236		236	
8' Wide							
	Demolish	Ea.		13.45		13.45	Includes material and labor to install
	Install	Ea.	41.50	15.70		57.20	header, double studs each side,
	Demolish and Install	Ea.	41.50	29.15		70.65	cripples, blocking and nails, up to 8'
	Reinstall	Ea.		12.57		12.57	opening in 2" x 4" stud wall 8' high.
	Minimum Charge	Job		236		236	
10' Wide							
	Demolish	Ea.		13.45		13.45	Includes material and labor to install
	Install	Ea.	56	15.70		71.70	header, double studs each side,
	Demolish and Install	Ea.	56	29.15		85.15	cripples, blocking and nails, up to 10'
	Reinstall	Ea.		12.57		12.57	opening in 2" x 4" stud wall 8' high.
	Minimum Charge	Job		236		236	
12' Wide							
	Demolish	Ea.		13.45		13.45	Includes material and labor to install
	Install	Ea.	77	15.70		92.70	header, double studs each side,
	Demolish and Install	Ea.	77	29.15		106.15	cripples, blocking and nails, up to 12'
	Reinstall	Ea.		12.57		12.57	opening in 2" x 4" stud wall 8' high.
	Minimum Charge	Job		236		236	

Door w / 2" x 6" Lumber

		Unit	Material	Labor	Equip.	Total	Specification
3' Wide							
	Demolish	Ea.		12.65		12.65	Includes material and labor to install
	Install	Ea.	31	14.75		45.75	header, double studs each side,
	Demolish and Install	Ea.	31	27.40		58.40	cripples, blocking and nails, up to 3'
	Reinstall	Ea.		11.78		11.78	opening in 2" x 6" stud wall 8' high.
	Minimum Charge	Job		236		236	
4' Wide							
	Demolish	Ea.		12.65		12.65	Includes material and labor to install
	Install	Ea.	32	14.75		46.75	header, double studs each side,
	Demolish and Install	Ea.	32	27.40		59.40	cripples, blocking and nails, up to 4'
	Reinstall	Ea.		11.78		11.78	opening in 2" x 6" stud wall 8' high.
	Minimum Charge	Job		236		236	
5' Wide							
	Demolish	Ea.		12.65		12.65	Includes material and labor to install
	Install	Ea.	37	14.75		51.75	header, double studs each side,
	Demolish and Install	Ea.	37	27.40		64.40	cripples, blocking and nails up to 5'
	Reinstall	Ea.		11.78		11.78	opening in 2" x 6" stud wall 8' high.
	Minimum Charge	Job		236		236	
6' Wide							
	Demolish	Ea.		12.65		12.65	Includes material and labor to install
	Install	Ea.	38.50	14.75		53.25	header, double studs each side,
	Demolish and Install	Ea.	38.50	27.40		65.90	cripples, blocking and nails up to 6'
	Reinstall	Ea.		11.78		11.78	opening in 2" x 6" stud wall 8' high.
	Minimum Charge	Job		236		236	
8' Wide							
	Demolish	Ea.		13.45		13.45	Includes material and labor to install
	Install	Ea.	51.50	15.70		67.20	header, double studs each side,
	Demolish and Install	Ea.	51.50	29.15		80.65	cripples, blocking and nails up to 8'
	Reinstall	Ea.		12.57		12.57	opening in 2" x 6" stud wall 8' high.
	Minimum Charge	Job		236		236	
10' Wide							
	Demolish	Ea.		13.45		13.45	Includes material and labor to install
	Install	Ea.	66.50	15.70		82.20	header, double studs each side,
	Demolish and Install	Ea.	66.50	29.15		95.65	cripples, blocking and nails up to 10'
	Reinstall	Ea.		12.57		12.57	opening in 2" x 6" stud wall 8' high.
	Minimum Charge	Job		236		236	

39

Rough Frame / Structure

Rough-in Opening	Unit	Material	Labor	Equip.	Total	Specification
12' Wide						Includes material and labor to install header, double studs each side, cripples, blocking and nails, up to 12' opening in 2" x 6" stud wall 8' high.
Demolish	Ea.		13.45		13.45	
Install	Ea.	87.50	15.70		103.20	
Demolish and Install	Ea.	87.50	29.15		116.65	
Reinstall	Ea.		12.57		12.57	
Minimum Charge	Job		236		236	

Rough-in Opening	Unit	Material	Labor	Equip.	Total	Specification
Window w / 2" x 4" Lumber						
2' Wide						Includes material and labor to install header, double studs each side of opening, cripples, blocking, nails and sub-sills for opening up to 2' in a 2" x 4" stud wall 8' high.
Demolish	Ea.		16.85		16.85	
Install	Ea.	21	19.65		40.65	
Demolish and Install	Ea.	21	36.50		57.50	
Reinstall	Ea.		15.71		15.71	
Minimum Charge	Job		236		236	
3' Wide						Includes material and labor to install header, double studs each side of opening, cripples, blocking, nails and sub-sills for opening up to 3' in a 2" x 4" stud wall 8' high.
Demolish	Ea.		16.85		16.85	
Install	Ea.	25	19.65		44.65	
Demolish and Install	Ea.	25	36.50		61.50	
Reinstall	Ea.		15.71		15.71	
Minimum Charge	Job		236		236	
4' Wide						Includes material and labor to install header, double studs each side of opening, cripples, blocking, nails and sub-sills for opening up to 4' in a 2" x 4" stud wall 8' high.
Demolish	Ea.		16.85		16.85	
Install	Ea.	27.50	19.65		47.15	
Demolish and Install	Ea.	27.50	36.50		64	
Reinstall	Ea.		15.71		15.71	
Minimum Charge	Job		236		236	
5' Wide						Includes material and labor to install header, double studs each side of opening, cripples, blocking, nails and sub-sills for opening up to 5' in a 2" x 4" stud wall 8' high.
Demolish	Ea.		16.85		16.85	
Install	Ea.	32	19.65		51.65	
Demolish and Install	Ea.	32	36.50		68.50	
Reinstall	Ea.		15.71		15.71	
Minimum Charge	Job		236		236	
6' Wide						Includes material and labor to install header, double studs each side of opening, cripples, blocking, nails and sub-sills for opening up to 6' in a 2" x 4" stud wall 8' high.
Demolish	Ea.		16.85		16.85	
Install	Ea.	35	19.65		54.65	
Demolish and Install	Ea.	35	36.50		71.50	
Reinstall	Ea.		15.71		15.71	
Minimum Charge	Job		236		236	
7' Wide						Includes material and labor to install header, double studs each side of opening, cripples, blocking, nails and sub-sills for opening up to 7' in a 2" x 4" stud wall 8' high.
Demolish	Ea.		16.85		16.85	
Install	Ea.	46	19.65		65.65	
Demolish and Install	Ea.	46	36.50		82.50	
Reinstall	Ea.		15.71		15.71	
Minimum Charge	Job		236		236	
8' Wide						Includes material and labor to install header, double studs each side of opening, cripples, blocking, nails and sub-sills for opening up to 8' in a 2" x 4" stud wall 8' high.
Demolish	Ea.		18.35		18.35	
Install	Ea.	51	21.50		72.50	
Demolish and Install	Ea.	51	39.85		90.85	
Reinstall	Ea.		17.13		17.13	
Minimum Charge	Job		236		236	

Rough Frame / Structure

Rough-in Opening

	Unit	Material	Labor	Equip.	Total	Specification
10' Wide						
Demolish	Ea.		18.35		18.35	Includes material and labor to install
Install	Ea.	67	21.50		88.50	header, double studs each side of
Demolish and Install	Ea.	67	39.85		106.85	opening, cripples, blocking, nails and
Reinstall	Ea.		17.13		17.13	sub-sills for opening up to 10' in a 2"
Minimum Charge	Job		236		236	x 4" stud wall 8' high.
12' Wide						
Demolish	Ea.		18.35		18.35	Includes material and labor to install
Install	Ea.	91.50	21.50		113	header, double studs each side of
Demolish and Install	Ea.	91.50	39.85		131.35	opening, cripples, blocking, nails and
Reinstall	Ea.		17.13		17.13	sub-sills for opening up to 12' in a 2"
Minimum Charge	Job		236		236	x 4" stud wall 8' high.

Window w / 2" x 6" Lumber

	Unit	Material	Labor	Equip.	Total	Specification
2' Wide						
Demolish	Ea.		16.85		16.85	Includes material and labor to install
Install	Ea.	34	19.65		53.65	header, double studs each side of
Demolish and Install	Ea.	34	36.50		70.50	opening, cripples, blocking, nails and
Reinstall	Ea.		15.71		15.71	sub-sills for opening up to 2' in a 2" x
Minimum Charge	Job		236		236	6" stud wall 8' high.
3' Wide						
Demolish	Ea.		16.85		16.85	Includes material and labor to install
Install	Ea.	38	19.65		57.65	header, double studs each side of
Demolish and Install	Ea.	38	36.50		74.50	opening, cripples, blocking, nails and
Reinstall	Ea.		15.71		15.71	sub-sills for opening up to 3' in a 2" x
Minimum Charge	Job		236		236	6" stud wall 8' high.
4' Wide						
Demolish	Ea.		16.85		16.85	Includes material and labor to install
Install	Ea.	41.50	19.65		61.15	header, double studs each side of
Demolish and Install	Ea.	41.50	36.50		78	opening, cripples, blocking, nails and
Reinstall	Ea.		15.71		15.71	sub-sills for opening up to 4' in a 2" x
Minimum Charge	Job		236		236	6" stud wall 8' high.
5' Wide						
Demolish	Ea.		16.85		16.85	Includes material and labor to install
Install	Ea.	46	19.65		65.65	header, double studs each side of
Demolish and Install	Ea.	46	36.50		82.50	opening, cripples, blocking, nails and
Reinstall	Ea.		15.71		15.71	sub-sills for opening up to 5' in a 2" x
Minimum Charge	Job		236		236	6" stud wall 8' high.
6' Wide						
Demolish	Ea.		16.85		16.85	Includes material and labor to install
Install	Ea.	50.50	19.65		70.15	header, double studs each side of
Demolish and Install	Ea.	50.50	36.50		87	opening, cripples, blocking, nails and
Reinstall	Ea.		15.71		15.71	sub-sills for opening up to 6' in a 2" x
Minimum Charge	Job		236		236	6" stud wall 8' high.
7' Wide						
Demolish	Ea.		16.85		16.85	Includes material and labor to install
Install	Ea.	62.50	19.65		82.15	header, double studs each side of
Demolish and Install	Ea.	62.50	36.50		99	opening, cripples, blocking, nails and
Reinstall	Ea.		15.71		15.71	sub-sills for opening up to 7' in a 2" x
Minimum Charge	Job		236		236	6" stud wall 8' high.

For customer support on your Contractor's Pricing Guide: Residential Repair & Remodeling, call 800.448.8182.

Rough Frame / Structure

Rough-in Opening

	Unit	Material	Labor	Equip.	Total	Specification
8' Wide						
Demolish	Ea.		18.35		18.35	Includes material and labor to install
Install	Ea.	68	21.50		89.50	header, double studs each side of
Demolish and Install	Ea.	68	39.85		107.85	opening, cripples, blocking, nails and
Reinstall	Ea.		17.13		17.13	sub-sills for opening up to 8' in a 2" x
Minimum Charge	Job		236		236	6" stud wall 8' high.
10' Wide						
Demolish	Ea.		18.35		18.35	Includes material and labor to install
Install	Ea.	85.50	21.50		107	header, double studs each side of
Demolish and Install	Ea.	85.50	39.85		125.35	opening, cripples, blocking, nails and
Reinstall	Ea.		17.13		17.13	sub-sills for opening up to 10' in a 2"
Minimum Charge	Job		236		236	x 6" stud wall 8' high.
12' Wide						
Demolish	Ea.		18.35		18.35	Includes material and labor to install
Install	Ea.	111	21.50		132.50	header, double studs each side of
Demolish and Install	Ea.	111	39.85		150.85	opening, cripples, blocking, nails and
Reinstall	Ea.		17.13		17.13	sub-sills for opening up to 12' in a 2"
Minimum Charge	Job		236		236	x 6" stud wall 8' high.

Glue-Laminated Beams

	Unit	Material	Labor	Equip.	Total	Specification
3-1/2" x 6"						
Demolish	L.F.		3.01		3.01	Includes material and labor to install
Install	L.F.	6.05	3.62	.86	10.53	glue-laminated wood beam.
Demolish and Install	L.F.	6.05	6.63	.86	13.54	
Reinstall	L.F.		2.90	.69	3.59	
Clean	L.F.	.06	.45		.51	
Paint	L.F.	.39	1.15		1.54	
Minimum Charge	Job		236		236	
3-1/2" x 9"						
Demolish	L.F.		3.01		3.01	Includes material and labor to install
Install	L.F.	8.10	3.62	.86	12.58	glue-laminated wood beam.
Demolish and Install	L.F.	8.10	6.63	.86	15.59	
Reinstall	L.F.		2.90	.69	3.59	
Clean	L.F.	.07	.47		.54	
Paint	L.F.	.50	1.49		1.99	
Minimum Charge	Job		236		236	
3-1/2" x 12"						
Demolish	L.F.		4.51		4.51	Includes material and labor to install
Install	L.F.	10.80	3.62	.86	15.28	glue-laminated wood beam.
Demolish and Install	L.F.	10.80	8.13	.86	19.79	
Reinstall	L.F.		2.90	.69	3.59	
Clean	L.F.	.08	.49		.57	
Paint	L.F.	.61	1.86		2.47	
Minimum Charge	Job		236		236	
3-1/2" x 15"						
Demolish	L.F.		4.51		4.51	Includes material and labor to install
Install	L.F.	13.55	3.75	.89	18.19	glue-laminated wood beam.
Demolish and Install	L.F.	13.55	8.26	.89	22.70	
Reinstall	L.F.		3	.72	3.72	
Clean	L.F.	.10	.50		.60	
Paint	L.F.	.72	2.17		2.89	
Minimum Charge	Job		236		236	

For customer support on your Contractor's Pricing Guide: Residential Repair & Remodeling, call 800.448.8182.

Glue-Laminated Beams		Unit	Material	Labor	Equip.	Total	Specification
3-1/2" x 18"							
	Demolish	L.F.		5.65		5.65	Includes material and labor to install
	Install	L.F.	18.15	3.75	.89	22.79	glue-laminated wood beam.
	Demolish and Install	L.F.	18.15	9.40	.89	28.44	
	Reinstall	L.F.		3	.72	3.72	
	Clean	L.F.	.11	.52		.63	
	Paint	L.F.	.85	2.52		3.37	
	Minimum Charge	Job		236		236	
5-1/8" x 6"							
	Demolish	L.F.		4.51		4.51	Includes material and labor to install
	Install	L.F.	8.85	3.62	.86	13.33	glue-laminated wood beam.
	Demolish and Install	L.F.	8.85	8.13	.86	17.84	
	Reinstall	L.F.		2.90	.69	3.59	
	Clean	L.F.	.06	.45		.51	
	Paint	L.F.	.41	1.26		1.67	
	Minimum Charge	Job		236		236	
5-1/8" x 9"							
	Demolish	L.F.		5.65		5.65	Includes material and labor to install
	Install	L.F.	13.30	3.62	.86	17.78	glue-laminated wood beam.
	Demolish and Install	L.F.	13.30	9.27	.86	23.43	
	Reinstall	L.F.		2.90	.69	3.59	
	Clean	L.F.	.07	.47		.54	
	Paint	L.F.	.53	1.62		2.15	
	Minimum Charge	Job		236		236	
5-1/8" x 12"							
	Demolish	L.F.		6.70		6.70	Includes material and labor to install
	Install	L.F.	17.70	3.62	.86	22.18	glue-laminated wood beam.
	Demolish and Install	L.F.	17.70	10.32	.86	28.88	
	Reinstall	L.F.		2.90	.69	3.59	
	Clean	L.F.	.09	.49		.58	
	Paint	L.F.	.65	1.96		2.61	
	Minimum Charge	Job		236		236	
5-1/8" x 18"							
	Demolish	L.F.		11.30		11.30	Includes material and labor to install
	Install	L.F.	26.50	3.75	.89	31.14	glue-laminated wood beam.
	Demolish and Install	L.F.	26.50	15.05	.89	42.44	
	Reinstall	L.F.		3	.72	3.72	
	Clean	L.F.	.11	.52		.63	
	Paint	L.F.	.76	2.64		3.40	
	Minimum Charge	Job		236		236	
6-3/4" x 12"							
	Demolish	L.F.		6.70		6.70	Includes material and labor to install
	Install	L.F.	23	3.75	.89	27.64	glue-laminated wood beam.
	Demolish and Install	L.F.	23	10.45	.89	34.34	
	Reinstall	L.F.		3	.72	3.72	
	Clean	L.F.	.10	.50		.60	
	Paint	L.F.	.72	2.17		2.89	
	Minimum Charge	Job		236		236	
6-3/4" x 15"							
	Demolish	L.F.		11.30		11.30	Includes material and labor to install
	Install	L.F.	29	3.75	.89	33.64	glue-laminated wood beam.
	Demolish and Install	L.F.	29	15.05	.89	44.94	
	Reinstall	L.F.		3	.72	3.72	
	Clean	L.F.	.11	.52		.63	
	Paint	L.F.	.85	2.52		3.37	
	Minimum Charge	Job		236		236	

Glue-Laminated Beams

Glue-Laminated Beams	Unit	Material	Labor	Equip.	Total	Specification
6-3/4" x 18"						
Demolish	L.F.		11.30		11.30	Includes material and labor to install
Install	L.F.	35	3.88	.93	39.81	glue-laminated wood beam.
Demolish and Install	L.F.	35	15.18	.93	51.11	
Reinstall	L.F.		3.10	.74	3.84	
Clean	L.F.	.12	.54		.66	
Paint	L.F.	.96	2.88		3.84	
Minimum Charge	Job		236		236	
Hardware						
5-1/4" Glue-Lam Seat						
Install	Ea.	197	2.62		199.62	Includes labor and material to install
Minimum Charge	Job		236		236	beam hangers.
6-3/4" Glue-Lam Seat						
Install	Ea.	231	2.62		233.62	Includes labor and material to install
Minimum Charge	Job		236		236	beam hangers.
Earthquake Strapping						
Install	Ea.	3	2.95		5.95	Includes labor and material to replace
Minimum Charge	Job		236		236	an earthquake strap.
Hurricane Clips						
Install	Ea.	1.12	3.25		4.37	Includes labor and material to install a
Minimum Charge	Job		236		236	hurricane clip.

Metal Stud Framing

Metal Stud Framing	Unit	Material	Labor	Equip.	Total	Specification
16" O.C. System						
4", 16 Ga.						
Demolish	S.F.		1.48		1.48	Includes material and labor to install
Install	S.F.	1.34	1.45		2.79	load bearing cold rolled metal stud
Demolish and Install	S.F.	1.34	2.93		4.27	walls, to 10' high, including studs, top
Reinstall	S.F.		1.16		1.16	and bottom track and screws.
Minimum Charge	Job		236		236	
6", 16 Ga.						
Demolish	S.F.		1.48		1.48	Includes material and labor to install
Install	S.F.	1.69	1.47		3.16	load bearing cold rolled metal stud
Demolish and Install	S.F.	1.69	2.95		4.64	walls, to 10' high, including studs, top
Reinstall	S.F.		1.18		1.18	and bottom track and screws.
Minimum Charge	Job		236		236	
4", 25 Ga.						
Demolish	S.F.		1.03		1.03	Includes material and labor to install
Install	S.F.	.46	.99		1.45	cold rolled metal stud walls, to 10'
Demolish and Install	S.F.	.46	2.02		2.48	high, including studs, top and bottom
Reinstall	S.F.		.79		.79	track and screws.
Minimum Charge	Job		236		236	
6", 25 Ga.						
Demolish	S.F.		1.03		1.03	Includes material and labor to install
Install	S.F.	.57	1		1.57	cold rolled metal stud walls, to 10'
Demolish and Install	S.F.	.57	2.03		2.60	high, including studs, top and bottom
Reinstall	S.F.		.80		.80	track and screws.
Minimum Charge	Job		236		236	
24" O.C. System						
4", 25 Ga.						
Demolish	S.F.		.87		.87	Includes material and labor to install
Install	S.F.	.34	.64		.98	cold rolled metal stud walls, to 10'
Demolish and Install	S.F.	.34	1.51		1.85	high, including studs, top and bottom
Reinstall	S.F.		.51		.51	track and screws.
Minimum Charge	Job		236		236	

Metal Stud Framing

Metal Stud Framing	Unit	Material	Labor	Equip.	Total	Specification
6", 25 Ga.						
Demolish	S.F.		.87		.87	Includes material and labor to install
Install	S.F.	.42	.65		1.07	cold rolled metal stud walls, to 10'
Demolish and Install	S.F.	.42	1.52		1.94	high, including studs, top and bottom
Reinstall	S.F.		.52		.52	track and screws.
Minimum Charge	Job		236		236	
4", 16 Ga.						
Demolish	S.F.		1.01		1.01	Includes material and labor to install
Install	S.F.	.98	1.05		2.03	load bearing cold rolled metal stud
Demolish and Install	S.F.	.98	2.06		3.04	walls, to 10' high, including studs, top
Reinstall	S.F.		.84		.84	and bottom track and screws.
Minimum Charge	Job		236		236	
6", 16 Ga.						
Demolish	S.F.		1.01		1.01	Includes material and labor to install
Install	S.F.	1.24	1.07		2.31	load bearing cold rolled metal stud
Demolish and Install	S.F.	1.24	2.08		3.32	walls, to 10' high, including studs, top
Reinstall	S.F.		.86		.86	and bottom track and screws.
Minimum Charge	Job		236		236	

Metal Joist

Metal Joist	Unit	Material	Labor	Equip.	Total	Specification
Bar Joist						
18K9						
Demolish	L.F.		1.18	.32	1.50	Includes material, labor and
Install	L.F.	9.35	2.47	1.04	12.86	equipment to install open web joist.
Demolish and Install	L.F.	9.35	3.65	1.36	14.36	
Reinstall	L.F.		1.98	.83	2.81	
Minimum Charge	Job		2175	520	2695	
16K6						
Demolish	L.F.		1.18	.32	1.50	Includes material, labor and
Install	L.F.	5.15	1.66	.08	6.89	equipment to install open web joist.
Demolish and Install	L.F.	5.15	2.84	.40	8.39	
Reinstall	L.F.		1.32	.06	1.38	
Minimum Charge	Job		2175	520	2695	

Exterior Sheathing

Exterior Sheathing	Unit	Material	Labor	Equip.	Total	Specification
CDX Plywood						
5/16"						
Demolish	S.F.		.69		.69	Cost includes material and labor to
Install	S.F.	.65	.59		1.24	install 5/16" CDX plywood sheathing.
Demolish and Install	S.F.	.65	1.28		1.93	
Minimum Charge	Job		236		236	
3/8"						
Demolish	S.F.		.70		.70	Cost includes material and labor to
Install	S.F.	.67	.79		1.46	install 3/8" CDX plywood sheathing.
Demolish and Install	S.F.	.67	1.49		2.16	
Minimum Charge	Job		236		236	
1/2"						
Demolish	S.F.		.72		.72	Cost includes material and labor to
Install	S.F.	.69	.84		1.53	install 1/2" CDX plywood sheathing.
Demolish and Install	S.F.	.69	1.56		2.25	
Minimum Charge	Job		236		236	

Rough Frame / Structure

Exterior Sheathing

	Unit	Material	Labor	Equip.	Total	Specification
5/8"						
Demolish	S.F.		.73		.73	Cost includes material and labor to
Install	S.F.	.86	.90		1.76	install 5/8" CDX plywood sheathing.
Demolish and Install	S.F.	.86	1.63		2.49	
Minimum Charge	Job		236		236	
3/4"						
Demolish	S.F.		.75		.75	Cost includes material and labor to
Install	S.F.	1.03	.97		2	install 3/4" CDX plywood sheathing.
Demolish and Install	S.F.	1.03	1.72		2.75	
Minimum Charge	Job		236		236	
OSB						
1/2"						
Demolish	S.F.		.72		.72	Cost includes material and labor to
Install	S.F.	.52	.67		1.19	install 4' x 8' x 1/2" OSB sheathing.
Demolish and Install	S.F.	.52	1.39		1.91	
Minimum Charge	Job		236		236	
5/8"						
Demolish	S.F.		.73		.73	Cost includes material and labor to
Install	S.F.	.52	.54		1.06	install 4' x 8' x 5/8" OSB sheathing.
Demolish and Install	S.F.	.52	1.27		1.79	
Minimum Charge	Job		236		236	
Vapor Barrier						
Black Paper						
Install	S.F.	.06	.13		.19	Includes material and labor to install
Minimum Charge	Job		236		236	#15 felt building paper.

Plywood Sheathing

	Unit	Material	Labor	Equip.	Total	Specification
Finish Plywood						
5/16"						
Demolish	S.F.		.69		.69	Includes material and labor to install
Install	S.F.	1.16	.29		1.45	exterior 5/16" AC plywood on walls.
Demolish and Install	S.F.	1.16	.98		2.14	
Minimum Charge	Job		236		236	
3/8"						
Demolish	S.F.		.70		.70	Includes material and labor to install
Install	S.F.	1.16	.39		1.55	exterior 3/8" AC plywood on walls.
Demolish and Install	S.F.	1.16	1.09		2.25	
Minimum Charge	Job		236		236	
1/2"						
Demolish	S.F.		.72		.72	Includes material and labor to install
Install	S.F.	1.36	.42		1.78	exterior 1/2" AC plywood on walls.
Demolish and Install	S.F.	1.36	1.14		2.50	
Minimum Charge	Job		236		236	
5/8"						
Demolish	S.F.		.73		.73	Includes material and labor to install
Install	S.F.	1.50	.42		1.92	exterior 5/8" AC plywood on walls.
Demolish and Install	S.F.	1.50	1.15		2.65	
Reinstall	S.F.		.42		.42	
Minimum Charge	Job		236		236	
3/4"						
Demolish	S.F.		.75		.75	Includes material and labor to install
Install	S.F.	1.62	.42		2.04	exterior 3/4" AC plywood on walls.
Demolish and Install	S.F.	1.62	1.17		2.79	
Reinstall	S.F.		.42		.42	
Minimum Charge	Job		236		236	

For customer support on your Contractor's Pricing Guide: Residential Repair & Remodeling, call 800.448.8182.

Stairs

	Unit	Material	Labor	Equip.	Total	Specification
Job-built						
Treads and Risers						
Demolish	Ea.		50.50		50.50	Includes material and labor to install
Install	Ea.	780	470		1250	three 2" x 12" stringers, treads and
Demolish and Install	Ea.	780	520.50		1300.50	risers of 3/4" CDX plywood, installed
Clean	Flight	1.50	30.50		32	in a straight or "L" shaped run
Paint	Ea.	.12	78		78.12	including carpet.
Minimum Charge	Job		236		236	
Landing						
Demolish	S.F.		.40		.40	Includes material and labor to install
Install	S.F.	80	3.27		83.27	landing framing, 3/4" CDX plywood
Demolish and Install	S.F.	80	3.67		83.67	surface and carpet.
Clean	S.F.		.45		.45	
Paint	S.F.	.50	.97		1.47	
Minimum Charge	Job		236		236	

Ceiling Framing

	Unit	Material	Labor	Equip.	Total	Specification
16" O.C. System						
2" x 6" Joists						
Demolish	S.F.		.62		.62	Cost includes material and labor to
Install	S.F.	.61	.78		1.39	install 2" x 6" joists including end and
Demolish and Install	S.F.	.61	1.40		2.01	header joist installed 16" O.C. Does
Reinstall	S.F.		.62		.62	not include beams, ledger strips,
Minimum Charge	Job		236		236	blocking or bridging.
2" x 8" Joists						
Demolish	S.F.		.65		.65	Cost includes material and labor to
Install	S.F.	.81	.95		1.76	install 2" x 8" joists including end and
Demolish and Install	S.F.	.81	1.60		2.41	header joist installed 16" O.C. Does
Reinstall	S.F.		.76		.76	not include beams, ledger strips,
Minimum Charge	Job		236		236	blocking or bridging.
2" x 10" Joists						
Demolish	S.F.		.66		.66	Cost includes material and labor to
Install	S.F.	1.21	1.13		2.34	install 2" x 10" joists including end
Demolish and Install	S.F.	1.21	1.79		3	and header joist installed 16" O.C.
Reinstall	S.F.		.90		.90	Does not include beams, ledger strips,
Minimum Charge	Job		236		236	blocking or bridging.
2" x 12" Joists						
Demolish	S.F.		.69		.69	Cost includes material and labor to
Install	S.F.	1.53	1.38		2.91	install 2" x 12" joists including end
Demolish and Install	S.F.	1.53	2.07		3.60	and header joist installed 16" O.C.
Reinstall	S.F.		1.10		1.10	Does not include beams, ledger strips,
Minimum Charge	Job		236		236	blocking or bridging.
Block / Bridge						
Demolish	Ea.		1.26		1.26	Includes material and labor to install
Install	Ea.	2.13	2.65		4.78	set of cross bridging or per block of
Demolish and Install	Ea.	2.13	3.91		6.04	solid bridging for 2" x 10" joists cut to
Minimum Charge	Job		236		236	size on site.

Rough Frame / Structure

Ceiling Framing		Unit	Material	Labor	Equip.	Total	Specification
24" O.C. System							
2" x 6" Joists							
	Demolish	S.F.		.42		.42	Cost includes material and labor to
	Install	S.F.	.41	.52		.93	install 2" x 6" joists including end and
	Demolish and Install	S.F.	.41	.94		1.35	header joist installed 24" O.C. Does
	Reinstall	S.F.		.41		.41	not include beams, ledger strips,
	Minimum Charge	Job		236		236	blocking or bridging.
2" x 8" Joists							
	Demolish	S.F.		.43		.43	Cost includes material and labor to
	Install	S.F.	.55	.64		1.19	install 2" x 8" joists including end and
	Demolish and Install	S.F.	.55	1.07		1.62	header joist installed 24" O.C. Does
	Reinstall	S.F.		.51		.51	not include beams, ledger strips,
	Minimum Charge	Job		236		236	blocking or bridging.
2" x 10" Joists							
	Demolish	S.F.		.44		.44	Cost includes material and labor to
	Install	S.F.	.81	.75		1.56	install 2" x 10" joists including end
	Demolish and Install	S.F.	.81	1.19		2	and header joist installed 24" O.C.
	Reinstall	S.F.		.60		.60	Does not include beams, ledger strips,
	Minimum Charge	Job		236		236	blocking or bridging.
2" x 12" Joists							
	Demolish	S.F.		.46		.46	Cost includes material and labor to
	Install	S.F.	1.02	.92		1.94	install 2" x 12" joists including end
	Demolish and Install	S.F.	1.02	1.38		2.40	and header joist installed 24" O.C.
	Reinstall	S.F.		.73		.73	Does not include beams, ledger strips,
	Minimum Charge	Job		236		236	blocking or bridging.
Block / Bridge							
	Demolish	Ea.		1.26		1.26	Includes material and labor to install
	Install	Ea.	2.13	2.65		4.78	set of cross bridging or per block of
	Demolish and Install	Ea.	2.13	3.91		6.04	solid bridging for 2" x 10" joists cut to
	Minimum Charge	Job		236		236	size on site.
Ledger Strips							
1" x 2"							
	Demolish	L.F.		.34		.34	Cost includes material and labor to
	Install	L.F.	.30	1.37		1.67	install 1" x 2" ledger strip nailed to the
	Demolish and Install	L.F.	.30	1.71		2.01	face of studs, beams or joist.
	Minimum Charge	Job		236		236	
1" x 3"							
	Demolish	L.F.		.34		.34	Includes material and labor to install
	Install	L.F.	.58	1.57		2.15	up to 1" x 4" ledger strip nailed to the
	Demolish and Install	L.F.	.58	1.91		2.49	face of studs, beams or joist.
	Minimum Charge	Job		236		236	
1" x 4"							
	Demolish	L.F.		.34		.34	Includes material and labor to install
	Install	L.F.	.58	1.57		2.15	up to 1" x 4" ledger strip nailed to the
	Demolish and Install	L.F.	.58	1.91		2.49	face of studs, beams or joist.
	Minimum Charge	Job		236		236	
2" x 2"							
	Demolish	L.F.		.37		.37	Cost includes material and labor to
	Install	L.F.	.46	1.43		1.89	install 2" x 2" ledger strip nailed to the
	Demolish and Install	L.F.	.46	1.80		2.26	face of studs, beams or joist.
	Minimum Charge	Job		236		236	
2" x 4"							
	Demolish	L.F.		.40		.40	Cost includes material and labor to
	Install	L.F.	.47	1.88		2.35	install 2" x 4" ledger strip nailed to the
	Demolish and Install	L.F.	.47	2.28		2.75	face of studs, beams or joist.
	Reinstall	L.F.		1.51		1.51	
	Minimum Charge	Job		236		236	

Rough Frame / Structure

Ceiling Framing

Ceiling Framing	Unit	Material	Labor	Equip.	Total	Specification
Ledger Boards						
2" x 4"						
Demolish	L.F.		.40		.40	Cost includes material and labor to
Install	L.F.	5.30	2.62		7.92	install 2" x 4" ledger board fastened to
Demolish and Install	L.F.	5.30	3.02		8.32	a wall, joists or studs.
Reinstall	L.F.		2.09		2.09	
Minimum Charge	Job		236		236	
4" x 6"						
Demolish	L.F.		.67		.67	Cost includes material and labor to
Install	L.F.	9.80	3.37		13.17	install 4" x 6" ledger board fastened to
Demolish and Install	L.F.	9.80	4.04		13.84	a wall, joists or studs.
Reinstall	L.F.		2.69		2.69	
Minimum Charge	Job		236		236	
4" x 8"						
Demolish	L.F.		.90		.90	Cost includes material and labor to
Install	L.F.	11.05	3.93		14.98	install 4" x 8" ledger board fastened to
Demolish and Install	L.F.	11.05	4.83		15.88	a wall, joists or studs.
Reinstall	L.F.		3.14		3.14	
Minimum Charge	Job		236		236	
Earthquake Strapping						
Install	Ea.	3	2.95		5.95	Includes labor and material to replace
Minimum Charge	Job		236		236	an earthquake strap.
Hurricane Clips						
Install	Ea.	1.12	3.25		4.37	Includes labor and material to install a
Minimum Charge	Job		236		236	hurricane clip.

Roof Framing

Roof Framing	Unit	Material	Labor	Equip.	Total	Specification
16" O.C. System						
2" x 4" Rafters						
Demolish	S.F.		.92		.92	Cost includes material and labor to
Install	S.F.	.35	.84		1.19	install 2" x 4", 16" O.C. rafter framing
Demolish and Install	S.F.	.35	1.76		2.11	for flat, shed or gable roofs, 25' span,
Reinstall	S.F.		.67		.67	up to 5/12 slope, per S.F. of roof.
Minimum Charge	Job		236		236	
2" x 6" Rafters						
Demolish	S.F.		.96		.96	Cost includes material and labor to
Install	S.F.	.59	.88		1.47	install 2" x 6", 16" O.C. rafter framing
Demolish and Install	S.F.	.59	1.84		2.43	for flat, shed or gable roofs, 25' span,
Reinstall	S.F.		.71		.71	up to 5/12 slope, per S.F. of roof.
Minimum Charge	Job		236		236	
2" x 8" Rafters						
Demolish	S.F.		.99		.99	Cost includes material and labor to
Install	S.F.	.80	.94		1.74	install 2" x 8", 16" O.C. rafter framing
Demolish and Install	S.F.	.80	1.93		2.73	for flat, shed or gable roofs, 25' span,
Reinstall	S.F.		.75		.75	up to 5/12 slope, per S.F. of roof.
Minimum Charge	Job		236		236	
2" x 10" Rafters						
Demolish	S.F.		.99		.99	Cost includes material and labor to
Install	S.F.	1.20	1.43		2.63	install 2" x 10", 16" O.C. rafter
Demolish and Install	S.F.	1.20	2.42		3.62	framing for flat, shed or gable roofs,
Reinstall	S.F.		1.14		1.14	25' span, up to 5/12 slope, per S.F.
Minimum Charge	Job		236		236	of roof.

For customer support on your Contractor's Pricing Guide: Residential Repair & Remodeling, call 800.448.8182.

Roof Framing	Unit	Material	Labor	Equip.	Total	Specification
2" x 12" Rafters						
Demolish	S.F.		1		1	Cost includes material and labor to
Install	S.F.	1.51	1.55		3.06	install 2" x 12", 16" O.C. rafter
Demolish and Install	S.F.	1.51	2.55		4.06	framing for flat, shed or gable roofs,
Reinstall	S.F.		1.24		1.24	25' span, up to 5/12 slope, per S.F.
Minimum Charge	Job		236		236	of roof.
24" O.C. System						
2" x 4" Rafters						
Demolish	S.F.		.69		.69	Cost includes material and labor to
Install	S.F.	.23	.63		.86	install 2" x 4", 24" O.C. rafter framing
Demolish and Install	S.F.	.23	1.32		1.55	for flat, shed or gable roofs, 25' span,
Reinstall	S.F.		.51		.51	up to 5/12 slope, per S.F. of roof.
Minimum Charge	Job		236		236	
2" x 6" Rafters						
Demolish	S.F.		.72		.72	Cost includes material and labor to
Install	S.F.	.40	.66		1.06	install 2" x 6", 24" O.C. rafter framing
Demolish and Install	S.F.	.40	1.38		1.78	for flat, shed or gable roofs, 25' span,
Reinstall	S.F.		.53		.53	up to 5/12 slope, per S.F. of roof.
Minimum Charge	Job		236		236	
2" x 8" Rafters						
Demolish	S.F.		.74		.74	Cost includes material and labor to
Install	S.F.	.54	.71		1.25	install 2" x 8", 24" O.C. rafter framing
Demolish and Install	S.F.	.54	1.45		1.99	for flat, shed or gable roofs, 25' span,
Reinstall	S.F.		.57		.57	up to 5/12 slope, per S.F. of roof.
Minimum Charge	Job		236		236	
2" x 10" Rafters						
Demolish	S.F.		.74		.74	Cost includes material and labor to
Install	S.F.	.80	1.07		1.87	install 2" x 10", 24" O.C. rafter
Demolish and Install	S.F.	.80	1.81		2.61	framing for flat, shed or gable roofs,
Reinstall	S.F.		.86		.86	25' span, up to 5/12 slope, per S.F.
Minimum Charge	Job		236		236	of roof.
2" x 12" Rafters						
Demolish	S.F.		.75		.75	Cost includes material and labor to
Install	S.F.	1.01	1.17		2.18	install 2" x 12", 24" O.C. rafter
Demolish and Install	S.F.	1.01	1.92		2.93	framing for flat, shed or gable roofs,
Reinstall	S.F.		.93		.93	25' span, up to 5/12 slope, per S.F.
Minimum Charge	Job		236		236	of roof.
Rafter						
2" x 4"						
Demolish	L.F.		.94		.94	Cost includes material and labor to
Install	L.F.	.47	1.18		1.65	install 2" x 4" rafters for flat, shed or
Demolish and Install	L.F.	.47	2.12		2.59	gable roofs, up to 5/12 slope, 25'
Reinstall	L.F.		.94		.94	span, per L.F.
Minimum Charge	Job		236		236	
2" x 6"						
Demolish	L.F.		.95		.95	Cost includes material and labor to
Install	L.F.	.79	1.18		1.97	install 2" x 6" rafters for flat, shed or
Demolish and Install	L.F.	.79	2.13		2.92	gable roofs, up to 5/12 slope, 25'
Reinstall	L.F.		.94		.94	span, per L.F.
Minimum Charge	Job		236		236	
2" x 8"						
Demolish	L.F.		.97		.97	Cost includes material and labor to
Install	L.F.	1.07	1.26		2.33	install 2" x 8" rafters for flat, shed or
Demolish and Install	L.F.	1.07	2.23		3.30	gable roofs, up to 5/12 slope, 25'
Reinstall	L.F.		1.01		1.01	span, per L.F.
Minimum Charge	Job		236		236	

Roof Framing

	Unit	Material	Labor	Equip.	Total	Specification
2" x 10"						
Demolish	L.F.		.98		.98	Cost includes material and labor to
Install	L.F.	1.60	1.90		3.50	install 2" x 10" rafters for flat, shed or
Demolish and Install	L.F.	1.60	2.88		4.48	gable roofs, up to 5/12 slope, 25'
Reinstall	L.F.		1.52		1.52	span, per L.F.
Minimum Charge	Job		236		236	
2" x 12"						
Demolish	L.F.		.99		.99	Cost includes material and labor to
Install	L.F.	2.01	2.07		4.08	install 2" x 12" rafters for flat, shed or
Demolish and Install	L.F.	2.01	3.06		5.07	gable roofs, up to 5/12 slope, 25'
Reinstall	L.F.		1.66		1.66	span, per L.F.
Minimum Charge	Job		236		236	
2"x 4" Valley / Jack						
Demolish	L.F.		.94		.94	Includes material and labor to install
Install	L.F.	.79	1.98		2.77	up to 2" x 6" valley/jack rafters for
Demolish and Install	L.F.	.79	2.92		3.71	flat, shed or gable roofs, up to 5/12
Reinstall	L.F.		1.59		1.59	slope, 25' span, per L.F.
Minimum Charge	Job		236		236	
2"x 6" Valley / Jack						
Demolish	L.F.		.95		.95	Includes material and labor to install
Install	L.F.	.79	1.98		2.77	up to 2" x 6" valley/jack rafters for
Demolish and Install	L.F.	.79	2.93		3.72	flat, shed or gable roofs, up to 5/12
Reinstall	L.F.		1.59		1.59	slope, 25' span, per L.F.
Minimum Charge	Job		236		236	

Ridgeboard

	Unit	Material	Labor	Equip.	Total	Specification
2" x 4"						
Demolish	L.F.		.90		.90	Cost includes material and labor to
Install	L.F.	.47	1.71		2.18	install 2" x 4" ridgeboard for flat, shed
Demolish and Install	L.F.	.47	2.61		3.08	or gable roofs, up to 5/12 slope, 25'
Reinstall	L.F.		1.37		1.37	span, per L.F.
Minimum Charge	Job		236		236	
2" x 6"						
Demolish	L.F.		.92		.92	Cost includes material and labor to
Install	L.F.	.79	1.88		2.67	install 2" x 6" ridgeboard for flat, shed
Demolish and Install	L.F.	.79	2.80		3.59	or gable roofs, up to 5/12 slope, 25'
Reinstall	L.F.		1.51		1.51	span, per L.F.
Minimum Charge	Job		236		236	
2" x 8"						
Demolish	L.F.		.95		.95	Cost includes material and labor to
Install	L.F.	1.07	2.09		3.16	install 2" x 8" ridgeboard for flat, shed
Demolish and Install	L.F.	1.07	3.04		4.11	or gable roofs, up to 5/12 slope, 25'
Reinstall	L.F.		1.68		1.68	span, per L.F.
Minimum Charge	Job		236		236	
2" x 10"						
Demolish	L.F.		.98		.98	Cost includes material and labor to
Install	L.F.	1.60	2.36		3.96	install 2" x 10" ridgeboard for flat,
Demolish and Install	L.F.	1.60	3.34		4.94	shed or gable roofs, up to 5/12
Reinstall	L.F.		1.88		1.88	slope, 25' span, per L.F.
Minimum Charge	Job		236		236	
2" x 12"						
Demolish	L.F.		1.01		1.01	Cost includes material and labor to
Install	L.F.	2.01	1.35		3.36	install 2" x 12" ridgeboard for flat,
Demolish and Install	L.F.	2.01	2.36		4.37	shed or gable roofs, up to 5/12
Reinstall	L.F.		1.08		1.08	slope, 25' span, per L.F.
Minimum Charge	Job		236		236	

Roof Framing		Unit	Material	Labor	Equip.	Total	Specification
Collar Beam							
1" x 6"							
	Demolish	L.F.		.71		.71	Cost includes material and labor to
	Install	L.F.	.87	.59		1.46	install 1" x 6" collar beam or tie for
	Demolish and Install	L.F.	.87	1.30		2.17	roof framing.
	Reinstall	L.F.		.47		.47	
	Minimum Charge	Job		236		236	
2" x 6"							
	Demolish	L.F.		.83		.83	Cost includes material and labor to
	Install	L.F.	.79	.59		1.38	install 2" x 6" collar beam or tie for
	Demolish and Install	L.F.	.79	1.42		2.21	roof framing.
	Reinstall	L.F.		.47		.47	
	Minimum Charge	Job		236		236	
Purlins							
2" x 6"							
	Demolish	L.F.		.81		.81	Cost includes material and labor to
	Install	L.F.	.79	.52		1.31	install 2" x 6" purlins below roof
	Demolish and Install	L.F.	.79	1.33		2.12	rafters.
	Reinstall	L.F.		.42		.42	
	Minimum Charge	Job		236		236	
2" x 8"							
	Demolish	L.F.		.82		.82	Cost includes material and labor to
	Install	L.F.	1.07	.53		1.60	install 2" x 8" purlins below roof
	Demolish and Install	L.F.	1.07	1.35		2.42	rafters.
	Reinstall	L.F.		.42		.42	
	Minimum Charge	Job		236		236	
2" x 10"							
	Demolish	L.F.		.82		.82	Cost includes material and labor to
	Install	L.F.	1.60	.54		2.14	install 2" x 10" purlins below roof
	Demolish and Install	L.F.	1.60	1.36		2.96	rafters.
	Reinstall	L.F.		.43		.43	
	Minimum Charge	Job		236		236	
2" x 12"							
	Demolish	L.F.		.83		.83	Cost includes material and labor to
	Install	L.F.	2.01	.54		2.55	install 2" x 12" purlins below roof
	Demolish and Install	L.F.	2.01	1.37		3.38	rafters.
	Reinstall	L.F.		.43		.43	
	Minimum Charge	Job		236		236	
4" x 6"							
	Demolish	L.F.		.83		.83	Cost includes material and labor to
	Install	L.F.	3.47	.54		4.01	install 4" x 6" purlins below roof
	Demolish and Install	L.F.	3.47	1.37		4.84	rafters.
	Reinstall	L.F.		.43		.43	
	Minimum Charge	Job		236		236	
4" x 8"							
	Demolish	L.F.		.84		.84	Cost includes material and labor to
	Install	L.F.	4.72	.55		5.27	install 4" x 8" purlins below roof
	Demolish and Install	L.F.	4.72	1.39		6.11	rafters.
	Reinstall	L.F.		.44		.44	
	Minimum Charge	Job		236		236	
Ledger Board							
2" x 6"							
	Demolish	L.F.		.40		.40	Cost includes material and labor to
	Install	L.F.	.79	2.12		2.91	install 2" x 6" ledger board nailed to
	Demolish and Install	L.F.	.79	2.52		3.31	the face of a rafter.
	Reinstall	L.F.		1.70		1.70	
	Minimum Charge	Job		236		236	

Rough Frame / Structure

Roof Framing

Roof Framing	Unit	Material	Labor	Equip.	Total	Specification
4" x 6"						
Demolish	L.F.		.67		.67	Cost includes material and labor to
Install	L.F.	3.47	3.12		6.59	install 4" x 6" ledger board nailed to
Demolish and Install	L.F.	3.47	3.79		7.26	the face of a rafter.
Reinstall	L.F.		2.50		2.50	
Minimum Charge	Job		236		236	
4" x 8"						
Demolish	L.F.		.90		.90	Cost includes material and labor to
Install	L.F.	4.72	3.60		8.32	install 4" x 8" ledger board nailed to
Demolish and Install	L.F.	4.72	4.50		9.22	the face of a rafter.
Reinstall	L.F.		2.88		2.88	
Minimum Charge	Job		236		236	

Outriggers

Outriggers	Unit	Material	Labor	Equip.	Total	Specification
2" x 4"						
Demolish	L.F.		.28		.28	Cost includes material and labor to
Install	L.F.	.47	1.88		2.35	install 2" x 4" outrigger rafters for flat,
Demolish and Install	L.F.	.47	2.16		2.63	shed or gabled roofs, up to 5/12
Reinstall	L.F.		1.51		1.51	slope, per L.F.
Minimum Charge	Job		236		236	
2" x 6"						
Demolish	L.F.		.42		.42	Cost includes material and labor to
Install	L.F.	.79	1.88		2.67	install 2" x 6" outrigger rafters for flat,
Demolish and Install	L.F.	.79	2.30		3.09	shed or gabled roofs, up to 5/12
Reinstall	L.F.		1.51		1.51	slope, per L.F.
Minimum Charge	Job		236		236	

Lookout Rafter

Lookout Rafter	Unit	Material	Labor	Equip.	Total	Specification
2" x 4"						
Demolish	L.F.		.94		.94	Cost includes material and labor to
Install	L.F.	.47	.79		1.26	install 2" x 4" lookout rafters for flat,
Demolish and Install	L.F.	.47	1.73		2.20	shed or gabled roofs, up to 5/12
Reinstall	L.F.		.63		.63	slope, per L.F.
Minimum Charge	Job		236		236	
2" x 6"						
Demolish	L.F.		.95		.95	Cost includes material and labor to
Install	L.F.	.79	.80		1.59	install 2" x 6" lookout rafters for flat,
Demolish and Install	L.F.	.79	1.75		2.54	shed or gabled roofs, up to 5/12
Reinstall	L.F.		.64		.64	slope, per L.F.
Minimum Charge	Job		236		236	

Fly Rafter

Fly Rafter	Unit	Material	Labor	Equip.	Total	Specification
2" x 4"						
Demolish	L.F.		.94		.94	Cost includes material and labor to
Install	L.F.	.47	.79		1.26	install 2" x 4" fly rafters for flat, shed
Demolish and Install	L.F.	.47	1.73		2.20	or gabled roofs, up to 5/12 slope,
Reinstall	L.F.		.63		.63	per L.F.
Minimum Charge	Job		236		236	
2" x 6"						
Demolish	L.F.		.95		.95	Cost includes material and labor to
Install	L.F.	.78	.80		1.58	install 2" x 6" fly rafters for flat, shed
Demolish and Install	L.F.	.78	1.75		2.53	or gabled roofs, up to 5/12 slope,
Reinstall	L.F.		.64		.64	per L.F.
Minimum Charge	Job		236		236	

Rough Frame / Structure

Residential Trusses	Unit	Material	Labor	Equip.	Total	Specification
W or Fink Truss						
24' Span						
Demolish	Ea.		36.50	8.75	45.25	Includes material, labor and
Install	Ea.	86	36	8.65	130.65	equipment to install wood gang-nailed
Demolish and Install	Ea.	86	72.50	17.40	175.90	residential truss, up to 24' span.
Reinstall	Ea.		36.12	8.65	44.77	
Minimum Charge	Job		236		236	
28' Span						
Demolish	Ea.		39	9.30	48.30	Includes material, labor and
Install	Ea.	108	41	9.80	158.80	equipment to install wood gang-nailed
Demolish and Install	Ea.	108	80	19.10	207.10	residential truss, up to 28' span.
Reinstall	Ea.		40.89	9.79	50.68	
Minimum Charge	Job		236		236	
32' Span						
Demolish	Ea.		43	10.30	53.30	Includes material, labor and
Install	Ea.	125	43.50	10.40	178.90	equipment to install wood gang-nailed
Demolish and Install	Ea.	125	86.50	20.70	232.20	residential truss, up to 32' span.
Reinstall	Ea.		43.34	10.38	53.72	
Minimum Charge	Job		236		236	
36' Span						
Demolish	Ea.		46.50	11.10	57.60	Includes material, labor and
Install	Ea.	152	47	11.30	210.30	equipment to install wood gang-nailed
Demolish and Install	Ea.	152	93.50	22.40	267.90	residential truss, up to 36' span.
Reinstall	Ea.		47.11	11.28	58.39	
Minimum Charge	Job		236		236	
Gable End						
24' Span						
Demolish	Ea.		36.50	8.75	45.25	Includes material, labor and
Install	Ea.	94.50	38.50	9.25	142.25	equipment to install wood gang-nailed
Demolish and Install	Ea.	94.50	75	18	187.50	residential gable end truss, up to 24'
Reinstall	Ea.		38.70	9.26	47.96	span.
Clean	Ea.	.63	48.50		49.13	
Paint	Ea.	4.77	52		56.77	
Minimum Charge	Job		236		236	
28' Span						
Demolish	Ea.		39	9.30	48.30	Includes material, labor and
Install	Ea.	112	46	11.05	169.05	equipment to install wood gang-nailed
Demolish and Install	Ea.	112	85	20.35	217.35	residential gable end truss, up to 30'
Reinstall	Ea.		46.11	11.04	57.15	span.
Clean	Ea.	.65	66		66.65	
Paint	Ea.	4.98	71		75.98	
Minimum Charge	Job		236		236	
32' Span						
Demolish	Ea.		43	10.30	53.30	Includes material, labor and
Install	Ea.	133	47	11.30	191.30	equipment to install wood gang-nailed
Demolish and Install	Ea.	133	90	21.60	244.60	residential gable end truss, up to 32'
Reinstall	Ea.		47.11	11.28	58.39	span.
Clean	Ea.	.68	86.50		87.18	
Paint	Ea.	5.20	93		98.20	
Minimum Charge	Job		236		236	
36' Span						
Demolish	Ea.		46.50	11.10	57.60	Includes material, labor and
Install	Ea.	154	54	12.95	220.95	equipment to install wood gang-nailed
Demolish and Install	Ea.	154	100.50	24.05	278.55	residential gable end truss, up to 36'
Reinstall	Ea.		54.18	12.97	67.15	span.
Clean	Ea.	.72	109		109.72	
Paint	Ea.	5.45	117		122.45	
Minimum Charge	Job		236		236	

Rough Frame / Structure

Residential Trusses

	Unit	Material	Labor	Equip.	Total	Specification
Truss Hardware						
5-1/4″ Glue-Lam Seat						
Install	Ea.	197	2.62		199.62	Includes labor and material to install
Minimum Charge	Job		236		236	beam hangers.
6-3/4″ Glue-Lam Seat						
Install	Ea.	231	2.62		233.62	Includes labor and material to install
Minimum Charge	Job		236		236	beam hangers.
2″ x 4″ Joist Hanger						
Install	Ea.	.85	2.69		3.54	Includes labor and material to install
Minimum Charge	Job		236		236	joist and beam hangers.
2″ x 12″ Joist Hanger						
Install	Ea.	1.61	2.86		4.47	Cost includes labor and material to
Minimum Charge	Job		236		236	install 2″ x 12″ joist hanger.

Commercial Trusses

	Unit	Material	Labor	Equip.	Total	Specification
Engineered Lumber, Truss / Joist						
9-1/2″						
Demolish	S.F.		.66		.66	Includes material and labor to install
Install	S.F.	1.75	.47		2.22	engineered lumber truss / joists per
Demolish and Install	S.F.	1.75	1.13		2.88	S.F. of floor area based on joists 16″
Reinstall	S.F.		.38		.38	O.C. Beams, supports and bridging
Minimum Charge	Job		236		236	are not included.
11-7/8″						
Demolish	S.F.		.69		.69	Includes material and labor to install
Install	S.F.	2	.48		2.48	engineered lumber truss / joists per
Demolish and Install	S.F.	2	1.17		3.17	S.F. of floor area based on joists 16″
Reinstall	S.F.		.39		.39	O.C. Beams, supports and bridging
Minimum Charge	Job		236		236	are not included.
14″						
Demolish	S.F.		.71		.71	Includes material and labor to install
Install	S.F.	2.56	.52		3.08	engineered lumber truss / joists per
Demolish and Install	S.F.	2.56	1.23		3.79	S.F. of floor area based on joists 16″
Reinstall	S.F.		.41		.41	O.C. Beams, supports and bridging
Minimum Charge	Job		236		236	are not included.
16″						
Demolish	S.F.		.74		.74	Includes material and labor to install
Install	S.F.	4.08	.54		4.62	engineered lumber truss / joists per
Demolish and Install	S.F.	4.08	1.28		5.36	S.F. of floor area based on joists 16″
Reinstall	S.F.		.43		.43	O.C. Beams, supports and bridging
Minimum Charge	Job		236		236	are not included.
Bowstring Truss						
100′ Clear Span						
Demolish	S.F. Flr.		.67	.45	1.12	Includes material, labor and
Install	S.F. Flr.	8.15	.54	.27	8.96	equipment to install bowstring truss
Demolish and Install	S.F. Flr.	8.15	1.21	.72	10.08	system up to 100′ clear span per S.F.
Minimum Charge	Job		236		236	of floor area.
120′ Clear Span						
Demolish	S.F. Flr.		.74	.50	1.24	Includes material, labor and
Install	S.F. Flr.	8.65	.60	.30	9.55	equipment to install bowstring truss
Demolish and Install	S.F. Flr.	8.65	1.34	.80	10.79	system up to 120′ clear span per S.F.
Minimum Charge	Job		236		236	of floor area.

55

Roof Sheathing		Unit	Material	Labor	Equip.	Total	Specification
Plywood							
3/8"							
	Demolish	S.F.		.53		.53	Cost includes material and labor to
	Install	S.F.	.67	.62		1.29	install 3/8" CDX plywood roof
	Demolish and Install	S.F.	.67	1.15		1.82	sheathing.
	Reinstall	S.F.		.62		.62	
	Minimum Charge	Job		236		236	
1/2"							
	Demolish	S.F.		.53		.53	Cost includes material and labor to
	Install	S.F.	.69	.67		1.36	install 1/2" CDX plywood roof
	Demolish and Install	S.F.	.69	1.20		1.89	sheathing.
	Minimum Charge	Job		236		236	
5/8"							
	Demolish	S.F.		.53		.53	Cost includes material and labor to
	Install	S.F.	.86	.73		1.59	install 5/8" CDX plywood roof
	Demolish and Install	S.F.	.86	1.26		2.12	sheathing.
	Minimum Charge	Job		236		236	
3/4"							
	Demolish	S.F.		.53		.53	Cost includes material and labor to
	Install	S.F.	1.03	.79		1.82	install 3/4" CDX plywood roof
	Demolish and Install	S.F.	1.03	1.32		2.35	sheathing.
	Minimum Charge	Job		236		236	
OSB							
1/2"							
	Demolish	S.F.		.53		.53	Cost includes material and labor to
	Install	S.F.	.52	.67		1.19	install 4' x 8' x 1/2" OSB sheathing.
	Demolish and Install	S.F.	.52	1.20		1.72	
	Minimum Charge	Job		236		236	
5/8"							
	Demolish	S.F.		.53		.53	Cost includes material and labor to
	Install	S.F.	.52	.54		1.06	install 4' x 8' x 5/8" OSB sheathing.
	Demolish and Install	S.F.	.52	1.07		1.59	
	Minimum Charge	Job		236		236	
Plank							
	Demolish	S.F.		.53		.53	Cost includes material and labor to
	Install	S.F.	2.37	1.30		3.67	install 1" x 6" or 1" x 8" utility T&G
	Demolish and Install	S.F.	2.37	1.83		4.20	board sheathing laid diagonal.
	Reinstall	S.F.		1.30		1.30	
	Minimum Charge	Job		236		236	
Skip-type							
1" x 4"							
	Demolish	S.F.		.81		.81	Includes material and labor to install
	Install	S.F.	.73	.39		1.12	sheathing board material 1" x 4", 7"
	Demolish and Install	S.F.	.73	1.20		1.93	O.C.
	Minimum Charge	Job		236		236	
1" x 6"							
	Demolish	S.F.		.81		.81	Includes material and labor to install
	Install	S.F.	.87	.33		1.20	sheathing board material 1" x 6", 7"
	Demolish and Install	S.F.	.87	1.14		2.01	or 9" O.C.
	Minimum Charge	Job		236		236	

Rough Frame / Structure

Wood Deck

	Unit	Material	Labor	Equip.	Total	Specification
Stairs						
Demolish	Ea.		465		465	Includes material and labor to install
Install	Ea.	160	835		995	wood steps. 5 - 7 risers included.
Demolish and Install	Ea.	160	1300		1460	
Reinstall	Ea.		668.80		668.80	
Clean	Flight	1.50	30.50		32	
Paint	Flight	21.50	52		73.50	
Minimum Charge	Job		236		236	
Bench Seating						
Demolish	L.F.		6.75		6.75	Includes material and labor to install
Install	L.F.	2.33	23.50		25.83	pressure treated bench seating.
Demolish and Install	L.F.	2.33	30.25		32.58	
Reinstall	L.F.		18.85		18.85	
Clean	L.F.	.06	1.82		1.88	
Paint	L.F.	.68	3.13		3.81	
Minimum Charge	Job		236		236	
Privacy Wall						
Demolish	S.F. Flr.		1.62		1.62	Includes material, labor and
Install	S.F. Flr.	1.75	.87	.17	2.79	equipment to install privacy wall.
Demolish and Install	S.F. Flr.	1.75	2.49	.17	4.41	
Reinstall	S.F. Flr.		.69	.13	.82	
Clean	S.F. Flr.	.03	.37		.40	
Paint	S.F.	.11	.30		.41	
Minimum Charge	Job		236		236	
Treated Column or Post						
4" x 4"						
Demolish	L.F.		1.01		1.01	Includes material and labor to install
Install	L.F.	1.38	2.42		3.80	4" x 4" treated column or post.
Demolish and Install	L.F.	1.38	3.43		4.81	
Reinstall	L.F.		1.93		1.93	
Clean	L.F.	.06	.32		.38	
Paint	L.F.	.18	.98		1.16	
Minimum Charge	Job		236		236	
4" x 6"						
Demolish	L.F.		1.47		1.47	Includes material and labor to install
Install	L.F.	2.15	3.43		5.58	4" x 6" treated column or post.
Demolish and Install	L.F.	2.15	4.90		7.05	
Reinstall	L.F.		2.74		2.74	
Clean	L.F.	.07	.45		.52	
Paint	L.F.	.18	.98		1.16	
Minimum Charge	Job		236		236	
Cedar Columns or Posts						
4" x 4"						
Demolish	L.F.		1.01		1.01	Includes material and labor to install
Install	L.F.	4.14	2.42		6.56	4" x 4" cedar columns or posts.
Demolish and Install	L.F.	4.14	3.43		7.57	
Reinstall	L.F.		1.93		1.93	
Clean	L.F.	.06	.32		.38	
Paint	L.F.	.18	.98		1.16	
Minimum Charge	Job		236		236	
4" x 6"						
Demolish	L.F.		1.47		1.47	Includes material and labor to install
Install	L.F.	6.45	3.43		9.88	4" x 6" cedar columns or posts.
Demolish and Install	L.F.	6.45	4.90		11.35	
Reinstall	L.F.		2.74		2.74	
Clean	L.F.	.07	.45		.52	
Paint	L.F.	.18	.98		1.16	
Minimum Charge	Job		236		236	

Wood Deck	Unit	Material	Labor	Equip.	Total	Specification
Redwood Columns or Posts						
4″ x 4″						
Demolish	L.F.		1.01		1.01	Includes material and labor to install
Install	L.F.	7.20	2.42		9.62	4″ x 4″ redwood columns or posts.
Demolish and Install	L.F.	7.20	3.43		10.63	
Reinstall	L.F.		1.93		1.93	
Clean	L.F.	.06	.32		.38	
Paint	L.F.	.18	.98		1.16	
Minimum Charge	Job		236		236	
4″ x 6″						
Demolish	L.F.		1.47		1.47	Includes material and labor to install
Install	L.F.	14.10	3.43		17.53	4″ x 6″ redwood columns or posts.
Demolish and Install	L.F.	14.10	4.90		19	
Reinstall	L.F.		2.74		2.74	
Clean	L.F.	.07	.45		.52	
Paint	L.F.	.18	.98		1.16	
Minimum Charge	Job		236		236	
Treated Girder - Single						
4″ x 8″						
Demolish	L.F.		3.01		3.01	Includes material and labor to install a
Install	L.F.	4.47	1.80		6.27	treated lumber girder consisting of one
Demolish and Install	L.F.	4.47	4.81		9.28	4″ x 8″ for porch or deck construction.
Reinstall	L.F.		1.44		1.44	
Clean	L.F.	.06	.39		.45	
Minimum Charge	Job		236		236	
Treated Girder - Double						
2″ x 8″						
Demolish	L.F.		3.01		3.01	Includes material and labor to install a
Install	L.F.	2.55	1.64		4.19	treated lumber girder consisting of two
Demolish and Install	L.F.	2.55	4.65		7.20	2″ x 8″s for porch or deck
Reinstall	L.F.		1.31		1.31	construction.
Clean	L.F.	.06	.39		.45	
Minimum Charge	Job		236		236	
2″ x 10″						
Demolish	L.F.		3.76		3.76	Includes material and labor to install a
Install	L.F.	3.18	1.71		4.89	treated lumber girder consisting of two
Demolish and Install	L.F.	3.18	5.47		8.65	2″ x 10″s for porch or deck
Reinstall	L.F.		1.37		1.37	construction.
Clean	L.F.	.07	.45		.52	
Minimum Charge	Job		236		236	
2″ x 12″						
Demolish	L.F.		4.51		4.51	Includes material and labor to install a
Install	L.F.	4.71	1.80		6.51	treated lumber girder consisting of two
Demolish and Install	L.F.	4.71	6.31		11.02	2″ x 12″s for porch or deck
Reinstall	L.F.		1.44		1.44	construction.
Clean	L.F.	.08	.52		.60	
Minimum Charge	Job		236		236	
Treated Girder - Triple						
2″ x 8″						
Demolish	L.F.		4.51		4.51	Includes material and labor to install a
Install	L.F.	3.83	1.80		5.63	treated lumber girder consisting of
Demolish and Install	L.F.	3.83	6.31		10.14	three 2″ x 8″s for porch or deck
Reinstall	L.F.		1.44		1.44	construction.
Clean	L.F.	.07	.45		.52	
Minimum Charge	Job		236		236	

Rough Frame / Structure

Wood Deck

		Unit	Material	Labor	Equip.	Total	Specification
2" x 10"							
	Demolish	L.F.		5.65		5.65	Includes material and labor to install a
	Install	L.F.	4.77	1.88		6.65	treated lumber girder consisting of
	Demolish and Install	L.F.	4.77	7.53		12.30	three 2" x 10"s for porch or deck
	Reinstall	L.F.		1.51		1.51	construction.
	Clean	L.F.	.08	.52		.60	
	Minimum Charge	Job		236		236	
2" x 12"							
	Demolish	L.F.		6.70		6.70	Includes material and labor to install a
	Install	L.F.	7.05	1.98		9.03	treated lumber girder consisting of
	Demolish and Install	L.F.	7.05	8.68		15.73	three 2" x 12"s for porch or deck
	Reinstall	L.F.		1.59		1.59	construction.
	Clean	L.F.	.09	.58		.67	
	Minimum Charge	Job		236		236	

Treated Ledger - Bolted

		Unit	Material	Labor	Equip.	Total	Specification
2" x 8"							
	Demolish	L.F.		.50		.50	Includes material and labor to install a
	Install	L.F.	1.42	2.42		3.84	treated lumber 2" x 8" ledger with
	Demolish and Install	L.F.	1.42	2.92		4.34	carriage bolts 4' on center for porch
	Reinstall	L.F.		1.93		1.93	or deck construction.
	Clean	L.F.	.01	.26		.27	
	Minimum Charge	Job		236		236	
2" x 10"							
	Demolish	L.F.		.50		.50	Includes material and labor to install a
	Install	L.F.	1.72	2.45		4.17	treated lumber 2" x 10" ledger with
	Demolish and Install	L.F.	1.72	2.95		4.67	carriage bolts 4' on center for porch
	Reinstall	L.F.		1.96		1.96	or deck construction.
	Clean	L.F.	.02	.30		.32	
	Minimum Charge	Job		236		236	
2" x 12"							
	Demolish	L.F.		.50		.50	Includes material and labor to install a
	Install	L.F.	2.46	2.48		4.94	treated lumber 2" x 12" ledger with
	Demolish and Install	L.F.	2.46	2.98		5.44	carriage bolts 4' on center for porch
	Reinstall	L.F.		1.98		1.98	or deck construction.
	Clean	L.F.	.02	.30		.32	
	Minimum Charge	Job		236		236	

Treated Joists

		Unit	Material	Labor	Equip.	Total	Specification
2" x 8"							
	Demolish	L.F.		.86		.86	Includes material and labor to install
	Install	L.F.	1.28	.86		2.14	treated lumber 2" x 8" deck or floor
	Demolish and Install	L.F.	1.28	1.72		3	joists for porch or deck construction.
	Reinstall	L.F.		.69		.69	
	Clean	L.F.	.06	.32		.38	
	Minimum Charge	Job		236		236	
2" x 10"							
	Demolish	L.F.		.89		.89	Includes material and labor to install
	Install	L.F.	1.60	1.05		2.65	treated lumber 2" x 10" deck or floor
	Demolish and Install	L.F.	1.60	1.94		3.54	joists for porch or deck construction.
	Reinstall	L.F.		.84		.84	
	Clean	L.F.	.06	.39		.45	
	Minimum Charge	Job		236		236	
2" x 12"							
	Demolish	L.F.		.92		.92	Includes material and labor to install
	Install	L.F.	1.90	1.08		2.98	treated lumber 2" x 12" deck or floor
	Demolish and Install	L.F.	1.90	2		3.90	joists for porch or deck construction.
	Reinstall	L.F.		.86		.86	
	Clean	L.F.	.07	.45		.52	
	Minimum Charge	Job		236		236	

For customer support on your Contractor's Pricing Guide: Residential Repair & Remodeling, call 800.448.8182.

Wood Deck

	Unit	Material	Labor	Equip.	Total	Specification
Treated Railings and Trim						
2" x 2"						
Demolish	L.F.		.34		.34	Includes material and labor to install
Install	L.F.	.46	1.57		2.03	treated lumber 2" x 2" railings and
Demolish and Install	L.F.	.46	1.91		2.37	trim for porch or deck construction.
Reinstall	L.F.		1.26		1.26	
Clean	L.F.		.24		.24	
Paint	L.F.	.18	.98		1.16	
Minimum Charge	Job		236		236	
2" x 4"						
Demolish	L.F.		.34		.34	Includes material and labor to install
Install	L.F.	.67	1.57		2.24	treated lumber 2" x 4" railings and
Demolish and Install	L.F.	.67	1.91		2.58	trim for porch or deck construction.
Reinstall	L.F.		1.26		1.26	
Clean	L.F.	.01	.26		.27	
Paint	L.F.	.18	.98		1.16	
Minimum Charge	Job		236		236	
2" x 6"						
Demolish	L.F.		.34		.34	Includes material and labor to install
Install	L.F.	.85	1.57		2.42	treated lumber 2" x 6" railings and
Demolish and Install	L.F.	.85	1.91		2.76	trim for porch or deck construction.
Reinstall	L.F.		1.26		1.26	
Clean	L.F.	.01	.30		.31	
Paint	L.F.	.18	.98		1.16	
Minimum Charge	Job		236		236	
Cedar Railings and Trim						
1" x 4"						
Demolish	L.F.		.34		.34	Includes material and labor to install
Install	L.F.	3.12	1.57		4.69	cedar 1" x 4" railings and trim for
Demolish and Install	L.F.	3.12	1.91		5.03	porch or deck construction.
Reinstall	L.F.		1.26		1.26	
Clean	L.F.		.24		.24	
Paint	L.F.	.18	.98		1.16	
Minimum Charge	Job		236		236	
2" x 4"						
Demolish	L.F.		.34		.34	Includes material and labor to install
Install	L.F.	6.40	1.57		7.97	cedar 2" x 4" railings and trim for
Demolish and Install	L.F.	6.40	1.91		8.31	porch or deck construction.
Reinstall	L.F.		1.26		1.26	
Clean	L.F.	.01	.26		.27	
Paint	L.F.	.18	.98		1.16	
Minimum Charge	Job		236		236	
2" x 6"						
Demolish	L.F.		.34		.34	Includes material and labor to install
Install	L.F.	11.60	1.57		13.17	cedar 2" x 6" railings and trim for
Demolish and Install	L.F.	11.60	1.91		13.51	porch or deck construction.
Reinstall	L.F.		1.26		1.26	
Clean	L.F.	.01	.30		.31	
Paint	L.F.	.18	.98		1.16	
Minimum Charge	Job		236		236	
Redwood Railings and Trim						
1" x 4"						
Demolish	L.F.		.34		.34	Includes material and labor to install
Install	L.F.	1.31	1.57		2.88	redwood 1" x 4" railings and trim for
Demolish and Install	L.F.	1.31	1.91		3.22	porch or deck construction.
Reinstall	L.F.		1.26		1.26	
Clean	L.F.		.24		.24	
Paint	L.F.	.18	.98		1.16	
Minimum Charge	Job		236		236	

Rough Frame / Structure

Wood Deck		Unit	Material	Labor	Equip.	Total	Specification
2" x 6"							
	Demolish	L.F.		.34		.34	Includes material and labor to install
	Install	L.F.	8.35	1.57		9.92	redwood 2" x 6" railings and trim for
	Demolish and Install	L.F.	8.35	1.91		10.26	porch or deck construction.
	Reinstall	L.F.		1.26		1.26	
	Clean	L.F.	.01	.30		.31	
	Paint	L.F.	.18	.98		1.16	
	Minimum Charge	Job		236		236	
Treated Decking							
1" x 4"							
	Demolish	L.F.		.49		.49	Includes material and labor to install
	Install	S.F.	3.21	1.71		4.92	treated lumber 1" x 4" decking for
	Demolish and Install	S.F.	3.21	2.20		5.41	porch or deck construction.
	Reinstall	S.F.		1.37		1.37	
	Clean	S.F.	.03	.37		.40	
	Paint	S.F.	.23	.78		1.01	
	Minimum Charge	Job		236		236	
2" x 4"							
	Demolish	L.F.		.49		.49	Includes material and labor to install
	Install	S.F.	2.26	1.57		3.83	treated lumber 2" x 4" decking for
	Demolish and Install	S.F.	2.26	2.06		4.32	porch or deck construction.
	Reinstall	S.F.		1.26		1.26	
	Clean	S.F.	.03	.37		.40	
	Paint	S.F.	.23	.78		1.01	
	Minimum Charge	Job		236		236	
2" x 6"							
	Demolish	L.F.		.49		.49	Includes material and labor to install
	Install	S.F.	1.82	1.47		3.29	treated lumber 2" x 6" decking for
	Demolish and Install	S.F.	1.82	1.96		3.78	porch or deck construction.
	Reinstall	S.F.		1.18		1.18	
	Clean	S.F.	.03	.37		.40	
	Paint	S.F.	.23	.78		1.01	
	Minimum Charge	Job		236		236	
5/4" x 6"							
	Demolish	L.F.		.49		.49	Includes material and labor to install
	Install	S.F.	2.39	1.47		3.86	treated lumber 5/4" x 6" decking for
	Demolish and Install	S.F.	2.39	1.96		4.35	porch or deck construction.
	Reinstall	S.F.		1.18		1.18	
	Clean	S.F.	.03	.37		.40	
	Paint	S.F.	.23	.78		1.01	
	Minimum Charge	Job		236		236	
Wood / Plastic Composite Decking							
5/4" x 6"							
	Demolish	L.F.		.49		.49	Includes material and labor to install
	Install	L.F.	3.77	1.47		5.24	wood/plastic composite 5/4" x 6"
	Demolish and Install	L.F.	3.77	1.96		5.73	decking for porch or deck
	Reinstall	L.F.		1.18		1.18	construction.
	Clean	S.F.	.03	.37		.40	
	Paint	S.F.	.23	.78		1.01	
	Minimum Charge	Job		236		236	
Square Edge Fir Decking							
1" x 4"							
	Demolish	L.F.		.49		.49	Includes material and labor to install
	Install	L.F.	3.22	1.71		4.93	square edge fir 1" x 4" decking for
	Demolish and Install	L.F.	3.22	2.20		5.42	porch or deck construction.
	Reinstall	L.F.		1.37		1.37	
	Clean	S.F.	.03	.37		.40	
	Paint	S.F.	.23	.78		1.01	
	Minimum Charge	Job		236		236	

For customer support on your Contractor's Pricing Guide: Residential Repair & Remodeling, call 800.448.8182.

Wood Deck	Unit	Material	Labor	Equip.	Total	Specification
Tongue and Groove Fir Decking						
1″ x 4″						
Demolish	L.F.		.49		.49	Includes material and labor to install
Install	L.F.	1.69	2.09		3.78	treated lumber 1″ x 4″ decking for
Demolish and Install	L.F.	1.69	2.58		4.27	porch or deck construction.
Reinstall	L.F.		1.68		1.68	
Clean	S.F.	.03	.37		.40	
Paint	S.F.	.23	.78		1.01	
Minimum Charge	Job		236		236	
Mahogany Decking						
1″ x 4″						
Demolish	L.F.		.49		.49	Includes material and labor to install
Install	L.F.	2.30	1.71		4.01	treated lumber 1″ x 4″ decking for
Demolish and Install	L.F.	2.30	2.20		4.50	porch or deck construction.
Reinstall	L.F.		1.37		1.37	
Clean	S.F.	.03	.37		.40	
Paint	S.F.	.23	.78		1.01	
Minimum Charge	Job		236		236	
PVC Decking						
5/4″ x 6″						
Demolish	L.F.		.49		.49	Includes material and labor to install
Install	L.F.	3.69	1.71		5.40	PVC 5/4″ x 6″ decking for porch or
Demolish and Install	L.F.	3.69	2.20		5.89	deck construction.
Reinstall	L.F.		1.37		1.37	
Clean	S.F.	.03	.37		.40	
Minimum Charge	Job		236		236	
PVC Handrails and Balusters						
Straight						
Demolish	L.F.		3.37		3.37	Includes material and labor to install
Install	L.F.	32.50	4.91		37.41	straight PVC handrails and balusters
Demolish and Install	L.F.	32.50	8.28		40.78	for porch or deck construction.
Reinstall	L.F.		3.93		3.93	
Clean	L.F.	.30	2.84		3.14	
Minimum Charge	Job		236		236	
Angled						
Demolish	L.F.		3.37		3.37	Includes material and labor to install
Install	L.F.	22	6.55		28.55	angled PVC handrails and balusters
Demolish and Install	L.F.	22	9.92		31.92	for porch or deck construction.
Reinstall	L.F.		5.24		5.24	
Clean	L.F.	.30	2.84		3.14	
Minimum Charge	Job		236		236	
PVC Post Sleeve						
for 4″ x 4″ post						
Demolish	L.F.		1.01		1.01	Includes material and labor to install
Install	L.F.	17.25	4.91		22.16	PVC post sleeves over 4″ x 4″ porch
Demolish and Install	L.F.	17.25	5.92		23.17	or deck posts.
Reinstall	L.F.		3.93		3.93	
Clean	L.F.	.06	.32		.38	
Minimum Charge	Job		236		236	
Post Cap						
Demolish	Ea.		5.05		5.05	Includes material and labor to install
Install	Ea.	29	9.80		38.80	PVC post caps for porch or deck
Demolish and Install	Ea.	29	14.85		43.85	construction.
Reinstall	Ea.		7.85		7.85	
Clean	L.F.	.06	.32		.38	
Minimum Charge	Job		236		236	

Porch	Unit	Material	Labor	Equip.	Total	Specification
Stairs						
Demolish	Ea.		465		465	Includes material and labor to install
Install	Ea.	160	835		995	wood steps. 5 - 7 risers included.
Demolish and Install	Ea.	160	1300		1460	
Reinstall	Ea.		668.80		668.80	
Clean	Flight	1.50	30.50		32	
Paint	Flight	21.50	52		73.50	
Minimum Charge	Job		236		236	
Bench Seating						
Demolish	L.F.		6.75		6.75	Includes material and labor to install
Install	L.F.	2.33	23.50		25.83	pressure treated bench seating.
Demolish and Install	L.F.	2.33	30.25		32.58	
Reinstall	L.F.		18.85		18.85	
Clean	L.F.	.06	1.82		1.88	
Paint	L.F.	.68	3.13		3.81	
Minimum Charge	Job		236		236	
Privacy Wall						
Demolish	S.F. Flr.		1.62		1.62	Includes material, labor and
Install	S.F. Flr.	1.75	.87	.17	2.79	equipment to install privacy wall.
Demolish and Install	S.F. Flr.	1.75	2.49	.17	4.41	
Reinstall	S.F. Flr.		.69	.13	.82	
Clean	S.F. Flr.	.03	.37		.40	
Paint	S.F.	.11	.30		.41	
Minimum Charge	Job		236		236	
Wood Board Ceiling						
Demolish	S.F.		.67		.67	Includes material and labor to install
Install	S.F.	3.74	2.09		5.83	wood board on ceiling.
Demolish and Install	S.F.	3.74	2.76		6.50	
Clean	S.F.	.01	.21		.22	
Paint	S.F.	.20	.49		.69	
Minimum Charge	Job		236		236	
Re-screen						
Install	S.F.	1.85	1.88		3.73	Includes labor and material to
Minimum Charge	Job		236		236	re-screen wood frame.

For customer support on your Contractor's Pricing Guide: Residential Repair & Remodeling, call 800.448.8182.

Exterior Trim

Exterior Trim	Unit	Material	Labor	Equip.	Total	Specification
Stock Lumber						
1" x 2"						
Demolish	L.F.		.34		.34	Cost includes material and labor to
Install	L.F.	.32	1.43		1.75	install 1" x 2" pine trim.
Demolish and Install	L.F.	.32	1.77		2.09	
Clean	L.F.		.24		.24	
Paint	L.F.	.02	.61		.63	
Minimum Charge	Job		236		236	
1" x 3"						
Demolish	L.F.		.34		.34	Cost includes material and labor to
Install	L.F.	.34	1.63		1.97	install 1" x 3" pine trim.
Demolish and Install	L.F.	.34	1.97		2.31	
Clean	L.F.		.24		.24	
Paint	L.F.	.02	.61		.63	
Minimum Charge	Job		236		236	
1" x 4"						
Demolish	L.F.		.34		.34	Cost includes material and labor to
Install	L.F.	.62	1.88		2.50	install 1" x 4" pine trim.
Demolish and Install	L.F.	.62	2.22		2.84	
Clean	L.F.		.24		.24	
Paint	L.F.	.02	.61		.63	
Minimum Charge	Job		236		236	
1" x 6"						
Demolish	L.F.		.34		.34	Cost includes material and labor to
Install	L.F.	.90	1.88		2.78	install 1" x 6" pine trim.
Demolish and Install	L.F.	.90	2.22		3.12	
Clean	L.F.	.01	.24		.25	
Paint	L.F.	.09	.61		.70	
Minimum Charge	Job		236		236	
1" x 8"						
Demolish	L.F.		.37		.37	Cost includes material and labor to
Install	L.F.	1.52	2.36		3.88	install 1" x 8" pine trim.
Demolish and Install	L.F.	1.52	2.73		4.25	
Reinstall	L.F.		1.88		1.88	
Clean	L.F.	.01	.26		.27	
Paint	L.F.	.09	.61		.70	
Minimum Charge	Job		236		236	

Exterior Trim		Unit	Material	Labor	Equip.	Total	Specification
Stock Lumber							
1" x 10"							
	Demolish	L.F.		.56		.56	Cost includes material and labor to
	Install	L.F.	1.94	2.62		4.56	install 1" x 10" pine trim.
	Demolish and Install	L.F.	1.94	3.18		5.12	
	Minimum Charge	Job		236		236	
1" x 12"							
	Demolish	L.F.		.56		.56	Cost includes material and labor to
	Install	L.F.	2.42	2.62		5.04	install 1" x 12" pine trim.
	Demolish and Install	L.F.	2.42	3.18		5.60	
	Minimum Charge	Job		236		236	
PVC							
1" x 4"							
	Demolish	L.F.		.34		.34	Includes material and labor to install
	Install	L.F.	1.79	1.88		3.67	1" x 4" PVC exterior trim.
	Demolish and Install	L.F.	1.79	2.22		4.01	
	Reinstall	L.F.		1.51		1.51	
	Clean	L.F.		.24		.24	
	Paint	L.F.	.02	.61		.63	
	Minimum Charge	Job		236		236	
1" x 6"							
	Demolish	L.F.		.34		.34	Includes material and labor to install
	Install	L.F.	2.32	1.88		4.20	1" x 6" PVC exterior trim.
	Demolish and Install	L.F.	2.32	2.22		4.54	
	Reinstall	L.F.		1.51		1.51	
	Clean	L.F.	.01	.24		.25	
	Paint	L.F.	.09	.61		.70	
	Minimum Charge	Job		236		236	
1" x 8"							
	Demolish	L.F.		.37		.37	Includes material and labor to install
	Install	L.F.	3.28	2.09		5.37	1" x 8" PVC exterior trim.
	Demolish and Install	L.F.	3.28	2.46		5.74	
	Reinstall	L.F.		1.68		1.68	
	Clean	L.F.	.01	.26		.27	
	Paint	L.F.	.09	.61		.70	
	Minimum Charge	Job		236		236	
1" x 10"							
	Demolish	L.F.		.38		.38	Includes material and labor to install
	Install	L.F.	4.22	2.09		6.31	1" x 10" PVC exterior trim.
	Demolish and Install	L.F.	4.22	2.47		6.69	
	Reinstall	L.F.		1.68		1.68	
	Clean	L.F.	.02	.30		.32	
	Paint	L.F.	.09	.61		.70	
	Minimum Charge	Job		236		236	
1" x 12"							
	Demolish	L.F.		.40		.40	Includes material and labor to install
	Install	L.F.	4.62	2.36		6.98	1" x 12" PVC exterior trim.
	Demolish and Install	L.F.	4.62	2.76		7.38	
	Reinstall	L.F.		1.88		1.88	
	Clean	L.F.	.02	.30		.32	
	Paint	L.F.	.09	.61		.70	
	Minimum Charge	Job		236		236	

Fascia		Unit	Material	Labor	Equip.	Total	Specification
Aluminum							
	Demolish	L.F.		.70		.70	Includes material and labor to install
	Install	S.F.	2.29	2.24		4.53	aluminum fascia.
	Demolish and Install	S.F.	2.29	2.94		5.23	
	Reinstall	S.F.		1.80		1.80	
	Clean	L.F.		.45		.45	
	Minimum Charge	Job		236		236	
Vinyl							
	Demolish	L.F.		.70		.70	Cost includes material and labor to
	Install	L.F.	5.05	2.69		7.74	install 6″ vinyl fascia with 12″ vinyl
	Demolish and Install	L.F.	5.05	3.39		8.44	soffit and J-channel.
	Reinstall	L.F.		2.15		2.15	
	Clean	L.F.		.45		.45	
	Minimum Charge	Job		236		236	
Plywood							
	Demolish	L.F.		.70		.70	Cost includes material and labor to
	Install	L.F.	1.68	1.05		2.73	install 12″ wide plywood fascia.
	Demolish and Install	L.F.	1.68	1.75		3.43	
	Reinstall	L.F.		.84		.84	
	Clean	L.F.	.02	.30		.32	
	Paint	L.F.	.18	.98		1.16	
	Minimum Charge	Job		236		236	
Cedar							
	Demolish	L.F.		.70		.70	Cost includes material and labor to
	Install	L.F.	4.86	1.05		5.91	install 12″ wide cedar fascia.
	Demolish and Install	L.F.	4.86	1.75		6.61	
	Reinstall	L.F.		.84		.84	
	Clean	L.F.	.02	.30		.32	
	Paint	L.F.	.18	.98		1.16	
	Minimum Charge	Job		236		236	
Redwood							
1″ x 6″							
	Demolish	L.F.		.54		.54	Cost includes material and labor to
	Install	L.F.	1.99	1.88		3.87	install 1″ x 6″ redwood fascia board.
	Demolish and Install	L.F.	1.99	2.42		4.41	
	Reinstall	L.F.		1.51		1.51	
	Minimum Charge	Job		236		236	
1″ x 8″							
	Demolish	L.F.		.61		.61	Cost includes material and labor to
	Install	L.F.	2.65	2.05		4.70	install 1″ x 8″ redwood fascia board.
	Demolish and Install	L.F.	2.65	2.66		5.31	
	Reinstall	L.F.		1.64		1.64	
	Minimum Charge	Job		236		236	
2″ x 6″							
	Demolish	L.F.		.54		.54	Cost includes material and labor to
	Install	L.F.	.79	3.77		4.56	install 2″ x 6″ exterior trim board.
	Demolish and Install	L.F.	.79	4.31		5.10	
	Reinstall	L.F.		3.02		3.02	
	Minimum Charge	Job		236		236	
2″ x 8″							
	Demolish	L.F.		.61		.61	Cost includes material and labor to
	Install	L.F.	1.07	4.19		5.26	install 2″ x 8″ exterior trim board.
	Demolish and Install	L.F.	1.07	4.80		5.87	
	Reinstall	L.F.		3.35		3.35	
	Minimum Charge	Job		236		236	

Fascia	Unit	Material	Labor	Equip.	Total	Specification
Hem-fir Std & Better						
2" x 6"						
Demolish	L.F.		.54		.54	Cost includes material and labor to
Install	L.F.	.79	3.77		4.56	install 2" x 6" exterior trim board.
Demolish and Install	L.F.	.79	4.31		5.10	
Reinstall	L.F.		3.02		3.02	
Clean	L.F.	.01	.30		.31	
Paint	L.F.	.18	.98		1.16	
Minimum Charge	Job		236		236	
2" x 8"						
Demolish	L.F.		.61		.61	Cost includes material and labor to
Install	L.F.	1.07	4.19		5.26	install 2" x 8" exterior trim board.
Demolish and Install	L.F.	1.07	4.80		5.87	
Reinstall	L.F.		3.35		3.35	
Clean	L.F.	.01	.30		.31	
Paint	L.F.	.18	.98		1.16	
Minimum Charge	Job		236		236	
2" x 10"						
Demolish	L.F.		.70		.70	Cost includes material and labor to
Install	L.F.	1.60	5.25		6.85	install 2" x 10" exterior trim board.
Demolish and Install	L.F.	1.60	5.95		7.55	
Reinstall	L.F.		4.19		4.19	
Clean	L.F.	.02	.33		.35	
Paint	L.F.	.18	.98		1.16	
Minimum Charge	Job		236		236	
Barge Rafter						
2" x 6"						
Demolish	L.F.		1.08		1.08	Cost includes material and labor to
Install	L.F.	.79	.52		1.31	install 2" x 6" barge rafters.
Demolish and Install	L.F.	.79	1.60		2.39	
Reinstall	L.F.		.42		.42	
Clean	L.F.	.01	.30		.31	
Paint	L.F.	.18	.98		1.16	
Minimum Charge	Job		236		236	
2" x 8"						
Demolish	L.F.		1.10		1.10	Cost includes material and labor to
Install	L.F.	1.07	.55		1.62	install 2" x 8" barge rafters.
Demolish and Install	L.F.	1.07	1.65		2.72	
Reinstall	L.F.		.44		.44	
Clean	L.F.	.01	.30		.31	
Paint	L.F.	.18	.98		1.16	
Minimum Charge	Job		236		236	
2" x 10"						
Demolish	L.F.		1.11		1.11	Cost includes material and labor to
Install	L.F.	1.60	.57		2.17	install 2" x 10" barge rafters.
Demolish and Install	L.F.	1.60	1.68		3.28	
Reinstall	L.F.		.46		.46	
Clean	L.F.	.02	.33		.35	
Paint	L.F.	.18	.98		1.16	
Minimum Charge	Job		236		236	

Soffit	Unit	Material	Labor	Equip.	Total	Specification
Aluminum						
12" w / Fascia						
Demolish	L.F.		.70		.70	Cost includes material and labor to
Install	L.F.	5.15	2.69		7.84	install 12" aluminum soffit with 6"
Demolish and Install	L.F.	5.15	3.39		8.54	fascia and J-channel.
Reinstall	L.F.		2.15		2.15	
Clean	L.F.		.45		.45	
Paint	L.F.	.22	.98		1.20	
Minimum Charge	Job		236		236	
18" w / Fascia						
Demolish	L.F.		.70		.70	Cost includes material and labor to
Install	L.F.	6.20	3.14		9.34	install 18" aluminum soffit with 6"
Demolish and Install	L.F.	6.20	3.84		10.04	fascia and J-channel.
Reinstall	L.F.		2.51		2.51	
Clean	L.F.		.45		.45	
Paint	L.F.	.22	.98		1.20	
Minimum Charge	Job		236		236	
24" w / Fascia						
Demolish	L.F.		.70		.70	Cost includes material and labor to
Install	L.F.	7.30	3.49		10.79	install 24" aluminum soffit with 6"
Demolish and Install	L.F.	7.30	4.19		11.49	fascia and J-channel.
Reinstall	L.F.		2.79		2.79	
Clean	L.F.		.66		.66	
Paint	L.F.	.22	.98		1.20	
Minimum Charge	Job		236		236	
Vinyl						
Demolish	L.F.		.70		.70	Cost includes material and labor to
Install	L.F.	5.05	2.69		7.74	install 6" vinyl fascia with 12" vinyl
Demolish and Install	L.F.	5.05	3.39		8.44	soffit and J-channel.
Reinstall	L.F.		2.15		2.15	
Clean	L.F.		.45		.45	
Paint	L.F.	.22	.98		1.20	
Minimum Charge	Job		236		236	
Plywood						
Demolish	L.F.		.78		.78	Cost includes material and labor to
Install	L.F.	1.62	1.35		2.97	install 12" wide plywood soffit.
Demolish and Install	L.F.	1.62	2.13		3.75	
Reinstall	L.F.		1.08		1.08	
Clean	L.F.	.02	.30		.32	
Paint	L.F.	.18	.98		1.16	
Minimum Charge	Job		236		236	
Redwood						
Demolish	L.F.		.78		.78	Cost includes material and labor to
Install	L.F.	5.25	1.35		6.60	install 12" wide redwood soffit.
Demolish and Install	L.F.	5.25	2.13		7.38	
Reinstall	L.F.		1.08		1.08	
Clean	L.F.	.02	.30		.32	
Paint	L.F.	.18	.98		1.16	
Minimum Charge	Job		236		236	
Cedar						
Demolish	L.F.		.78		.78	Cost includes material and labor to
Install	L.F.	4.80	1.35		6.15	install 12" wide cedar soffit.
Demolish and Install	L.F.	4.80	2.13		6.93	
Reinstall	L.F.		1.08		1.08	
Clean	L.F.	.02	.30		.32	
Paint	L.F.	.18	.98		1.16	
Minimum Charge	Job		236		236	

Exterior Trim

Soffit

Douglas Fir

	Unit	Material	Labor	Equip.	Total	Specification
Demolish	L.F.		.78		.78	Cost includes material and labor to
Install	L.F.	2.35	1.35		3.70	install 12″ wide Douglas fir soffit.
Demolish and Install	L.F.	2.35	2.13		4.48	
Reinstall	L.F.		1.08		1.08	
Clean	L.F.	.02	.30		.32	
Paint	L.F.	.18	.98		1.16	
Minimum Charge	Job		236		236	

For customer support on your Contractor's Pricing Guide: Residential Repair & Remodeling, call 800.448.8182.

Roofing and Flashing

Tile Roofing

Metal Flashing		Unit	Material	Labor	Equip.	Total	Specification
Galvanized							
6"							
	Demolish	L.F.		.27		.27	Includes material and labor to install
	Install	L.F.	.68	1.11		1.79	flat galvanized steel flashing, 6" wide.
	Demolish and Install	L.F.	.68	1.38		2.06	
	Paint	L.F.	.12	.49		.61	
	Minimum Charge	Job		221		221	
14"							
	Demolish	L.F.		.27		.27	Includes material and labor to install
	Install	L.F.	1.56	1.11		2.67	flat galvanized steel flashing, 14"
	Demolish and Install	L.F.	1.56	1.38		2.94	wide.
	Paint	L.F.	.12	.49		.61	
	Minimum Charge	Job		221		221	
20"							
	Demolish	L.F.		.27		.27	Includes material and labor to install
	Install	L.F.	2.33	1.11		3.44	flat galvanized steel flashing, 20"
	Demolish and Install	L.F.	2.33	1.38		3.71	wide.
	Paint	L.F.	.12	.49		.61	
	Minimum Charge	Job		221		221	
Aluminum							
6"							
	Demolish	L.F.		.27		.27	Includes material and labor to install
	Install	L.F.	.43	1.11		1.54	flat aluminum flashing, 6" wide.
	Demolish and Install	L.F.	.43	1.38		1.81	
	Paint	L.F.	.12	.49		.61	
	Minimum Charge	Job		221		221	
14"							
	Demolish	L.F.		.27		.27	Includes material and labor to install
	Install	L.F.	.85	1.11		1.96	flat aluminum flashing, 14" wide.
	Demolish and Install	L.F.	.85	1.38		2.23	
	Paint	L.F.	.12	.49		.61	
	Minimum Charge	Job		221		221	
20"							
	Demolish	L.F.		.27		.27	Includes material and labor to install
	Install	L.F.	1.16	1.11		2.27	flat aluminum flashing, 20" wide.
	Demolish and Install	L.F.	1.16	1.38		2.54	
	Paint	L.F.	.12	.49		.61	
	Minimum Charge	Job		221		221	

Metal Flashing	Unit	Material	Labor	Equip.	Total	Specification
Copper						
6"						
Demolish	L.F.		.27		.27	Includes material and labor to install
Install	L.F.	4.46	1.48		5.94	flat copper flashing, 6" wide.
Demolish and Install	L.F.	4.46	1.75		6.21	
Minimum Charge	Job		259		259	
14"						
Demolish	L.F.		.27		.27	Includes material and labor to install
Install	L.F.	10.40	1.48		11.88	flat copper flashing, 14" wide.
Demolish and Install	L.F.	10.40	1.75		12.15	
Minimum Charge	Job		259		259	
18"						
Demolish	L.F.		.27		.27	Includes material and labor to install
Install	L.F.	14.85	1.48		16.33	flat copper flashing, 18" wide.
Demolish and Install	L.F.	14.85	1.75		16.60	
Minimum Charge	Job		259		259	
Aluminum Valley						
Demolish	L.F.		.27		.27	Includes material and labor to install
Install	L.F.	4.43	2.77		7.20	aluminum valley flashing, .024" thick.
Demolish and Install	L.F.	4.43	3.04		7.47	
Paint	L.F.	.12	.49		.61	
Minimum Charge	Job		236		236	
Gravel Stop						
Demolish	L.F.		.43		.43	Includes material and labor to install
Install	L.F.	7.20	3.57		10.77	aluminum gravel stop, .050" thick, 4"
Demolish and Install	L.F.	7.20	4		11.20	high, mill finish.
Reinstall	L.F.		2.86		2.86	
Paint	L.F.	.20	.49		.69	
Minimum Charge	Job		236		236	
Drip Edge						
Demolish	L.F.		.40		.40	Includes material and labor to install
Install	L.F.	.65	1.18		1.83	aluminum drip edge, .016" thick, 5"
Demolish and Install	L.F.	.65	1.58		2.23	wide, mill finish.
Reinstall	L.F.		.94		.94	
Paint	L.F.	.20	.49		.69	
Minimum Charge	Job		236		236	
End Wall						
Demolish	L.F.		1.24		1.24	Includes material and labor to install
Install	L.F.	2.53	2.77		5.30	aluminum end wall flashing, .024"
Demolish and Install	L.F.	2.53	4.01		6.54	thick.
Minimum Charge	Job		236		236	
Side Wall						
Demolish	L.F.		.24		.24	Includes material and labor to install
Install	L.F.	2.46	2.77		5.23	aluminum side wall flashing, .024"
Demolish and Install	L.F.	2.46	3.01		5.47	thick.
Minimum Charge	Job		236		236	
Coping and Wall Cap						
Demolish	L.F.		1.68		1.68	Includes material and labor to install
Install	L.F.	2.53	3.05		5.58	aluminum flashing, mill finish, 0.40"
Demolish and Install	L.F.	2.53	4.73		7.26	thick, 12" wide.
Clean	L.F.	.03	1.14		1.17	
Minimum Charge	Job		221		221	

Roof Shingles

Roof Shingles	Unit	Material	Labor	Equip.	Total	Specification
Underlayment						
#15						
Demolish	Sq.		13.45		13.45	Includes material and labor to install
Install	Sq.	5.95	6.90		12.85	#15 felt underlayment.
Demolish and Install	Sq.	5.95	20.35		26.30	
Minimum Charge	Job		221		221	
#30						
Demolish	Sq.		13.45		13.45	Includes material and labor to install
Install	Sq.	11.60	7.65		19.25	#30 felt underlayment.
Demolish and Install	Sq.	11.60	21.10		32.70	
Minimum Charge	Job		221		221	
Self Adhering						
Demolish	Sq.		33.50		33.50	Includes material and labor to install
Install	Sq.	87	20		107	self adhering ice barrier roofing
Demolish and Install	Sq.	87	53.50		140.50	underlayment.
Minimum Charge	Job		221		221	

Rolled Roofing

Rolled Roofing	Unit	Material	Labor	Equip.	Total	Specification
90 lb.						
Demolish	Sq.		33.50		33.50	Cost includes material and labor to
Install	Sq.	39	37		76	install 90 lb. mineral surface rolled
Demolish and Install	Sq.	39	70.50		109.50	roofing.
Minimum Charge	Job		221		221	

Composition Shingles

Composition Shingles	Unit	Material	Labor	Equip.	Total	Specification
Fiberglass (3-tab)						
25 Year						
Demolish	Sq.		36.50		36.50	Cost includes material and labor to
Install	Sq.	81	49.50		130.50	install 25 year fiberglass shingles.
Demolish and Install	Sq.	81	86		167	
Minimum Charge	Job		221		221	
30 Year						
Demolish	Sq.		36.50		36.50	Cost includes material and labor to
Install	Sq.	98	55.50		153.50	install 30 year fiberglass shingles.
Demolish and Install	Sq.	98	92		190	
Minimum Charge	Job		221		221	
Emergency Repairs						
Install	Ea.	174	221		395	Includes labor and material for
Minimum Charge	Job		221		221	emergency repairs.
Architectural						
Laminated, 25 Year						
Demolish	Sq.		36.50		36.50	Cost includes material and labor to
Install	Sq.	113	61.50		174.50	install 25 year fiberglass roof shingles.
Demolish and Install	Sq.	113	98		211	
Minimum Charge	Job		221		221	
Laminated, 30 Year						
Demolish	Sq.		36.50		36.50	Cost includes material and labor to
Install	Sq.	148	79		227	install 30 year laminated roof
Demolish and Install	Sq.	148	115.50		263.50	shingles.
Minimum Charge	Job		221		221	

73

Composition Shingles		Unit	Material	Labor	Equip.	Total	Specification
Laminated, 40 Year							
	Demolish	Sq.		36.50		36.50	Cost includes material and labor to
	Install	Sq.	280	92		372	install 40 year laminated roof
	Demolish and Install	Sq.	280	128.50		408.50	shingles.
	Minimum Charge	Job		221		221	
Laminated Shake, 40 Year							
	Demolish	Sq.		36.50		36.50	Cost includes material and labor to
	Install	Sq.	280	92		372	install 40 year laminated roof
	Demolish and Install	Sq.	280	128.50		408.50	shingles.
	Minimum Charge	Job		221		221	
Emergency Repairs							
	Install	Ea.	174	221		395	Includes labor and material for
	Minimum Charge	Job		221		221	emergency repairs.
Roof Jack							
	Demolish	Ea.		7.35		7.35	Includes material and labor to install
	Install	Ea.	22.50	47		69.50	residential roof jack, w/bird screen,
	Demolish and Install	Ea.	22.50	54.35		76.85	backdraft damper, 3" & 4" diam.
	Paint	Ea.	.07	4.89		4.96	round duct.
	Minimum Charge	Job		270		270	
Ridge Vent							
	Demolish	L.F.		1.30		1.30	Includes material and labor to install a
	Install	L.F.	4.33	3.34		7.67	mill finish aluminum ridge vent strip.
	Demolish and Install	L.F.	4.33	4.64		8.97	
	Minimum Charge	Job		236		236	
Shingle Molding							
1" x 2"							
	Demolish	L.F.		.25		.25	Cost includes material and labor to
	Install	L.F.	.32	1.43		1.75	install 1" x 2" pine trim.
	Demolish and Install	L.F.	.32	1.68		2	
	Paint	L.F.	.02	.61		.63	
	Minimum Charge	Job		236		236	
1" x 3"							
	Demolish	L.F.		.25		.25	Cost includes material and labor to
	Install	L.F.	.34	1.63		1.97	install 1" x 3" pine trim.
	Demolish and Install	L.F.	.34	1.88		2.22	
	Paint	L.F.	.02	.61		.63	
	Minimum Charge	Job		236		236	
1" x 4"							
	Demolish	L.F.		.25		.25	Cost includes material and labor to
	Install	L.F.	.62	1.88		2.50	install 1" x 4" pine trim.
	Demolish and Install	L.F.	.62	2.13		2.75	
	Paint	L.F.	.02	.61		.63	
	Minimum Charge	Job		236		236	
Add for Steep Pitch							
	Install	Sq.		27.50		27.50	Includes labor and material added for
Add for Additional Story							2nd story or steep roofs.
	Install	Sq.		27.50		27.50	Includes labor and material added for
							2nd story or steep roofs.

Wood Shingles	Unit	Material	Labor	Equip.	Total	Specification
Cedar Wood Shingles						
16" Red Label (#2)						
Demolish	Sq.		90		90	Includes material and labor to install
Install	Sq.	175	188		363	wood shingles/shakes.
Demolish and Install	Sq.	175	278		453	
Minimum Charge	Job		236		236	
16" Blue Label (#1)						
Demolish	Sq.		90		90	Includes material and labor to install
Install	Sq.	325	145		470	wood shingles/shakes.
Demolish and Install	Sq.	325	235		560	
Minimum Charge	Job		236		236	
18" Red Label (#2)						
Demolish	Sq.		90		90	Includes material and labor to install
Install	Sq.	315	121		436	wood shingles/shakes.
Demolish and Install	Sq.	315	211		526	
Minimum Charge	Job		236		236	
18" Blue Label (#1)						
Demolish	Sq.		90		90	Includes material and labor to install
Install	Sq.	279	132		411	wood shingles/shakes.
Demolish and Install	Sq.	279	222		501	
Minimum Charge	Job		236		236	
Replace Shingle						
Install	Ea.	81	11.80		92.80	Includes labor and material to replace
Minimum Charge	Job		236		236	individual wood shingle/shake. Cost
						is per shingle.
Cedar Shake						
18" Medium Handsplit						
Demolish	Sq.		90		90	Includes material and labor to install
Install	Sq.	305	181		486	wood shingles/shakes.
Demolish and Install	Sq.	305	271		576	
Minimum Charge	Job		236		236	
18" Heavy Handsplit						
Demolish	Sq.		90		90	Includes material and labor to install
Install	Sq.	305	201		506	wood shingles/shakes.
Demolish and Install	Sq.	305	291		596	
Minimum Charge	Job		236		236	
24" Heavy Handsplit						
Demolish	Sq.		90		90	Includes material and labor to install
Install	Sq.	330	161		491	wood shingles/shakes.
Demolish and Install	Sq.	330	251		581	
Minimum Charge	Job		236		236	
24" Medium Handsplit						
Demolish	Sq.		90		90	Includes material and labor to install
Install	Sq.	330	145		475	wood shingles/shakes.
Demolish and Install	Sq.	330	235		565	
Minimum Charge	Job		236		236	
24" Straight Split						
Demolish	Sq.		90		90	Includes material and labor to install
Install	Sq.	345	214		559	wood shingles/shakes.
Demolish and Install	Sq.	345	304		649	
Minimum Charge	Job		236		236	
24" Tapersplit						
Demolish	Sq.		90		90	Includes material and labor to install
Install	Sq.	298	214		512	wood shingles/shakes.
Demolish and Install	Sq.	298	304		602	
Minimum Charge	Job		236		236	

Roofing

Wood Shingles

Wood Shingles		Unit	Material	Labor	Equip.	Total	Specification
18" Straight Split							
	Demolish	Sq.		90		90	Includes material and labor to install
	Install	Sq.	330	214		544	wood shingles/shakes.
	Demolish and Install	Sq.	330	304		634	
	Minimum Charge	Job		236		236	
Replace Shingle							
	Install	Ea.	81	11.80		92.80	Includes labor and material to replace
	Minimum Charge	Job		236		236	individual wood shingle/shake. Cost is per shingle.

Slate Tile Roofing

Slate Tile Roofing		Unit	Material	Labor	Equip.	Total	Specification
Unfading Green							
	Demolish	Sq.		81		81	Includes material and labor to install
	Install	Sq.	560	254		814	clear Vermont slate tile.
	Demolish and Install	Sq.	560	335		895	
	Reinstall	Sq.		203.52		203.52	
	Minimum Charge	Job		223		223	
Unfading Purple							
	Demolish	Sq.		81		81	Includes material and labor to install
	Install	Sq.	480	254		734	clear Vermont slate tile.
	Demolish and Install	Sq.	480	335		815	
	Reinstall	Sq.		203.52		203.52	
	Minimum Charge	Job		223		223	
Replace Slate Tiles							
	Install	Ea.	7.80	23.50		31.30	Includes labor and material to replace
	Minimum Charge	Job		223		223	individual slate tiles. Cost is per tile.
Variegated Purple							
	Demolish	Sq.		81		81	Includes material and labor to install
	Install	Sq.	505	254		759	clear Vermont slate tile.
	Demolish and Install	Sq.	505	335		840	
	Reinstall	Sq.		203.52		203.52	
	Minimum Charge	Job		223		223	
Unfading Grey / Black							
	Demolish	Sq.		81		81	Includes material and labor to install
	Install	Sq.	540	254		794	clear Vermont slate tile.
	Demolish and Install	Sq.	540	335		875	
	Reinstall	Sq.		203.52		203.52	
	Minimum Charge	Job		223		223	
Unfading Red							
	Demolish	Sq.		81		81	Includes material and labor to install
	Install	Sq.	1300	254		1554	clear Vermont slate tile.
	Demolish and Install	Sq.	1300	335		1635	
	Reinstall	Sq.		203.52		203.52	
	Minimum Charge	Job		223		223	
Weathering Black							
	Demolish	Sq.		81		81	Includes material and labor to install
	Install	Sq.	545	254		799	clear Pennsylvania slate tile.
	Demolish and Install	Sq.	545	335		880	
	Reinstall	Sq.		203.52		203.52	
	Minimum Charge	Job		223		223	
Weathering Green							
	Demolish	Sq.		81		81	Includes material and labor to install
	Install	Sq.	445	254		699	clear Vermont slate tile.
	Demolish and Install	Sq.	445	335		780	
	Reinstall	Sq.		203.52		203.52	
	Minimum Charge	Job		223		223	

Roofing

Flat Clay Tile Roofing

	Unit	Material	Labor	Equip.	Total	Specification
Glazed						
Demolish	Sq.		90		90	Includes material and labor to install
Install	Sq.	545	223		768	tile roofing. Cost based on terra cotta
Demolish and Install	Sq.	545	313		858	red flat clay tile and includes ridge
Reinstall	Sq.		178.08		178.08	and rake tiles.
Minimum Charge	Job		223		223	
Terra Cotta Red						
Demolish	Sq.		90		90	Includes material and labor to install
Install	Sq.	545	223		768	tile roofing. Cost based on terra cotta
Demolish and Install	Sq.	545	313		858	red flat clay tile and includes ridge
Reinstall	Sq.		178.08		178.08	and rake tiles.
Minimum Charge	Job		223		223	

Mission Tile Roofing

	Unit	Material	Labor	Equip.	Total	Specification
Glazed Red						
Demolish	Sq.		90		90	Includes material and labor to install
Install	Sq.	605	243		848	tile roofing. Cost based on glazed red
Demolish and Install	Sq.	605	333		938	tile and includes birdstop, ridge and
Reinstall	Sq.		194.27		194.27	rake tiles.
Minimum Charge	Job		223		223	
Unglazed Red						
Demolish	Sq.		90		90	Includes material and labor to install
Install	Sq.	460	243		703	tile roofing. Cost based on unglazed
Demolish and Install	Sq.	460	333		793	red tile and includes birdstop, ridge
Reinstall	Sq.		194.27		194.27	and rake tiles.
Minimum Charge	Job		223		223	
Glazed Blue						
Demolish	Sq.		90		90	Includes material and labor to install
Install	Sq.	415	243		658	tile roofing. Cost based on glazed
Demolish and Install	Sq.	415	333		748	blue tile and includes birdstop, ridge
Reinstall	Sq.		194.27		194.27	and rake tiles.
Minimum Charge	Job		223		223	
Glazed White						
Demolish	Sq.		90		90	Includes material and labor to install
Install	Sq.	440	243		683	tile roofing. Cost based on glazed
Demolish and Install	Sq.	440	333		773	white tile and includes birdstop, ridge
Reinstall	Sq.		194.27		194.27	and rake tiles.
Minimum Charge	Job		223		223	
Unglazed White						
Demolish	Sq.		90		90	Includes material and labor to install
Install	Sq.	365	243		608	tile roofing. Cost based on unglazed
Demolish and Install	Sq.	365	333		698	white tile and includes birdstop, ridge
Reinstall	Sq.		194.27		194.27	and rake tiles.
Minimum Charge	Job		223		223	
Color Blend						
Demolish	Sq.		90		90	Includes material and labor to install
Install	Sq.	370	243		613	tile roofing. Cost based on color blend
Demolish and Install	Sq.	370	333		703	tile and includes birdstop, ridge and
Reinstall	Sq.		194.27		194.27	rake tiles.
Minimum Charge	Job		223		223	

For customer support on your Contractor's Pricing Guide: Residential Repair & Remodeling, call 800.448.8182.

Mission Tile Roofing

Mission Tile Roofing		Unit	Material	Labor	Equip.	Total	Specification
Accessories / Extras							
Add for Cap Furring Strips	Install	Sq.	98.50	278		376.50	Includes labor and material to install vertically placed wood furring strips under cap tiles.
Add for Tile Adhesive	Install	Sq.	23.50	16.25		39.75	Includes material and labor for adhesive material placed between tiles.
Add for Wire Attachment	Install	Sq.	15.35	194		209.35	Includes labor and material to install single wire anchors for attaching tile.
Add for Braided Runners	Install	Sq.	38.50	186		224.50	Includes labor and material to install braided wire runners for secure attachment in high wind areas.
Add for Stainless Steel Nails	Install	Sq.	25.50			25.50	Includes material only for stainless steel nails.
Add for Brass Nails	Install	Sq.	44			44	Includes material only for brass nails.
Add for Hurricane Clips	Install	Sq.	23.50	262		285.50	Includes material and labor for hurricane or wind clips for high wind areas.
Add for Copper Nails	Install	Sq.	44			44	Includes material only for copper nails.
Add for Vertical Furring Strips	Install	Sq.	19.80	52		71.80	Includes labor and material to install vertically placed wood furring strips under tiles.
Add for Horizontal Furring Strips	Install	Sq.	13.15	23		36.15	Includes labor and material to install horizontally placed wood furring strips under tiles.

Spanish Tile Roofing

Spanish Tile Roofing		Unit	Material	Labor	Equip.	Total	Specification
Glazed Red							
	Demolish	Sq.		90		90	Includes material and labor to install tile roofing. Cost based on glazed red tile and includes birdstop, ridge and rake tiles.
	Install	Sq.	620	186		806	
	Demolish and Install	Sq.	620	276		896	
	Reinstall	Sq.		148.40		148.40	
	Minimum Charge	Job		223		223	
Glazed White							
	Demolish	Sq.		90		90	Includes material and labor to install tile roofing. Cost based on glazed white tile and includes birdstop, ridge and rake tiles.
	Install	Sq.	395	247		642	
	Demolish and Install	Sq.	395	337		732	
	Reinstall	Sq.		197.87		197.87	
	Minimum Charge	Job		223		223	

Roofing

Spanish Tile Roofing		Unit	Material	Labor	Equip.	Total	Specification
Glazed Blue							
	Demolish	Sq.		90		90	Includes material and labor to install
	Install	Sq.	415	186		601	tile roofing. Cost based on glazed
	Demolish and Install	Sq.	415	276		691	blue tile and includes birdstop, ridge
	Reinstall	Sq.		148.40		148.40	and rake tiles.
	Minimum Charge	Job		223		223	
Unglazed Red							
	Demolish	Sq.		90		90	Includes material and labor to install
	Install	Sq.	615	186		801	tile roofing. Cost based on unglazed
	Demolish and Install	Sq.	615	276		891	red tile and includes birdstop, ridge
	Reinstall	Sq.		148.40		148.40	and rake tiles.
	Minimum Charge	Job		223		223	
Unglazed White							
	Demolish	Sq.		90		90	Includes material and labor to install
	Install	Sq.	545	186		731	tile roofing. Cost based on unglazed
	Demolish and Install	Sq.	545	276		821	white tile and includes birdstop, ridge
	Reinstall	Sq.		148.40		148.40	and rake tiles.
	Minimum Charge	Job		223		223	
Color Blend							
	Demolish	Sq.		90		90	Includes material and labor to install
	Install	Sq.	665	186		851	tile roofing. Cost based on color blend
	Demolish and Install	Sq.	665	276		941	tile and includes birdstop, ridge and
	Reinstall	Sq.		148.40		148.40	rake tiles.
	Minimum Charge	Job		223		223	
Accessories / Extras							
Add for Vertical Furring Strips							
	Install	Sq.	19.80	52		71.80	Includes labor and material to install vertically placed wood furring strips under tiles.
Add for Wire Attachment							
	Install	Sq.	8.25	148		156.25	Includes labor and material to install single wire ties.
Add for Braided Runners							
	Install	Sq.	30.50	127		157.50	Includes labor and material to install braided wire runners for secure attachment in high wind areas.
Add for Stainless Steel Nails							
	Install	Sq.	21			21	Includes material only for stainless steel nails.
Add for Copper Nails							
	Install	Sq.	37			37	Includes material only for copper nails.
Add for Brass Nails							
	Install	Sq.	33			33	Includes material only for brass nails.
Add for Hurricane Clips							
	Install	Sq.	12.30	154		166.30	Includes material and labor to install hurricane or wind clips for high wind areas.
Add for Mortar Set Tiles							
	Install	Sq.	51.50	223		274.50	Includes material and labor to install setting tiles by using mortar material placed between tiles.

Roofing

Spanish Tile Roofing		Unit	Material	Labor	Equip.	Total	Specification
Add for Tile Adhesive	Install	Sq.	13.40	14.45		27.85	Includes material and labor for adhesive placed between tiles.
Add for Cap Furring Strips	Install	Sq.	98.50	278		376.50	Includes labor and material to install vertically placed wood furring strips under cap tiles.
Add for Horizontal Furring Strips	Install	Sq.	13.15	23		36.15	Includes labor and material to install horizontally placed wood furring strips under tiles.

Concrete Tile Roofing		Unit	Material	Labor	Equip.	Total	Specification
Corrugated Red / Brown	Demolish	Sq.		90		90	Includes material and labor to install tile roofing. Cost includes birdstop, booster, ridge and rake tiles.
	Install	Sq.	114	330		444	
	Demolish and Install	Sq.	114	420		534	
	Reinstall	Sq.		263.82		263.82	
	Minimum Charge	Job		223		223	
Corrugated Gray	Demolish	Sq.		90		90	Includes material and labor to install tile roofing. Cost includes birdstop, booster, ridge and rake tiles.
	Install	Sq.	116	330		446	
	Demolish and Install	Sq.	116	420		536	
	Reinstall	Sq.		263.82		263.82	
	Minimum Charge	Job		223		223	
Corrugated Bright Red	Demolish	Sq.		90		90	Includes material and labor to install tile roofing. Cost includes birdstop, booster, ridge and rake tiles.
	Install	Sq.	116	330		446	
	Demolish and Install	Sq.	116	420		536	
	Reinstall	Sq.		263.82		263.82	
	Minimum Charge	Job		223		223	
Corrugated Black	Demolish	Sq.		90		90	Includes material and labor to install tile roofing. Cost includes birdstop, booster, ridge and rake tiles.
	Install	Sq.	116	330		446	
	Demolish and Install	Sq.	116	420		536	
	Reinstall	Sq.		263.82		263.82	
	Minimum Charge	Job		223		223	
Corrugated Green	Demolish	Sq.		90		90	Includes material and labor to install tile roofing. Cost includes birdstop, booster, ridge and rake tiles.
	Install	Sq.	116	330		446	
	Demolish and Install	Sq.	116	420		536	
	Reinstall	Sq.		263.82		263.82	
	Minimum Charge	Job		223		223	
Corrugated Blue	Demolish	Sq.		90		90	Includes material and labor to install tile roofing. Cost includes birdstop, booster, ridge and rake tiles.
	Install	Sq.	116	330		446	
	Demolish and Install	Sq.	116	420		536	
	Reinstall	Sq.		263.82		263.82	
	Minimum Charge	Job		223		223	
Flat Natural Gray	Demolish	Sq.		90		90	Includes material and labor to install tile roofing. Cost includes birdstop, booster, ridge and rake tiles.
	Install	Sq.	142	330		472	
	Demolish and Install	Sq.	142	420		562	
	Reinstall	Sq.		263.82		263.82	
	Minimum Charge	Job		223		223	

Roofing

Concrete Tile Roofing

	Unit	Material	Labor	Equip.	Total	Specification
Flat Red / Brown						
Demolish	Sq.		90		90	Includes material and labor to install
Install	Sq.	142	330		472	tile roofing. Cost includes birdstop,
Demolish and Install	Sq.	142	420		562	booster, ridge and rake tiles.
Reinstall	Sq.		263.82		263.82	
Minimum Charge	Job		223		223	
Flat Bright Red						
Demolish	Sq.		90		90	Includes material and labor to install
Install	Sq.	142	330		472	tile roofing. Cost includes birdstop,
Demolish and Install	Sq.	142	420		562	booster, ridge and rake tiles.
Reinstall	Sq.		263.82		263.82	
Minimum Charge	Job		223		223	
Flat Green						
Demolish	Sq.		90		90	Includes material and labor to install
Install	Sq.	142	330		472	tile roofing. Cost includes birdstop,
Demolish and Install	Sq.	142	420		562	booster, ridge and rake tiles.
Minimum Charge	Job		223		223	
Flat Blue						
Demolish	Sq.		90		90	Includes material and labor to install
Install	Sq.	142	330		472	tile roofing. Cost includes birdstop,
Demolish and Install	Sq.	142	420		562	booster, ridge and rake tiles.
Reinstall	Sq.		263.82		263.82	
Minimum Charge	Job		223		223	
Flat Black						
Demolish	Sq.		90		90	Includes material and labor to install
Install	Sq.	142	330		472	tile roofing. Cost includes birdstop,
Demolish and Install	Sq.	142	420		562	booster, ridge and rake tiles.
Reinstall	Sq.		263.82		263.82	
Minimum Charge	Job		223		223	

Aluminum Sheet

	Unit	Material	Labor	Equip.	Total	Specification
Corrugated						
.019″ Thick Natural						
Demolish	S.F.		.95		.95	Includes material and labor to install
Install	S.F.	1.39	2.16		3.55	aluminum sheet metal roofing.
Demolish and Install	S.F.	1.39	3.11		4.50	
Reinstall	S.F.		1.73		1.73	
Minimum Charge	Job		520		520	
.016″ Thick Natural						
Demolish	S.F.		.95		.95	Includes material and labor to install
Install	S.F.	1.12	2.16		3.28	aluminum sheet metal roofing.
Demolish and Install	S.F.	1.12	3.11		4.23	
Reinstall	S.F.		1.73		1.73	
Minimum Charge	Job		520		520	
.016″ Colored Finish						
Demolish	S.F.		.95		.95	Includes material and labor to install
Install	S.F.	1.63	2.16		3.79	aluminum sheet metal roofing.
Demolish and Install	S.F.	1.63	3.11		4.74	
Reinstall	S.F.		1.73		1.73	
Minimum Charge	Job		520		520	
.019″ Colored Finish						
Demolish	S.F.		.95		.95	Includes material and labor to install
Install	S.F.	1.74	2.16		3.90	aluminum sheet metal roofing.
Demolish and Install	S.F.	1.74	3.11		4.85	
Reinstall	S.F.		1.73		1.73	
Minimum Charge	Job		520		520	

Roofing

Aluminum Sheet	Unit	Material	Labor	Equip.	Total	Specification
Ribbed						
.019″ Thick Natural						
Demolish	S.F.		.95		.95	Includes material and labor to install
Install	S.F.	1.39	2.16		3.55	aluminum sheet metal roofing.
Demolish and Install	S.F.	1.39	3.11		4.50	
Reinstall	S.F.		1.73		1.73	
Minimum Charge	Job		520		520	
.016″ Thick Natural						
Demolish	S.F.		.95		.95	Includes material and labor to install
Install	S.F.	1.12	2.16		3.28	aluminum sheet metal roofing.
Demolish and Install	S.F.	1.12	3.11		4.23	
Reinstall	S.F.		1.73		1.73	
Minimum Charge	Job		520		520	
.050″ Colored Finish						
Demolish	S.F.		.95		.95	Includes material and labor to install
Install	S.F.	5.35	2.16		7.51	aluminum sheet metal roofing.
Demolish and Install	S.F.	5.35	3.11		8.46	
Reinstall	S.F.		1.73		1.73	
Minimum Charge	Job		520		520	
.032″ Thick Natural						
Demolish	S.F.		.95		.95	Includes material and labor to install
Install	S.F.	2.95	2.16		5.11	aluminum sheet metal roofing.
Demolish and Install	S.F.	2.95	3.11		6.06	
Reinstall	S.F.		1.73		1.73	
Minimum Charge	Job		520		520	
.040″ Thick Natural						
Demolish	S.F.		.95		.95	Includes material and labor to install
Install	S.F.	4.25	2.16		6.41	aluminum sheet metal roofing.
Demolish and Install	S.F.	4.25	3.11		7.36	
Reinstall	S.F.		1.73		1.73	
Minimum Charge	Job		520		520	
.050″ Thick Natural						
Demolish	S.F.		.95		.95	Includes material and labor to install
Install	S.F.	4.47	2.16		6.63	aluminum sheet metal roofing.
Demolish and Install	S.F.	4.47	3.11		7.58	
Reinstall	S.F.		1.73		1.73	
Minimum Charge	Job		520		520	
.016″ Colored Finish						
Demolish	S.F.		.95		.95	Includes material and labor to install
Install	S.F.	1.63	2.16		3.79	aluminum sheet metal roofing.
Demolish and Install	S.F.	1.63	3.11		4.74	
Reinstall	S.F.		1.73		1.73	
Minimum Charge	Job		520		520	
.019″ Colored Finish						
Demolish	S.F.		.95		.95	Includes material and labor to install
Install	S.F.	1.74	2.16		3.90	aluminum sheet metal roofing.
Demolish and Install	S.F.	1.74	3.11		4.85	
Reinstall	S.F.		1.73		1.73	
Minimum Charge	Job		520		520	
.032″ Colored Finish						
Demolish	S.F.		.95		.95	Includes material and labor to install
Install	S.F.	4.33	2.16		6.49	aluminum sheet metal roofing.
Demolish and Install	S.F.	4.33	3.11		7.44	
Reinstall	S.F.		1.73		1.73	
Minimum Charge	Job		520		520	

Roofing

Aluminum Sheet

.040" Colored Finish	Unit	Material	Labor	Equip.	Total	Specification
Demolish	S.F.		.95		.95	Includes material and labor to install
Install	S.F.	5.10	2.16		7.26	aluminum sheet metal roofing.
Demolish and Install	S.F.	5.10	3.11		8.21	
Reinstall	S.F.		1.73		1.73	
Minimum Charge	Job		520		520	

Fiberglass Sheet

12 Ounce Corrugated	Unit	Material	Labor	Equip.	Total	Specification
Demolish	S.F.		1.71		1.71	Cost includes material and labor to
Install	S.F.	5.10	1.76		6.86	install 12 ounce corrugated fiberglass
Demolish and Install	S.F.	5.10	3.47		8.57	sheet roofing.
Reinstall	S.F.		1.41		1.41	
Minimum Charge	Job		221		221	
8 Ounce Corrugated						
Demolish	S.F.		1.71		1.71	Cost includes material and labor to
Install	S.F.	2.86	1.76		4.62	install 8 ounce corrugated fiberglass
Demolish and Install	S.F.	2.86	3.47		6.33	roofing.
Reinstall	S.F.		1.41		1.41	
Minimum Charge	Job		221		221	

Galvanized Steel

Corrugated	Unit	Material	Labor	Equip.	Total	Specification
24 Gauge						
Demolish	S.F.		.95		.95	Includes material and labor to install
Install	S.F.	3.33	1.86		5.19	galvanized steel sheet metal roofing.
Demolish and Install	S.F.	3.33	2.81		6.14	Cost based on 24 gauge corrugated
Reinstall	S.F.		1.48		1.48	or ribbed steel roofing.
Minimum Charge	Job		221		221	
26 Gauge						
Demolish	S.F.		.95		.95	Includes material and labor to install
Install	S.F.	2.30	1.76		4.06	galvanized steel sheet metal roofing.
Demolish and Install	S.F.	2.30	2.71		5.01	Cost based on 26 gauge corrugated
Reinstall	S.F.		1.41		1.41	or ribbed steel roofing.
Minimum Charge	Job		221		221	
28 Gauge						
Demolish	S.F.		.95		.95	Includes material and labor to install
Install	S.F.	1.64	1.68		3.32	galvanized steel sheet metal roofing.
Demolish and Install	S.F.	1.64	2.63		4.27	Cost based on 28 gauge corrugated
Reinstall	S.F.		1.34		1.34	or ribbed steel roofing.
Minimum Charge	Job		221		221	
30 Gauge						
Demolish	S.F.		.95		.95	Includes material and labor to install
Install	S.F.	1.88	1.60		3.48	galvanized steel sheet metal roofing.
Demolish and Install	S.F.	1.88	2.55		4.43	Cost based on 30 gauge corrugated
Reinstall	S.F.		1.28		1.28	or ribbed steel roofing.
Minimum Charge	Job		221		221	

Roofing

Galvanized Steel

Galvanized Steel	Unit	Material	Labor	Equip.	Total	Specification
Ribbed						
24 Gauge						
Demolish	S.F.		.95		.95	Includes material and labor to install
Install	S.F.	3.33	1.86		5.19	galvanized steel sheet metal roofing.
Demolish and Install	S.F.	3.33	2.81		6.14	Cost based on 24 gauge corrugated
Reinstall	S.F.		1.48		1.48	or ribbed steel roofing.
Minimum Charge	Job		221		221	
26 Gauge						
Demolish	S.F.		.95		.95	Includes material and labor to install
Install	S.F.	2.30	1.76		4.06	galvanized steel sheet metal roofing.
Demolish and Install	S.F.	2.30	2.71		5.01	Cost based on 26 gauge corrugated
Reinstall	S.F.		1.41		1.41	or ribbed steel roofing.
Minimum Charge	Job		221		221	
28 Gauge						
Demolish	S.F.		.95		.95	Includes material and labor to install
Install	S.F.	1.64	1.68		3.32	galvanized steel sheet metal roofing.
Demolish and Install	S.F.	1.64	2.63		4.27	Cost based on 28 gauge corrugated
Reinstall	S.F.		1.34		1.34	or ribbed steel roofing.
Minimum Charge	Job		221		221	
30 Gauge						
Demolish	S.F.		.95		.95	Includes material and labor to install
Install	S.F.	1.88	1.60		3.48	galvanized steel sheet metal roofing.
Demolish and Install	S.F.	1.88	2.55		4.43	Cost based on 30 gauge corrugated
Reinstall	S.F.		1.28		1.28	or ribbed steel roofing.
Minimum Charge	Job		221		221	

Standing Seam

Standing Seam	Unit	Material	Labor	Equip.	Total	Specification
Copper						
16 Ounce						
Demolish	Sq.		95.50		95.50	Includes material and labor to install
Install	Sq.	1100	400		1500	copper standing seam metal roofing.
Demolish and Install	Sq.	1100	495.50		1595.50	Cost based on 16 ounce copper
Reinstall	Sq.		318.52		318.52	roofing.
Minimum Charge	Job		259		259	
18 Ounce						
Demolish	Sq.		95.50		95.50	Includes material and labor to install
Install	Sq.	1250	430		1680	copper standing seam metal roofing.
Demolish and Install	Sq.	1250	525.50		1775.50	Cost based on 18 ounce copper
Reinstall	Sq.		345.07		345.07	roofing.
Minimum Charge	Job		259		259	
20 Ounce						
Demolish	Sq.		95.50		95.50	Includes material and labor to install
Install	Sq.	1475	470		1945	copper standing seam metal roofing.
Demolish and Install	Sq.	1475	565.50		2040.50	Cost based on 20 ounce copper
Reinstall	Sq.		376.44		376.44	roofing.
Stainless Steel						
26 Gauge						
Demolish	Sq.		95.50		95.50	Cost includes material and labor to
Install	Sq.	865	450		1315	install 26 gauge stainless steel
Demolish and Install	Sq.	865	545.50		1410.50	standing seam metal roofing.
Reinstall	Sq.		360.07		360.07	
Minimum Charge	Job		259		259	

For customer support on your Contractor's Pricing Guide: Residential Repair & Remodeling, call 800.448.8182.

Roofing

Standing Seam

	Unit	Material	Labor	Equip.	Total	Specification
28 Gauge						Includes material and labor to install
Demolish	Sq.		95.50		95.50	standing seam metal roofing. Cost
Install	Sq.	730	430		1160	based on 28 gauge stainless steel
Demolish and Install	Sq.	730	525.50		1255.50	roofing.
Reinstall	Sq.		345.07		345.07	
Minimum Charge	Job		259		259	

Elastomeric Roofing

	Unit	Material	Labor	Equip.	Total	Specification
Neoprene						
1/16″						Includes material and labor to install
Demolish	Sq.		33		33	neoprene membrane, fully adhered,
Install	Sq.	250	212	41	503	1/16″ thick.
Demolish and Install	Sq.	250	245	41	536	
Minimum Charge	Job		1000	99.50	1099.50	
GRM Membrane						
1/16″						Includes material and labor to install
Demolish	Sq.		33		33	coal tar based elastomeric roofing
Install	Sq.	105	335	33	473	membrane, polyester fiber reinforced.
Demolish and Install	Sq.	105	368	33	506	
Minimum Charge	Job		1000	99.50	1099.50	
Cant Strips						
3″						Includes material and labor to install
Demolish	L.F.		1.02		1.02	cants 3″ x 3″, treated timber, cut
Install	L.F.	1	1.36		2.36	diagonally.
Demolish and Install	L.F.	1	2.38		3.38	
Minimum Charge	Job		111		111	
4″						Includes material and labor to install
Demolish	L.F.		1.02		1.02	cants 4″ x 4″, treated timber, cut
Install	L.F.	2.16	1.36		3.52	diagonally.
Demolish and Install	L.F.	2.16	2.38		4.54	
Minimum Charge	Job		111		111	
Polyvinyl Chloride (PVC)						
45 mil (loose laid)						Cost includes material and labor to
Demolish	Sq.		33		33	install 45 mil PVC elastomeric roofing,
Install	Sq.	107	39.50	3.91	150.41	loose-laid with stone.
Demolish and Install	Sq.	107	72.50	3.91	183.41	
Minimum Charge	Job		1000	99.50	1099.50	
48 mil (loose laid)						Cost includes material and labor to
Demolish	Sq.		33		33	install 48 mil PVC elastomeric roofing,
Install	Sq.	131	39.50	3.91	174.41	loose-laid with stone.
Demolish and Install	Sq.	131	72.50	3.91	207.41	
Minimum Charge	Job		1000	99.50	1099.50	
60 mil (loose laid)						Cost includes material and labor to
Demolish	Sq.		33		33	install 60 mil PVC elastomeric roofing,
Install	Sq.	133	39.50	3.91	176.41	loose-laid with stone.
Demolish and Install	Sq.	133	72.50	3.91	209.41	
Minimum Charge	Job		1000	99.50	1099.50	
45 mil (attached)						Cost includes material and labor to
Demolish	Sq.		52.50		52.50	install 45 mil PVC elastomeric roofing
Install	Sq.	89.50	78	7.65	175.15	fully attached to roof deck.
Demolish and Install	Sq.	89.50	130.50	7.65	227.65	
Minimum Charge	Job		1000	99.50	1099.50	

Elastomeric Roofing	Unit	Material	Labor	Equip.	Total	Specification
48 mil (attached)						
Demolish	Sq.		52.50		52.50	Cost includes material and labor to
Install	Sq.	166	78	7.65	251.65	install 48 mil PVC elastomeric roofing
Demolish and Install	Sq.	166	130.50	7.65	304.15	fully attached to roof deck.
Minimum Charge	Job		1000	99.50	1099.50	
60 mil (attached)						
Demolish	Sq.		52.50		52.50	Cost includes material and labor to
Install	Sq.	168	78	7.65	253.65	install 60 mil PVC elastomeric roofing
Demolish and Install	Sq.	168	130.50	7.65	306.15	fully attached to roof deck.
Minimum Charge	Job		1000	99.50	1099.50	

CSPE Type

	Unit	Material	Labor	Equip.	Total	Specification
45 mil (loose laid w / stone)						
Demolish	Sq.		33		33	Cost includes material and labor to
Install	Sq.	283	78	7.65	368.65	install 45 mil CSPE (chlorosulfonated
Demolish and Install	Sq.	283	111	7.65	401.65	polyethylene-hypalon) roofing material
Minimum Charge	Job		1000	99.50	1099.50	loose laid on deck and ballasted with stone.
45 mil (attached at seams)						
Demolish	Sq.		39.50		39.50	Cost includes material and labor to
Install	Sq.	345	58	5.70	408.70	install 45 mil CSPE (chlorosulfonated
Demolish and Install	Sq.	345	97.50	5.70	448.20	polyethylene-hypalon) roofing material
Minimum Charge	Job		1000	99.50	1099.50	attached to deck at seams with batten strips.
45 mil (fully attached)						
Demolish	Sq.		52.50		52.50	Cost includes material and labor to
Install	Sq.	265	58	5.70	328.70	install 45 mil CSPE (chlorosulfonated
Demolish and Install	Sq.	265	110.50	5.70	381.20	polyethylene-hypalon) roofing material
Minimum Charge	Job		1000	99.50	1099.50	fully attached to deck.

EPDM Type

	Unit	Material	Labor	Equip.	Total	Specification
45 mil (attached at seams)						
Demolish	Sq.		39.50		39.50	Cost includes material and labor to
Install	Sq.	87	58	5.70	150.70	install 45 mil EPDM (ethylene
Demolish and Install	Sq.	87	97.50	5.70	190.20	propylene diene monomer) roofing
Minimum Charge	Job		1000	99.50	1099.50	material attached to deck at seams.
45 mil (fully attached)						
Demolish	Sq.		52.50		52.50	Cost includes material and labor to
Install	Sq.	124	78	7.65	209.65	install 45 mil EPDM (ethylene
Demolish and Install	Sq.	124	130.50	7.65	262.15	propylene diene monomer) roofing
Minimum Charge	Job		1000	99.50	1099.50	material fully attached to deck.
45 mil (loose laid w / stone)						
Demolish	Sq.		33		33	Cost includes material and labor to
Install	Sq.	100	39.50	3.91	143.41	install 45 mil EPDM (ethylene
Demolish and Install	Sq.	100	72.50	3.91	176.41	propylene diene monomer) roofing
Minimum Charge	Job		1000	99.50	1099.50	material loose laid on deck and ballasted with stones (approx. 1/2 ton of stone per square).
60 mil (attached at seams)						
Demolish	Sq.		39.50		39.50	Cost includes material and labor to
Install	Sq.	104	58	5.70	167.70	install 60 mil EPDM (ethylene
Demolish and Install	Sq.	104	97.50	5.70	207.20	propylene diene monomer) roofing
Minimum Charge	Job		1000	99.50	1099.50	material attached to deck at seams.

Roofing

Elastomeric Roofing

	Unit	Material	Labor	Equip.	Total	Specification
60 mil (fully attached)						Cost includes material and labor to
Demolish	Sq.		52.50		52.50	install 60 mil EPDM (ethylene
Install	Sq.	141	78	7.65	226.65	propylene diene monomer) roofing
Demolish and Install	Sq.	141	130.50	7.65	279.15	material fully attached to deck.
Minimum Charge	Job		1000	99.50	1099.50	
60 mil (loose laid w / stone)						Cost includes material and labor to
Demolish	Sq.		33		33	install 60 mil EPDM (ethylene
Install	Sq.	119	39.50	3.91	162.41	propylene diene monomer) roofing
Demolish and Install	Sq.	119	72.50	3.91	195.41	material loose laid on deck and
Minimum Charge	Job		1000	99.50	1099.50	ballasted with stones (approx. 1/2 ton of stone per square).

Modified Bitumen Roofing

	Unit	Material	Labor	Equip.	Total	Specification
Fully Attached						
120 mil						
Demolish	Sq.		79		79	Cost includes material and labor to
Install	Sq.	72	9.85	1.92	83.77	install 120 mil modified bitumen
Demolish and Install	Sq.	72	88.85	1.92	162.77	roofing, fully attached to deck by
Minimum Charge	Job		111		111	torch.
150 mil						
Demolish	Sq.		79		79	Cost includes material and labor to
Install	Sq.	94	24	4.71	122.71	install 150 mil modified bitumen
Demolish and Install	Sq.	94	103	4.71	201.71	roofing, fully attached to deck by
Minimum Charge	Job		111		111	torch.
160 mil						
Demolish	Sq.		79		79	Cost includes material and labor to
Install	Sq.	105	25.50	4.91	135.41	install 160 mil modified bitumen
Demolish and Install	Sq.	105	104.50	4.91	214.41	roofing, fully attached to deck by
Minimum Charge	Job		111		111	torch.

Built-up Roofing

	Unit	Material	Labor	Equip.	Total	Specification
Emergency Repairs						
Install	Ea.	229	2075	405	2709	Minimum charge for work requiring
Minimum Charge	Job		1450	283	1733	hot asphalt.
Asphalt						
3-ply Asphalt						
Demolish	Sq.		82		82	Cost includes material and labor to
Install	Sq.	83	121	23.50	227.50	install 3-ply asphalt built-up roof
Demolish and Install	Sq.	83	203	23.50	309.50	including one 30 lb and two 15 lb
Minimum Charge	Job		1450	283	1733	layers of felt.
4-ply Asphalt						
Demolish	Sq.		97.50		97.50	Cost includes material and labor to
Install	Sq.	118	132	25.50	275.50	install 4-ply asphalt built-up roof
Demolish and Install	Sq.	118	229.50	25.50	373	including one 30 lb and three 15 lb
Minimum Charge	Job		1450	283	1733	layers of felt.

For customer support on your Contractor's Pricing Guide: Residential Repair & Remodeling, call 800.448.8182.

Built-up Roofing		Unit	Material	Labor	Equip.	Total	Specification
5-ply Asphalt							Cost includes material and labor to
	Demolish	Sq.		128		128	install 5-ply asphalt built-up roof
	Install	Sq.	160	145	28.50	333.50	including one 30 lb and four 15 lb
	Demolish and Install	Sq.	160	273	28.50	461.50	layers of felt.
	Minimum Charge	Job		1450	283	1733	
Components / Accessories							
Flood Coat							Includes labor and material to install
	Install	Sq.	32	68		100	flood coat.
R / R Gravel and Flood Coat							
	Install	Sq.	40.50	221		261.50	Includes labor and material to install
Aluminum UV Coat							flood coat and gravel.
	Install	Sq.	35	253	25	313	Includes labor and material to install
Gravel Coat							aluminum-based UV coating.
	Demolish	Sq.		41		41	Includes labor and material to install
	Install	Sq.	8.90	37		45.90	gravel coat.
	Demolish and Install	Sq.	8.90	78		86.90	
Mineral Surface Cap Sheet							Cost includes labor and material to
	Install	Sq.	92.50	7.65		100.15	install mineral surfaced cap sheet.
Fibrated Aluminum UV Coating							
	Install	Sq.	27	68		95	Includes labor and material to install
Impregnated Walk							fibered aluminum-based UV coating.
1″							
	Install	S.F.	3.27	2.36		5.63	Cost includes labor and material to
1/2″							install 1″ layer of asphalt coating.
	Install	S.F.	2.37	1.86		4.23	Cost includes labor and material to
3/4″							install 1/2″ layer of asphalt coating.
	Install	S.F.	2.86	1.97		4.83	Cost includes labor and material to
							install 3/4″ layer of asphalt coating.

| Doors | Roof Window | Double Hung Window |

Exterior Type Door	Unit	Material	Labor	Equip.	Total	Specification
Entry						
Flush (1-3/4")						
Demolish	Ea.		43.50		43.50	Cost includes material and labor to
Install	Ea.	263	29.50		292.50	install 3' x 6' 8" pre-hung 1-3/4"
Demolish and Install	Ea.	263	73		336	door, including casing and stop,
Reinstall	Ea.		29.45		29.45	hinges, jamb, aluminum sill,
Clean	Ea.	1.50	15.15		16.65	weatherstripped.
Paint	Ea.	10.85	56		66.85	
Minimum Charge	Job		236		236	
Raised Panel						
Demolish	Ea.		43.50		43.50	Cost includes material and labor to
Install	Ea.	505	26		531	install 3' x 6' 8" pre-hung 1-3/4"
Demolish and Install	Ea.	505	69.50		574.50	door, jamb, including casing and
Reinstall	Ea.		26.18		26.18	stop, hinges, aluminum sill,
Clean	Ea.	1.50	15.15		16.65	weatherstripping.
Paint	Ea.	10.85	56		66.85	
Minimum Charge	Job		236		236	
Entry w / Lights						
Good Quality						
Demolish	Ea.		43.50		43.50	Cost includes material and labor to
Install	Ea.	370	59		429	install 3' x 6' 8" pre-hung 1-3/4"
Demolish and Install	Ea.	370	102.50		472.50	door, solid wood lower panel, with up
Reinstall	Ea.		58.90		58.90	to 9-light glass panels, casing, sill,
Clean	Ea.	1.50	15.15		16.65	weatherstripped.
Paint	Ea.	7.10	26		33.10	
Minimum Charge	Job		236		236	
Better Quality						
Demolish	Ea.		43.50		43.50	Cost includes material and labor to
Install	Ea.	385	59		444	install 3' x 6' 8" pre-hung 1-3/4" Ash
Demolish and Install	Ea.	385	102.50		487.50	door, solid wood, raised panel door
Reinstall	Ea.		58.90		58.90	without glass panels.
Clean	Ea.	1.50	15.15		16.65	
Paint	Ea.	10.85	56		66.85	
Minimum Charge	Job		236		236	
Premium Quality						
Demolish	Ea.		43.50		43.50	Cost includes material and labor to
Install	Ea.	1000	27.50		1027.50	install 3' x 6' 8" pre-hung 1-3/4" Oak
Demolish and Install	Ea.	1000	71		1071	door, solid wood, raised panel door
Reinstall	Ea.		27.72		27.72	without glass panels.
Clean	Ea.	1.50	15.15		16.65	
Paint	Ea.	10.85	56		66.85	
Minimum Charge	Job		236		236	

89

For customer support on your Contractor's Pricing Guide: Residential Repair & Remodeling, call 800.448.8182.

Exterior Doors and Windows

Exterior Type Door		Unit	Material	Labor	Equip.	Total	Specification
Custom Quality							
	Demolish	Ea.		43.50		43.50	Cost includes material and labor to
	Install	Ea.	550	63		613	install 3' x 6' 8" pre-hung 1-3/4"
	Demolish and Install	Ea.	550	106.50		656.50	hand-made, solid wood, raised panel
	Reinstall	Ea.		62.83		62.83	door with custom stained glass.
	Clean	Ea.	.10	15.15		15.25	
	Paint	Ea.	2.72	130		132.72	
	Minimum Charge	Job		236		236	
Miami Style							
	Demolish	Ea.		43.50		43.50	Cost includes material and labor to
	Install	Ea.	242	27.50		269.50	install 3' x 6' 8" pre-hung 1-3/4"
	Demolish and Install	Ea.	242	71		313	lauan door of solid wood.
	Reinstall	Ea.		27.72		27.72	
	Clean	Ea.	.10	15.15		15.25	
	Paint	Ea.	10.85	56		66.85	
	Minimum Charge	Job		236		236	
French Style							
1-Light							
	Demolish	Ea.		43.50		43.50	Cost includes material and labor to
	Install	Ea.	700	78.50		778.50	install 3' x 6' 8" pre-hung 1-3/4"
	Demolish and Install	Ea.	700	122		822	French style door with 1 light.
	Reinstall	Ea.		78.53		78.53	
	Clean	Ea.	.94	22.50		23.44	
	Paint	Ea.	11.50	98		109.50	
	Minimum Charge	Job		236		236	
9-Light							
	Demolish	Ea.		43.50		43.50	Cost includes material and labor to
	Install	Ea.	780	33.50		813.50	install 3' x 6' 8" pre-hung 1-3/4"
	Demolish and Install	Ea.	780	77		857	French style door with 9 lights.
	Reinstall	Ea.		33.66		33.66	
	Clean	Ea.	.21	30.50		30.71	
	Paint	Ea.	11.50	98		109.50	
	Minimum Charge	Job		236		236	
15-Light							
	Demolish	Ea.		50.50		50.50	Cost includes material and labor to
	Install	Ea.	845	67.50		912.50	install 3' x 6' 8" pre-hung 1-3/4"
	Demolish and Install	Ea.	845	118		963	French style door with 15 lights.
	Reinstall	Ea.		67.31		67.31	
	Clean	Ea.	.21	30.50		30.71	
	Paint	Ea.	11.50	98		109.50	
	Minimum Charge	Job		236		236	
Jamb							
Wood							
	Demolish	Ea.		32.50		32.50	Includes material and labor to install
	Install	Ea.	116	12.75		128.75	flat pine jamb with square cut heads
	Demolish and Install	Ea.	116	45.25		161.25	and rabbeted sides for 6' 8" high and
	Clean	Ea.	.04	9.10		9.14	3-9/16" door including trim sets for
	Paint	Ea.	7.10	26		33.10	both sides.
	Minimum Charge	Job		236		236	
Metal 6' 8" High							
	Demolish	Ea.		26		26	Cost includes material and labor to
	Install	Ea.	152	29.50		181.50	install 18 gauge hollow metal door
	Demolish and Install	Ea.	152	55.50		207.50	frame to fit a 4-1/2" jamb 6' 8" high
	Reinstall	Ea.		29.45		29.45	and up to 3' 6" opening.
	Clean	Ea.	.31	7.60		7.91	
	Paint	Ea.	1.72	7.80		9.52	
	Minimum Charge	Job		236		236	

Exterior Doors and Windows

Exterior Type Door

Exterior Type Door	Unit	Material	Labor	Equip.	Total	Specification
Metal 7' High						
Demolish	Ea.		26		26	Cost includes material and labor to
Install	Ea.	157	31.50		188.50	install 18 gauge hollow metal door
Demolish and Install	Ea.	157	57.50		214.50	frame to fit a 4-1/2" jamb 7' high and
Reinstall	Ea.		31.41		31.41	up to 3' 6" opening.
Clean	Ea.	.31	7.60		7.91	
Paint	Ea.	1.72	7.80		9.52	
Minimum Charge	Job		236		236	
Metal 8' High						
Demolish	Ea.		26		26	Cost includes material and labor to
Install	Ea.	191	39.50		230.50	install 18 gauge hollow metal door
Demolish and Install	Ea.	191	65.50		256.50	frame to fit a 4-1/2" jamb 8' high and
Reinstall	Ea.		39.27		39.27	up to 3' 6" opening.
Clean	Ea.	.44	9.10		9.54	
Paint	Ea.	1.91	8.70		10.61	
Minimum Charge	Job		236		236	
Metal 9' High						
Demolish	Ea.		26		26	Cost includes material and labor to
Install	Ea.	242	47		289	install 18 gauge hollow metal door
Demolish and Install	Ea.	242	73		315	frame to fit a 4-1/2" jamb 9' high and
Reinstall	Ea.		47.12		47.12	up to 3' 6" opening.
Clean	Ea.	.44	9.10		9.54	
Paint	Ea.	1.91	8.70		10.61	
Minimum Charge	Job		236		236	
Casing Trim						
Single Width						
Demolish	Ea.		8.30		8.30	Cost includes material and labor to
Install	Opng.	26.50	39.50		66	install 11/16" x 2-1/2" pine ranch
Demolish and Install	Opng.	26.50	47.80		74.30	style casing for one side of a standard
Reinstall	Opng.		31.41		31.41	door opening.
Clean	Opng.	.21	3.64		3.85	
Paint	Ea.	5.05	7.80		12.85	
Minimum Charge	Job		236		236	
Double Width						
Demolish	Ea.		9.90		9.90	Cost includes material and labor to
Install	Opng.	31	47		78	install 11/16" x 2-1/2" pine ranch
Demolish and Install	Opng.	31	56.90		87.90	style door casing for one side of a
Reinstall	Opng.		37.70		37.70	double door opening.
Clean	Opng.	.25	7.25		7.50	
Paint	Ea.	5.05	8.90		13.95	
Minimum Charge	Job		236		236	

Residential Metal Door

Residential Metal Door	Unit	Material	Labor	Equip.	Total	Specification
Pre-hung, Steel-clad						
2' 8" x 6' 8"						
Demolish	Ea.		43.50		43.50	Cost includes material and labor to
Install	Ea.	455	55.50		510.50	install 2' 8" x 6' 8" pre-hung 1-3/4"
Demolish and Install	Ea.	455	99		554	24 gauge steel clad door with 4-1/2"
Reinstall	Ea.		55.44		55.44	wood frame, trims, hinges, aluminum
Clean	Ea.	1.50	15.15		16.65	sill, weatherstripped.
Paint	Ea.	11.50	56		67.50	
Minimum Charge	Job		236		236	

91

Residential Metal Door		Unit	Material	Labor	Equip.	Total	Specification
3' x 6' 8"							Cost includes material and labor to
	Demolish	Ea.		43.50		43.50	install 3' x 6' 8" pre-hung 1-3/4" 24
	Install	Ea.	325	63		388	gauge steel clad door with 4-1/2"
	Demolish and Install	Ea.	325	106.50		431.50	wood frame, trims, hinges, aluminum
	Reinstall	Ea.		62.83		62.83	sill, weatherstripped.
	Clean	Ea.	1.50	15.15		16.65	
	Paint	Ea.	11.50	56		67.50	
	Minimum Charge	Job		236		236	

Metal Door		Unit	Material	Labor	Equip.	Total	Specification
Steel, Door Only							
2' 8" x 6' 8", 18 Gauge							Cost includes material and labor to
	Demolish	Ea.		9.35		9.35	install 18 gauge 2' 8" x 6' 8" steel
	Install	Ea.	560	29.50		589.50	door only, 1-3/4" thick, hinges.
	Demolish and Install	Ea.	560	38.85		598.85	
	Reinstall	Ea.		29.45		29.45	
	Clean	Ea.	.63	11.35		11.98	
	Paint	Ea.	6.20	32.50		38.70	
	Minimum Charge	Job		236		236	
3' x 6' 8", 18 Gauge							Cost includes material and labor to
	Demolish	Ea.		9.35		9.35	install 18 gauge 3' x 6' 8" steel door
	Install	Ea.	555	31.50		586.50	only, 1-3/4" thick, hinges.
	Demolish and Install	Ea.	555	40.85		595.85	
	Reinstall	Ea.		31.41		31.41	
	Clean	Ea.	.63	11.35		11.98	
	Paint	Ea.	6.20	32.50		38.70	
	Minimum Charge	Job		236		236	
2' 8" x 7', 18 Gauge							Cost includes material and labor to
	Demolish	Ea.		9.35		9.35	install 18 gauge 2' 8" x 7' steel door
	Install	Ea.	505	31.50		536.50	only, 1-3/4" thick, hinges.
	Demolish and Install	Ea.	505	40.85		545.85	
	Reinstall	Ea.		31.41		31.41	
	Clean	Ea.	.63	11.35		11.98	
	Paint	Ea.	6.20	32.50		38.70	
	Minimum Charge	Job		236		236	
3' x 7', 18 Gauge							Cost includes material and labor to
	Demolish	Ea.		9.35		9.35	install 18 gauge 3' x 7' steel door
	Install	Ea.	560	33.50		593.50	only, 1-3/4" thick, hinges.
	Demolish and Install	Ea.	560	42.85		602.85	
	Reinstall	Ea.		33.66		33.66	
	Clean	Ea.	.63	11.35		11.98	
	Paint	Ea.	6.20	32.50		38.70	
	Minimum Charge	Job		236		236	
3' 6" x 7', 18 Gauge							Cost includes material and labor to
	Demolish	Ea.		9.35		9.35	install 18 gauge 3' 6" x 7' steel door
	Install	Ea.	655	36.50		691.50	only, 1-3/4" thick, hinges.
	Demolish and Install	Ea.	655	45.85		700.85	
	Reinstall	Ea.		36.25		36.25	
	Clean	Ea.	.63	11.35		11.98	
	Paint	Ea.	6.20	32.50		38.70	
	Minimum Charge	Job		236		236	

Exterior Doors and Windows

Metal Door

Metal Door	Unit	Material	Labor	Equip.	Total	Specification
Casing Trim						
Single Width						
Demolish	Ea.		8.30		8.30	Cost includes material and labor to
Install	Opng.	26.50	39.50		66	install 11/16" x 2-1/2" pine ranch
Demolish and Install	Opng.	26.50	47.80		74.30	style casing for one side of a standard
Reinstall	Opng.		31.41		31.41	door opening.
Clean	Opng.	.21	3.64		3.85	
Paint	Ea.	5.05	7.80		12.85	
Minimum Charge	Job		236		236	
Double Width						
Demolish	Ea.		9.90		9.90	Cost includes material and labor to
Install	Opng.	31	47		78	install 11/16" x 2-1/2" pine ranch
Demolish and Install	Opng.	31	56.90		87.90	style door casing for one side of a
Reinstall	Opng.		37.70		37.70	double door opening.
Clean	Opng.	.25	7.25		7.50	
Paint	Ea.	5.05	8.90		13.95	
Minimum Charge	Job		236		236	

Solid Core Door

Solid Core Door	Unit	Material	Labor	Equip.	Total	Specification
Exterior Type						
Flush (1-3/4")						
Demolish	Ea.		43.50		43.50	Cost includes material and labor to
Install	Ea.	440	59		499	install 3' x 6' 8" pre-hung 1-3/4"
Demolish and Install	Ea.	440	102.50		542.50	door, including casing and stop,
Reinstall	Ea.		58.90		58.90	hinges, jamb, aluminum sill,
Clean	Ea.	1.50	15.15		16.65	weatherstripped.
Paint	Ea.	11.50	56		67.50	
Minimum Charge	Job		236		236	
Raised Panel						
Demolish	Ea.		43.50		43.50	Cost includes material and labor to
Install	Ea.	505	26		531	install 3' x 6' 8" pre-hung 1-3/4"
Demolish and Install	Ea.	505	69.50		574.50	door, jamb, including casing and
Reinstall	Ea.		26.18		26.18	stop, hinges, aluminum sill,
Clean	Ea.	1.50	15.15		16.65	weatherstripping.
Paint	Ea.	11.50	130		141.50	
Minimum Charge	Job		236		236	
Entry w / Lights						
Demolish	Ea.		43.50		43.50	Cost includes material and labor to
Install	Ea.	370	59		429	install 3' x 6' 8" pre-hung 1-3/4"
Demolish and Install	Ea.	370	102.50		472.50	door, solid wood lower panel, with up
Reinstall	Ea.		58.90		58.90	to 9-light glass panels, casing, sill,
Clean	Ea.	1.50	15.15		16.65	weatherstripped.
Paint	Ea.	2.72	130		132.72	
Minimum Charge	Job		236		236	
Custom Entry						
Demolish	Ea.		43.50		43.50	Cost includes material and labor to
Install	Ea.	3600	94		3694	install prehung, fir, single glazed with
Demolish and Install	Ea.	3600	137.50		3737.50	lead caming, 1-3/4" x 6' 8" x 3'
Reinstall	Ea.		94.24		94.24	
Clean	Ea.	1.50	15.15		16.65	
Paint	Ea.	6.20	32.50		38.70	
Minimum Charge	Job		236		236	

Solid Core Door		Unit	Material	Labor	Equip.	Total	Specification
Dutch Style							
	Demolish	Ea.		43.50		43.50	Includes material and labor to install
	Install	Ea.	825	78.50		903.50	dutch door, pine, 1-3/4" x 6' 8" x 2'
	Demolish and Install	Ea.	825	122		947	8" wide.
	Reinstall	Ea.		78.53		78.53	
	Clean	Ea.	1.50	15.15		16.65	
	Paint	Ea.	2.72	130		132.72	
	Minimum Charge	Job		236		236	
Miami Style							
	Demolish	Ea.		43.50		43.50	Cost includes material and labor to
	Install	Ea.	242	27.50		269.50	install 3' x 6' 8" pre-hung 1-3/4"
	Demolish and Install	Ea.	242	71		313	lauan door of solid wood.
	Reinstall	Ea.		27.72		27.72	
	Clean	Ea.	1.50	15.15		16.65	
	Paint	Ea.	10.85	56		66.85	
	Minimum Charge	Job		236		236	
French Style							
	Demolish	Ea.		50.50		50.50	Cost includes material and labor to
	Install	Ea.	845	67.50		912.50	install 3' x 6' 8" pre-hung 1-3/4"
	Demolish and Install	Ea.	845	118		963	French style door with 15 lights.
	Reinstall	Ea.		67.31		67.31	
	Clean	Ea.	.10	15.15		15.25	
	Paint	Ea.	8.30	196		204.30	
	Minimum Charge	Job		236		236	
Double							
	Demolish	Ea.		33.50		33.50	Includes material and labor to install a
	Install	Ea.	1525	118		1643	pre-hung double entry door including
	Demolish and Install	Ea.	1525	151.50		1676.50	frame and exterior trim.
	Clean	Ea.	.06	54.50		54.56	
	Paint	Ea.	23	196		219	
	Minimum Charge	Job		118		118	
Commercial Grade							
3' x 6' 8"							
	Demolish	Ea.		43.50		43.50	Includes material and labor to install
	Install	Ea.	600	27.50		627.50	flush solid core 1-3/4" hardwood
	Demolish and Install	Ea.	600	71		671	veneer 3' x 6' 8" door only.
	Reinstall	Ea.		27.72		27.72	
	Clean	Ea.	1.50	15.15		16.65	
	Paint	Ea.	10.35	32.50		42.85	
	Minimum Charge	Job		236		236	
3' 6" x 6' 8"							
	Demolish	Ea.		43.50		43.50	Includes material and labor to install
	Install	Ea.	730	27.50		757.50	flush solid core 1-3/4" hardwood
	Demolish and Install	Ea.	730	71		801	veneer 3' 6" x 6' 8" door only.
	Reinstall	Ea.		27.72		27.72	
	Clean	Ea.	1.50	15.15		16.65	
	Paint	Ea.	10.35	32.50		42.85	
	Minimum Charge	Job		236		236	
3' x 7'							
	Demolish	Ea.		43.50		43.50	Includes material and labor to install
	Install	Ea.	345	67.50		412.50	flush solid core 1-3/4" hardwood
	Demolish and Install	Ea.	345	111		456	veneer 3' x 7' door only.
	Reinstall	Ea.		67.31		67.31	
	Clean	Ea.	1.50	15.15		16.65	
	Paint	Ea.	10.35	32.50		42.85	
	Minimum Charge	Job		236		236	

Exterior Doors and Windows

Solid Core Door		Unit	Material	Labor	Equip.	Total	Specification
3' 6" x 7'							
	Demolish	Ea.		43.50		43.50	Includes material and labor to install
	Install	Ea.	885	33.50		918.50	flush solid core 1-3/4" hardwood
	Demolish and Install	Ea.	885	77		962	veneer 3' 6" x 7' door only.
	Reinstall	Ea.		33.66		33.66	
	Clean	Ea.	1.50	15.15		16.65	
	Paint	Ea.	10.35	32.50		42.85	
	Minimum Charge	Job		236		236	
3' x 8'							
	Demolish	Ea.		43.50		43.50	Includes material and labor to install
	Install	Ea.	770	59		829	flush solid core 1-3/4" hardwood
	Demolish and Install	Ea.	770	102.50		872.50	veneer 3' x 8' door only.
	Reinstall	Ea.		58.90		58.90	
	Clean	Ea.	1.79	22.50		24.29	
	Paint	Ea.	10.35	32.50		42.85	
	Minimum Charge	Job		236		236	
3' 6" x 8'							
	Demolish	Ea.		43.50		43.50	Includes material and labor to install
	Install	Ea.	855	59		914	flush solid core 1-3/4" hardwood
	Demolish and Install	Ea.	855	102.50		957.50	veneer 3' 6" x 8' door only.
	Reinstall	Ea.		58.90		58.90	
	Clean	Ea.	1.79	22.50		24.29	
	Paint	Ea.	10.35	32.50		42.85	
	Minimum Charge	Job		236		236	
3' x 9'							
	Demolish	Ea.		43.50		43.50	Includes material and labor to install
	Install	Ea.	635	67.50		702.50	flush solid core 1-3/4" hardwood
	Demolish and Install	Ea.	635	111		746	veneer 3' x 9' door only.
	Reinstall	Ea.		67.31		67.31	
	Clean	Ea.	1.79	22.50		24.29	
	Paint	Ea.	10.35	32.50		42.85	
	Minimum Charge	Job		236		236	
Stain							
	Install	Ea.	12.60	43.50		56.10	Includes labor and material to stain
	Minimum Charge	Job		196		196	exterior type door and trim on both sides.
Door Only (SC)							
2' 4" x 6' 8"							
	Demolish	Ea.		9.35		9.35	Cost includes material and labor to
	Install	Ea.	231	26		257	install 2' 4" x 6' 8" lauan solid core
	Demolish and Install	Ea.	231	35.35		266.35	door slab.
	Reinstall	Ea.		26.18		26.18	
	Clean	Ea.	.63	11.35		11.98	
	Paint	Ea.	5	49		54	
	Minimum Charge	Job		236		236	
2' 8" x 6' 8"							
	Demolish	Ea.		9.35		9.35	Cost includes material and labor to
	Install	Ea.	241	26		267	install 2' 8" x 6' 8" lauan solid core
	Demolish and Install	Ea.	241	35.35		276.35	door slab.
	Reinstall	Ea.		26.18		26.18	
	Clean	Ea.	.63	11.35		11.98	
	Paint	Ea.	5	49		54	
	Minimum Charge	Job		236		236	

For customer support on your Contractor's Pricing Guide: Residential Repair & Remodeling, call 800.448.8182.

Exterior Doors and Windows

Solid Core Door		Unit	Material	Labor	Equip.	Total	Specification
3' x 6' 8"							
	Demolish	Ea.		9.35		9.35	Cost includes material and labor to
	Install	Ea.	244	27.50		271.50	install 3' x 6' 8" lauan solid core door
	Demolish and Install	Ea.	244	36.85		280.85	slab.
	Reinstall	Ea.		27.72		27.72	
	Clean	Ea.	.63	11.35		11.98	
	Paint	Ea.	5	49		54	
	Minimum Charge	Job		236		236	
3' 6" x 6' 8"							
	Demolish	Ea.		9.35		9.35	Cost includes material and labor to
	Install	Ea.	284	29.50		313.50	install 3' 6" x 6' 8" lauan solid core
	Demolish and Install	Ea.	284	38.85		322.85	door slab.
	Reinstall	Ea.		29.45		29.45	
	Clean	Ea.	.63	11.35		11.98	
	Paint	Ea.	5	49		54	
	Minimum Charge	Job		236		236	
Panel							
	Demolish	Ea.		9.35		9.35	Cost includes material and labor to
	Install	Ea.	310	55.50		365.50	install 3' x 6' 8" six (6) panel interior
	Demolish and Install	Ea.	310	64.85		374.85	door.
	Reinstall	Ea.		55.44		55.44	
	Clean	Ea.	.75	11.35		12.10	
	Paint	Ea.	5	49		54	
	Minimum Charge	Job		236		236	
Casing Trim							
Single Width							
	Demolish	Ea.		8.30		8.30	Cost includes material and labor to
	Install	Opng.	26.50	39.50		66	install 11/16" x 2-1/2" pine ranch
	Demolish and Install	Opng.	26.50	47.80		74.30	style casing for one side of a standard
	Reinstall	Opng.		31.41		31.41	door opening.
	Clean	Opng.	.21	3.64		3.85	
	Paint	Ea.	5.05	7.80		12.85	
	Minimum Charge	Job		236		236	
Double Width							
	Demolish	Ea.		9.90		9.90	Cost includes material and labor to
	Install	Opng.	31	47		78	install 11/16" x 2-1/2" pine ranch
	Demolish and Install	Opng.	31	56.90		87.90	style door casing for one side of a
	Reinstall	Opng.		37.70		37.70	double door opening.
	Clean	Opng.	.25	7.25		7.50	
	Paint	Ea.	5.05	8.90		13.95	
	Minimum Charge	Job		236		236	

Components		Unit	Material	Labor	Equip.	Total	Specification
Hardware							
Doorknob							
	Demolish	Ea.		18.70		18.70	Includes material and labor to install
	Install	Ea.	44.50	29.50		74	residential passage lockset, keyless
	Demolish and Install	Ea.	44.50	48.20		92.70	bored type.
	Reinstall	Ea.		23.56		23.56	
	Clean	Ea.	.15	11.35		11.50	
	Minimum Charge	Job		236		236	

For customer support on your Contractor's Pricing Guide: Residential Repair & Remodeling, call 800.448.8182.

Exterior Doors and Windows

Components		Unit	Material	Labor	Equip.	Total	Specification
Doorknob w / Lock							
	Demolish	Ea.		18.70		18.70	Includes material and labor to install
	Install	Ea.	42	29.50		71.50	privacy lockset.
	Demolish and Install	Ea.	42	48.20		90.20	
	Reinstall	Ea.		23.56		23.56	
	Clean	Ea.	.15	11.35		11.50	
	Minimum Charge	Job		236		236	
Lever Handle							
	Demolish	Ea.		18.70		18.70	Includes material and labor to install
	Install	Ea.	53.50	47		100.50	residential passage lockset, keyless
	Demolish and Install	Ea.	53.50	65.70		119.20	bored type with lever handle.
	Reinstall	Ea.		37.70		37.70	
	Clean	Ea.	.15	11.35		11.50	
	Minimum Charge	Job		236		236	
Deadbolt							
	Demolish	Ea.		16.35		16.35	Includes material and labor to install a
	Install	Ea.	81.50	33.50		115	deadbolt.
	Demolish and Install	Ea.	81.50	49.85		131.35	
	Reinstall	Ea.		26.93		26.93	
	Clean	Ea.	.15	11.35		11.50	
	Minimum Charge	Job		236		236	
Grip Handle Entry							
	Demolish	Ea.		14.95		14.95	Includes material and labor to install
	Install	Ea.	106	52.50		158.50	residential keyed entry lock set with
	Demolish and Install	Ea.	106	67.45		173.45	grip type handle.
	Reinstall	Ea.		41.88		41.88	
	Clean	Ea.	.15	11.35		11.50	
	Minimum Charge	Job		236		236	
Panic Device							
	Demolish	Ea.		40.50		40.50	Includes material and labor to install
	Install	Ea.	320	78.50		398.50	touch bar, low profile, exit only.
	Demolish and Install	Ea.	320	119		439	
	Reinstall	Ea.		62.83		62.83	
	Clean	Ea.	.08	20		20.08	
	Minimum Charge	Job		236		236	
Closer							
	Demolish	Ea.		9.80		9.80	Includes material and labor to install
	Install	Ea.	272	78.50		350.50	pneumatic heavy duty closer for
	Demolish and Install	Ea.	272	88.30		360.30	exterior type doors.
	Reinstall	Ea.		62.83		62.83	
	Clean	Ea.	.08	15.15		15.23	
	Minimum Charge	Job		236		236	
Peep-hole							
	Demolish	Ea.		9		9	Includes material and labor to install
	Install	Ea.	10.05	14.75		24.80	peep-hole for door.
	Demolish and Install	Ea.	10.05	23.75		33.80	
	Reinstall	Ea.		11.78		11.78	
	Clean	Ea.	.01	4.19		4.20	
	Minimum Charge	Job		236		236	
Exit Sign							
	Demolish	Ea.		27.50		27.50	Includes material and labor to install a
	Install	Ea.	113	67.50		180.50	wall mounted interior electric exit sign.
	Demolish and Install	Ea.	113	95		208	
	Reinstall	Ea.		54.04		54.04	
	Clean	Ea.	.01	8.50		8.51	
	Minimum Charge	Job		236		236	

Exterior Doors and Windows

Components		Unit	Material	Labor	Equip.	Total	Specification
Kickplate							
	Demolish	Ea.		14.55		14.55	Includes material and labor to install
	Install	Ea.	26.50	31.50		58	aluminum kickplate, 10" x 28".
	Demolish and Install	Ea.	26.50	46.05		72.55	
	Reinstall	Ea.		25.13		25.13	
	Clean	Ea.	.15	11.35		11.50	
	Minimum Charge	Job		236		236	
Threshold							
	Install	L.F.	3.93	4.71		8.64	Includes labor and material to install
	Minimum Charge	Job		118		118	oak threshold.
Weatherstripping							
	Install	L.F.	.75	1.18		1.93	Includes labor and material to install
	Minimum Charge	Job		118		118	rubber weatherstripping for door.
Metal Jamb							
6' 8" High							
	Demolish	Ea.		26		26	Cost includes material and labor to
	Install	Ea.	152	29.50		181.50	install 18 gauge hollow metal door
	Demolish and Install	Ea.	152	55.50		207.50	frame to fit a 4-1/2" jamb 6' 8" high
	Reinstall	Ea.		29.45		29.45	and up to 3' 6" opening.
	Clean	Ea.	.31	7.60		7.91	
	Paint	Ea.	1.72	7.80		9.52	
	Minimum Charge	Job		236		236	
7' High							
	Demolish	Ea.		26		26	Cost includes material and labor to
	Install	Ea.	157	31.50		188.50	install 18 gauge hollow metal door
	Demolish and Install	Ea.	157	57.50		214.50	frame to fit a 4-1/2" jamb 7' high and
	Reinstall	Ea.		31.41		31.41	up to 3' 6" opening.
	Clean	Ea.	.31	7.60		7.91	
	Paint	Ea.	1.72	7.80		9.52	
	Minimum Charge	Job		236		236	
8' High							
	Demolish	Ea.		26		26	Cost includes material and labor to
	Install	Ea.	191	39.50		230.50	install 18 gauge hollow metal door
	Demolish and Install	Ea.	191	65.50		256.50	frame to fit a 4-1/2" jamb 8' high and
	Reinstall	Ea.		39.27		39.27	up to 3' 6" opening.
	Clean	Ea.	.44	9.10		9.54	
	Paint	Ea.	1.91	8.70		10.61	
	Minimum Charge	Job		236		236	
9' High							
	Demolish	Ea.		26		26	Cost includes material and labor to
	Install	Ea.	242	47		289	install 18 gauge hollow metal door
	Demolish and Install	Ea.	242	73		315	frame to fit a 4-1/2" jamb 9' high and
	Reinstall	Ea.		47.12		47.12	up to 3' 6" opening.
	Clean	Ea.	.44	9.10		9.54	
	Paint	Ea.	1.91	8.70		10.61	
	Minimum Charge	Job		236		236	
Wood Jamb							
	Demolish	Ea.		32.50		32.50	Includes material and labor to install
	Install	Ea.	116	12.75		128.75	flat pine jamb with square cut heads
	Demolish and Install	Ea.	116	45.25		161.25	and rabbeted sides for 6' 8" high and
	Clean	Ea.	.04	9.10		9.14	3-9/16" door including trim sets for
	Paint	Ea.	7.10	26		33.10	both sides.
	Minimum Charge	Job		236		236	
Stain							
	Install	Ea.	8.70	32.50		41.20	Includes labor and material to stain
	Minimum Charge	Job		196		196	single door and trim on both sides.

For customer support on your Contractor's Pricing Guide: Residential Repair & Remodeling, call 800.448.8182.

Exterior Doors and Windows

Components		Unit	Material	Labor	Equip.	Total	Specification
Shave & Refit							
	Install	Ea.		39.50		39.50	Includes labor to shave and rework
	Minimum Charge	Job		118		118	door to fit opening at the job site.

Sliding Patio Door		Unit	Material	Labor	Equip.	Total	Specification
Aluminum							
6' Single Pane							
	Demolish	Ea.		65.50		65.50	Cost includes material and labor to
	Install	Ea.	515	236		751	install 6' x 6' 8" sliding anodized
	Demolish and Install	Ea.	515	301.50		816.50	aluminum patio door with single pane
	Reinstall	Ea.		188.48		188.48	tempered safety glass, anodized
	Clean	Ea.	.94	22.50		23.44	frame, screen, weatherstripped.
	Minimum Charge	Job		236		236	
8' Single Pane							
	Demolish	Ea.		65.50		65.50	Cost includes material and labor to
	Install	Ea.	655	315		970	install 8' x 6' 8" sliding aluminum
	Demolish and Install	Ea.	655	380.50		1035.50	patio door with single pane tempered
	Reinstall	Ea.		251.31		251.31	safety glass, anodized frame, screen,
	Clean	Ea.	.94	22.50		23.44	weatherstripped.
	Minimum Charge	Job		236		236	
10' Single Pane							
	Demolish	Ea.		65.50		65.50	Cost includes material and labor to
	Install	Ea.	765	470		1235	install 10' x 6' 8" sliding aluminum
	Demolish and Install	Ea.	765	535.50		1300.50	patio door with single pane tempered
	Reinstall	Ea.		376.96		376.96	safety glass, anodized frame, screen,
	Clean	Ea.	.94	22.50		23.44	weatherstripped.
	Minimum Charge	Job		236		236	
6' Insulated							
	Demolish	Ea.		65.50		65.50	Cost includes material and labor to
	Install	Ea.	1075	236		1311	install 6' x 6' 8" sliding aluminum
	Demolish and Install	Ea.	1075	301.50		1376.50	patio door with double insulated
	Reinstall	Ea.		188.48		188.48	tempered glass, baked-on enamel
	Clean	Ea.	.94	22.50		23.44	finish, screen, weatherstripped.
	Minimum Charge	Job		236		236	
8' Insulated							
	Demolish	Ea.		65.50		65.50	Cost includes material and labor to
	Install	Ea.	1275	315		1590	install 8' x 6' 8" sliding aluminum
	Demolish and Install	Ea.	1275	380.50		1655.50	patio door with insulated tempered
	Reinstall	Ea.		251.31		251.31	glass, baked-on enamel finish, screen,
	Clean	Ea.	.94	22.50		23.44	weatherstripped.
	Minimum Charge	Job		236		236	
10' Insulated							
	Demolish	Ea.		65.50		65.50	Cost includes material and labor to
	Install	Ea.	1600	470		2070	install 10' x 6' 8" sliding aluminum
	Demolish and Install	Ea.	1600	535.50		2135.50	patio door with double insulated
	Reinstall	Ea.		376.96		376.96	tempered glass, baked-on enamel
	Clean	Ea.	.94	22.50		23.44	finish, screen, weatherstripped.
	Minimum Charge	Job		236		236	

99

For customer support on your Contractor's Pricing Guide: Residential Repair & Remodeling, call 800.448.8182.

Sliding Patio Door	Unit	Material	Labor	Equip.	Total	Specification
Vinyl						
6' Single Pane						
Demolish	Ea.		65.50		65.50	Cost includes material and labor to
Install	Ea.	500	236		736	install 6' x 6' 8" sliding vinyl patio
Demolish and Install	Ea.	500	301.50		801.50	door with single pane tempered safety
Reinstall	Ea.		188.48		188.48	glass, anodized frame, screen,
Clean	Ea.	.94	22.50		23.44	weatherstripped.
Minimum Charge	Job		236		236	
8' Single Pane						
Demolish	Ea.		65.50		65.50	Cost includes material and labor to
Install	Ea.	665	315		980	install 8' x 6' 8" sliding vinyl patio
Demolish and Install	Ea.	665	380.50		1045.50	door with single pane tempered safety
Reinstall	Ea.		251.31		251.31	glass, anodized frame, screen,
Clean	Ea.	.94	22.50		23.44	weatherstripped.
Minimum Charge	Job		236		236	
10' Single Pane						
Demolish	Ea.		65.50		65.50	Cost includes material and labor to
Install	Ea.	720	470		1190	install 10' x 6' 8" sliding vinyl patio
Demolish and Install	Ea.	720	535.50		1255.50	door with single pane tempered safety
Reinstall	Ea.		376.96		376.96	glass, anodized frame, screen,
Clean	Ea.	.94	22.50		23.44	weatherstripped.
Minimum Charge	Job		236		236	
6' Insulated						
Demolish	Ea.		65.50		65.50	Cost includes material and labor to
Install	Ea.	1325	236		1561	install 6' x 6' 8" sliding vinyl patio
Demolish and Install	Ea.	1325	301.50		1626.50	door with double insulated tempered
Reinstall	Ea.		188.48		188.48	glass, anodized frame, screen,
Clean	Ea.	.94	22.50		23.44	weatherstripped.
Minimum Charge	Job		236		236	
8' Insulated						
Demolish	Ea.		65.50		65.50	Cost includes material and labor to
Install	Ea.	1100	315		1415	install 8' x 6' 8" sliding vinyl patio
Demolish and Install	Ea.	1100	380.50		1480.50	door with double insulated tempered
Reinstall	Ea.		251.31		251.31	glass, anodized frame, screen,
Clean	Ea.	.94	22.50		23.44	weatherstripped.
Minimum Charge	Job		236		236	
10' Insulated						
Demolish	Ea.		65.50		65.50	Cost includes material and labor to
Install	Ea.	1875	236		2111	install 10' x 6' 8" sliding vinyl patio
Demolish and Install	Ea.	1875	301.50		2176.50	door with double insulated tempered
Reinstall	Ea.		188.48		188.48	glass, anodized frame, screen,
Clean	Ea.	.94	22.50		23.44	weatherstripped.
Minimum Charge	Job		236		236	
Wood Framed						
6' Premium Insulated						
Demolish	Ea.		65.50		65.50	Cost includes material and labor to
Install	Ea.	1650	236		1886	install 6' x 6' 8" sliding wood patio
Demolish and Install	Ea.	1650	301.50		1951.50	door with pine frame, double insulated
Reinstall	Ea.		188.48		188.48	glass, screen, lock and dead bolt,
Clean	Ea.	.94	22.50		23.44	weatherstripped. Add for casing and
Minimum Charge	Job		236		236	finishing.

Exterior Doors and Windows

Sliding Patio Door		Unit	Material	Labor	Equip.	Total	Specification
8' Premium Insulated							Cost includes material and labor to
	Demolish	Ea.		65.50		65.50	install 8' x 6' 8" sliding wood patio
	Install	Ea.	2125	315		2440	door with pine frame, double insulated
	Demolish and Install	Ea.	2125	380.50		2505.50	glass, screen, lock and dead bolt,
	Reinstall	Ea.		251.31		251.31	weatherstripped. Add for casing and
	Clean	Ea.	.94	22.50		23.44	finishing.
	Minimum Charge	Job		236		236	
9' Premium Insulated							Cost includes material and labor to
	Demolish	Ea.		65.50		65.50	install 9' x 6' 8" sliding wood patio
	Install	Ea.	2125	315		2440	door with pine frame, double insulated
	Demolish and Install	Ea.	2125	380.50		2505.50	glass, screen, lock and dead bolt,
	Reinstall	Ea.		251.31		251.31	weatherstripped. Add for casing and
	Clean	Ea.	.94	22.50		23.44	finishing.
	Minimum Charge	Job		236		236	
Stain							Includes labor and material to stain
	Install	Ea.	12.60	43.50		56.10	exterior type door and trim on both
	Minimum Charge	Job		196		196	sides.
Casing Trim							Cost includes material and labor to
	Demolish	Ea.		9.90		9.90	install 11/16" x 2-1/2" pine ranch
	Install	Opng.	31	47		78	style door casing for one side of a
	Demolish and Install	Opng.	31	56.90		87.90	double door opening.
	Reinstall	Opng.		37.70		37.70	
	Clean	Opng.	.25	7.25		7.50	
	Paint	Ea.	5.05	8.90		13.95	
	Minimum Charge	Job		236		236	
Sliding Screen							Includes material and labor to install
	Demolish	Ea.		9.35		9.35	sliding screen door, 6' 8" x 3'.
	Install	Ea.	150	23.50		173.50	
	Demolish and Install	Ea.	150	32.85		182.85	
	Reinstall	Ea.		18.85		18.85	
	Clean	Ea.	.31	7.60		7.91	
	Minimum Charge	Job		236		236	

Screen Door		Unit	Material	Labor	Equip.	Total	Specification
Aluminum							Includes material and labor to install
	Demolish	Ea.		43.50		43.50	aluminum frame screen door with
	Install	Ea.	230	67.50		297.50	fiberglass screen cloth, closer, hinges,
	Demolish and Install	Ea.	230	111		341	latch.
	Reinstall	Ea.		53.85		53.85	
	Clean	Ea.	1.50	15.15		16.65	
	Paint	Ea.	4.84	15.65		20.49	
	Minimum Charge	Job		236		236	
Wood							Includes material and labor to install
	Demolish	Ea.		43.50		43.50	wood screen door with aluminum cloth
	Install	Ea.	340	78.50		418.50	screen. Frame and hardware not
	Demolish and Install	Ea.	340	122		462	included.
	Reinstall	Ea.		62.83		62.83	
	Clean	Ea.	1.50	15.15		16.65	
	Paint	Ea.	4.84	15.65		20.49	
	Minimum Charge	Job		236		236	
Shave & Refit							Includes labor to shave and rework
	Install	Ea.		39.50		39.50	door to fit opening at the job site.
	Minimum Charge	Job		236		236	

101

Exterior Doors and Windows

Storm Door

Storm Door		Unit	Material	Labor	Equip.	Total	Specification
Good Grade							
	Demolish	Ea.		43.50		43.50	Includes material and labor to install
	Install	Ea.	310	67.50		377.50	aluminum frame storm door with
	Demolish and Install	Ea.	310	111		421	fiberglass screen cloth, upper glass
	Reinstall	Ea.		53.85		53.85	panel, closer, hinges, latch.
	Clean	Ea.	1.50	15.15		16.65	
	Paint	Ea.	4.84	15.65		20.49	
	Minimum Charge	Job		236		236	
Better Grade							
	Demolish	Ea.		43.50		43.50	Includes material and labor to install
	Install	Ea.	282	67.50		349.50	aluminum frame storm door with
	Demolish and Install	Ea.	282	111		393	fiberglass screen cloth, upper and
	Reinstall	Ea.		53.85		53.85	lower glass panel, closer, latch.
	Clean	Ea.	1.50	15.15		16.65	
	Paint	Ea.	4.84	15.65		20.49	
	Minimum Charge	Job		236		236	
Premium Grade							
	Demolish	Ea.		43.50		43.50	Includes material and labor to install
	Install	Ea.	320	67.50		387.50	aluminum frame storm door with
	Demolish and Install	Ea.	320	111		431	fiberglass screen cloth, upper and
	Reinstall	Ea.		53.85		53.85	lower glass panel, closer, latch.
	Clean	Ea.	1.50	15.15		16.65	
	Paint	Ea.	4.84	15.65		20.49	
	Minimum Charge	Job		236		236	
Insulated							
	Demolish	Ea.		43.50		43.50	Includes material and labor to install
	Install	Ea.	310	33.50		343.50	aluminum frame storm door with
	Demolish and Install	Ea.	310	77		387	fiberglass screen cloth, upper and
	Reinstall	Ea.		26.93		26.93	lower glass panel, closer, latch.
	Clean	Ea.	1.50	15.15		16.65	
	Paint	Ea.	4.84	15.65		20.49	
	Minimum Charge	Job		236		236	

Garage Door

Garage Door		Unit	Material	Labor	Equip.	Total	Specification
9' One Piece Metal							
	Demolish	Ea.		110		110	Includes material and labor to install
	Install	Ea.	705	118		823	one piece steel garage door 9' x 7'
	Demolish and Install	Ea.	705	228		933	primed including hardware.
	Reinstall	Ea.		94.24		94.24	
	Clean	Ea.	1.87	45.50		47.37	
	Paint	Ea.	10.65	98		108.65	
	Minimum Charge	Job		236		236	
16' One Piece Metal							
	Demolish	Ea.		131		131	Includes material and labor to install
	Install	Ea.	925	157		1082	one piece steel garage door 16' x 7'
	Demolish and Install	Ea.	925	288		1213	primed including hardware.
	Reinstall	Ea.		125.65		125.65	
	Clean	Ea.	3.75	60.50		64.25	
	Paint	Ea.	21.50	112		133.50	
	Minimum Charge	Job		236		236	

For customer support on your Contractor's Pricing Guide: Residential Repair & Remodeling, call 800.448.8182.

Exterior Doors and Windows

Garage Door		Unit	Material	Labor	Equip.	Total	Specification
9' Metal Sectional							
	Demolish	Ea.		110		110	Includes material and labor to install
	Install	Ea.	1025	118		1143	sectional metal garage door 9' x 7'
	Demolish and Install	Ea.	1025	228		1253	primed including hardware.
	Reinstall	Ea.		94.24		94.24	
	Clean	Ea.	1.91	52		53.91	
	Paint	Ea.	21.50	112		133.50	
	Minimum Charge	Job		236		236	
16' Metal Sectional							
	Demolish	Ea.		131		131	Includes material and labor to install
	Install	Ea.	1200	157		1357	sectional metal garage door 16' x 7'
	Demolish and Install	Ea.	1200	288		1488	primed including hardware.
	Reinstall	Ea.		125.65		125.65	
	Clean	Ea.	3.75	60.50		64.25	
	Paint	Ea.	21.50	112		133.50	
	Minimum Charge	Job		236		236	
Metal / Steel Roll Up							
8' x 8' Steel							
	Demolish	Ea.		150		150	Cost includes material and labor to
	Install	Ea.	785	690		1475	install 8' x 8' steel overhead roll-up,
	Demolish and Install	Ea.	785	840		1625	chain hoist operated service door.
	Reinstall	Ea.		551.60		551.60	
	Clean	Ea.	1.91	52		53.91	
	Paint	Ea.	22.50	130		152.50	
	Minimum Charge	Job		1100		1100	
10' x 10' Steel							
	Demolish	Ea.		175		175	Cost includes material and labor to
	Install	Ea.	1200	790		1990	install 10' x 10' steel overhead roll-up,
	Demolish and Install	Ea.	1200	965		2165	chain hoist operated service door.
	Reinstall	Ea.		630.40		630.40	
	Clean	Ea.	2.99	72.50		75.49	
	Paint	Ea.	36	130		166	
	Minimum Charge	Job		1100		1100	
12' x 12' Steel							
	Demolish	Ea.		209		209	Cost includes material and labor to
	Install	Ea.	1300	920		2220	install 12' x 12' steel overhead roll-up,
	Demolish and Install	Ea.	1300	1129		2429	chain hoist operated service door.
	Reinstall	Ea.		735.47		735.47	
	Clean	Ea.	4.31	81		85.31	
	Paint	Ea.	51	156		207	
	Minimum Charge	Job		1100		1100	
14' x 14' Steel							
	Demolish	Ea.		262		262	Cost includes material and labor to
	Install	Ea.	3275	1375		4650	install 14' x 14' steel overhead roll-up,
	Demolish and Install	Ea.	3275	1637		4912	chain hoist operated service door.
	Reinstall	Ea.		1103.20		1103.20	
	Clean	Ea.	5.90	104		109.90	
	Paint	Ea.	60	196		256	
	Minimum Charge	Job		1100		1100	
18' x 18' Steel							
	Demolish	Ea.		291		291	Cost includes material and labor to
	Install	Opng.	3125	1850		4975	install 18' x 18' steel overhead roll-up,
	Demolish and Install	Opng.	3125	2141		5266	chain hoist operated service door.
	Reinstall	Opng.		1470.93		1470.93	
	Clean	Ea.	5.90	60.50		66.40	
	Paint	Ea.	116	390		506	
	Minimum Charge	Job		1100		1100	
Automatic Motor							
	Install	Ea.	650	47		697	Includes labor and material to install
	Minimum Charge	Job		236		236	electric opener motor, 1/3 HP.

Garage Door		Unit	Material	Labor	Equip.	Total	Specification
9' One Piece Wood							
	Demolish	Ea.		110		110	Includes material and labor to install
	Install	Ea.	820	118		938	one piece wood garage door 9' x 7'
	Demolish and Install	Ea.	820	228		1048	primed including hardware.
	Reinstall	Ea.		94.24		94.24	
	Clean	Ea.	1.87	45.50		47.37	
	Paint	Ea.	10.65	98		108.65	
	Minimum Charge	Job		236		236	
9' Wood Sectional							
	Demolish	Ea.		110		110	Includes material and labor to install
	Install	Ea.	1150	118		1268	sectional wood garage door 9' x 7'
	Demolish and Install	Ea.	1150	228		1378	unfinished including hardware.
	Reinstall	Ea.		94.24		94.24	
	Clean	Ea.	1.87	45.50		47.37	
	Paint	Ea.	10.65	98		108.65	
	Minimum Charge	Job		236		236	
16' Wood Sectional							
	Demolish	Ea.		131		131	Includes material and labor to install
	Install	Ea.	1825	157		1982	sectional fiberglass garage door 16' x
	Demolish and Install	Ea.	1825	288		2113	7' unfinished including hardware.
	Reinstall	Ea.		125.65		125.65	
	Clean	Ea.	3.75	60.50		64.25	
	Paint	Ea.	21.50	112		133.50	
	Minimum Charge	Job		236		236	
Electric Opener							
	Demolish	Ea.		81		81	Includes material and labor to install
	Install	Ea.	390	88.50		478.50	electric opener, 1/3 HP economy.
	Demolish and Install	Ea.	390	169.50		559.50	
	Reinstall	Ea.		70.72		70.72	
	Clean	Ea.	.08	20		20.08	
	Minimum Charge	Job		236		236	
Springs							
	Install	Pr.	200	118		318	Includes labor and material to replace
	Minimum Charge	Job		236		236	springs for garage door.

Commercial Metal Door		Unit	Material	Labor	Equip.	Total	Specification
Pre-hung Steel Door							
2' 8" x 6' 8"							
	Demolish	Ea.		65.50		65.50	Cost includes material and labor to
	Install	Ea.	715	29.50		744.50	install 2' 8" x 6' 8" pre-hung 1-3/4"
	Demolish and Install	Ea.	715	95		810	18 gauge steel door with steel frame,
	Reinstall	Ea.		29.45		29.45	hinges, aluminum sill, weatherstripped.
	Clean	Ea.	1.50	15.15		16.65	
	Paint	Ea.	6.20	32.50		38.70	
	Minimum Charge	Job		236		236	
3' x 6' 8"							
	Demolish	Ea.		65.50		65.50	Cost includes material and labor to
	Install	Ea.	720	31.50		751.50	install 3' x 6' 8" pre-hung 1-3/4" 18
	Demolish and Install	Ea.	720	97		817	gauge steel door with steel frame,
	Reinstall	Ea.		31.41		31.41	hinges, aluminum sill, weatherstripped.
	Clean	Ea.	1.50	15.15		16.65	
	Paint	Ea.	6.20	32.50		38.70	
	Minimum Charge	Job		236		236	

Exterior Doors and Windows

Commercial Metal Door	Unit	Material	Labor	Equip.	Total	Specification
3' 6" x 6' 8"						
Demolish	Ea.		65.50		65.50	Cost includes material and labor to
Install	Ea.	785	36.50		821.50	install 3' 6" x 6' 8" pre-hung 1-3/4"
Demolish and Install	Ea.	785	102		887	18 gauge steel door with steel frame,
Reinstall	Ea.		36.25		36.25	hinges, aluminum sill, weatherstripped.
Clean	Ea.	1.50	15.15		16.65	
Paint	Ea.	6.20	32.50		38.70	
Minimum Charge	Job		236		236	
4' x 6' 8"						
Demolish	Ea.		65.50		65.50	Cost includes material and labor to
Install	Ea.	835	47		882	install 4' x 6' 8" pre-hung 1-3/4" 18
Demolish and Install	Ea.	835	112.50		947.50	gauge steel door with steel frame,
Reinstall	Ea.		47.12		47.12	hinges, aluminum sill, weatherstripped.
Clean	Ea.	1.50	15.15		16.65	
Paint	Ea.	6.20	32.50		38.70	
Minimum Charge	Job		236		236	
Metal Jamb						
6' 8" High						
Demolish	Ea.		26		26	Cost includes material and labor to
Install	Ea.	152	29.50		181.50	install 18 gauge hollow metal door
Demolish and Install	Ea.	152	55.50		207.50	frame to fit a 4-1/2" jamb 6' 8" high
Reinstall	Ea.		29.45		29.45	and up to 3' 6" opening.
Clean	Ea.	.31	7.60		7.91	
Paint	Ea.	1.72	7.80		9.52	
Minimum Charge	Job		236		236	
7' High						
Demolish	Ea.		26		26	Cost includes material and labor to
Install	Ea.	157	31.50		188.50	install 18 gauge hollow metal door
Demolish and Install	Ea.	157	57.50		214.50	frame to fit a 4-1/2" jamb 7' high and
Reinstall	Ea.		31.41		31.41	up to 3' 6" opening.
Clean	Ea.	.31	7.60		7.91	
Paint	Ea.	1.72	7.80		9.52	
Minimum Charge	Job		236		236	
8' High						
Demolish	Ea.		26		26	Cost includes material and labor to
Install	Ea.	191	39.50		230.50	install 18 gauge hollow metal door
Demolish and Install	Ea.	191	65.50		256.50	frame to fit a 4-1/2" jamb 8' high and
Reinstall	Ea.		39.27		39.27	up to 3' 6" opening.
Clean	Ea.	.44	9.10		9.54	
Paint	Ea.	1.91	8.70		10.61	
Minimum Charge	Job		236		236	
9' High						
Demolish	Ea.		26		26	Cost includes material and labor to
Install	Ea.	242	47		289	install 18 gauge hollow metal door
Demolish and Install	Ea.	242	73		315	frame to fit a 4-1/2" jamb 9' high and
Reinstall	Ea.		47.12		47.12	up to 3' 6" opening.
Clean	Ea.	.44	9.10		9.54	
Paint	Ea.	1.91	8.70		10.61	
Minimum Charge	Job		236		236	
Hardware						
Doorknob w / Lock						
Demolish	Ea.		18.70		18.70	Includes material and labor to install
Install	Ea.	219	39.50		258.50	keyed entry lock set, bored type
Demolish and Install	Ea.	219	58.20		277.20	including knob trim, strike and strike
Reinstall	Ea.		31.41		31.41	box for standard commercial
Clean	Ea.	.15	11.35		11.50	application.
Minimum Charge	Job		236		236	

105

Commercial Metal Door	Unit	Material	Labor	Equip.	Total	Specification
Lever Handle						
Demolish	Ea.		18.70		18.70	Includes material and labor to install
Install	Ea.	179	47		226	passage lockset, keyless bored type
Demolish and Install	Ea.	179	65.70		244.70	with lever handle, non-locking
Reinstall	Ea.		37.70		37.70	standard commercial latchset.
Clean	Ea.	.15	11.35		11.50	
Minimum Charge	Job		236		236	
Deadbolt						
Demolish	Ea.		16.35		16.35	Includes material and labor to install a
Install	Ea.	81.50	33.50		115	deadbolt.
Demolish and Install	Ea.	81.50	49.85		131.35	
Reinstall	Ea.		26.93		26.93	
Clean	Ea.	.15	11.35		11.50	
Minimum Charge	Job		236		236	
Panic Device						
Demolish	Ea.		40.50		40.50	Includes material and labor to install
Install	Ea.	320	78.50		398.50	touch bar, low profile, exit only.
Demolish and Install	Ea.	320	119		439	
Reinstall	Ea.		62.83		62.83	
Clean	Ea.	.08	20		20.08	
Minimum Charge	Job		236		236	
Closer						
Demolish	Ea.		9.80		9.80	Includes material and labor to install
Install	Ea.	350	78.50		428.50	pneumatic heavy duty model for
Demolish and Install	Ea.	350	88.30		438.30	exterior type doors.
Reinstall	Ea.		62.83		62.83	
Clean	Ea.	.08	15.15		15.23	
Minimum Charge	Job		236		236	
Peep-Hole						
Demolish	Ea.		9		9	Includes material and labor to install
Install	Ea.	10.05	14.75		24.80	peep-hole for door.
Demolish and Install	Ea.	10.05	23.75		33.80	
Reinstall	Ea.		11.78		11.78	
Clean	Ea.	.01	4.19		4.20	
Minimum Charge	Job		236		236	
Exit Sign						
Demolish	Ea.		27.50		27.50	Includes material and labor to install a
Install	Ea.	113	67.50		180.50	wall mounted interior electric exit sign.
Demolish and Install	Ea.	113	95		208	
Reinstall	Ea.		54.04		54.04	
Clean	Ea.	.01	8.50		8.51	
Minimum Charge	Job		236		236	
Kickplate						
Demolish	Ea.		14.55		14.55	Includes material and labor to install
Install	Ea.	97	31.50		128.50	stainless steel kickplate, 10" x 36".
Demolish and Install	Ea.	97	46.05		143.05	
Reinstall	Ea.		25.13		25.13	
Clean	Ea.	.15	11.35		11.50	
Minimum Charge	Job		236		236	

Aluminum Window	Unit	Material	Labor	Equip.	Total	Specification
Single Hung Two Light						
2' x 2'						
Demolish	Ea.		25.50		25.50	Cost includes material and labor to
Install	Ea.	197	100		297	install 2' x 2' single hung, two light,
Demolish and Install	Ea.	197	125.50		322.50	double glazed aluminum window
Reinstall	Ea.		80.23		80.23	including screen.
Clean	Ea.	.02	9.75		9.77	
Minimum Charge	Job		276		276	

Exterior Doors and Windows

Aluminum Window		Unit	Material	Labor	Equip.	Total	Specification
2' x 2' 6"							
	Demolish	Ea.		25.50		25.50	Cost includes material and labor to
	Install	Ea.	73.50	100		173.50	install 2' x 2' 6" single hung, two
	Demolish and Install	Ea.	73.50	125.50		199	light, double glazed aluminum
	Reinstall	Ea.		80.23		80.23	window including screen.
	Clean	Ea.	.02	9.75		9.77	
	Minimum Charge	Job		276		276	
3' x 1' 6"							
	Demolish	Ea.		25.50		25.50	Cost includes material and labor to
	Install	Ea.	83	110		193	install 3' x 1' 6" single hung, two
	Demolish and Install	Ea.	83	135.50		218.50	light, double glazed aluminum
	Reinstall	Ea.		88.26		88.26	window including screen.
	Clean	Ea.	.01	8.50		8.51	
	Minimum Charge	Job		276		276	
3' x 2'							
	Demolish	Ea.		25.50		25.50	Cost includes material and labor to
	Install	Ea.	84.50	110		194.50	install 3' x 2' single hung, two light,
	Demolish and Install	Ea.	84.50	135.50		220	double glazed aluminum window
	Reinstall	Ea.		88.26		88.26	including screen.
	Clean	Ea.	.01	8.50		8.51	
	Minimum Charge	Job		276		276	
3' x 2' 6"							
	Demolish	Ea.		25.50		25.50	Cost includes material and labor to
	Install	Ea.	88	110		198	install 3' x 2' 6" single hung, two
	Demolish and Install	Ea.	88	135.50		223.50	light, double glazed aluminum
	Reinstall	Ea.		88.26		88.26	window including screen.
	Clean	Ea.	.02	9.75		9.77	
	Minimum Charge	Job		276		276	
3' x 3'							
	Demolish	Ea.		25.50		25.50	Cost includes material and labor to
	Install	Ea.	107	110		217	install 3' x 3' single hung, two light,
	Demolish and Install	Ea.	107	135.50		242.50	double glazed aluminum window
	Reinstall	Ea.		88.26		88.26	including screen.
	Clean	Ea.	.02	9.75		9.77	
	Minimum Charge	Job		276		276	
3' x 3' 6"							
	Demolish	Ea.		25.50		25.50	Cost includes material and labor to
	Install	Ea.	266	110		376	install 3' x 3' 6" single hung, two
	Demolish and Install	Ea.	266	135.50		401.50	light, double glazed aluminum
	Reinstall	Ea.		88.26		88.26	window including screen.
	Clean	Ea.	.02	11.35		11.37	
	Minimum Charge	Job		276		276	
3' x 4'							
	Demolish	Ea.		25.50		25.50	Cost includes material and labor to
	Install	Ea.	278	110		388	install 3' x 4' single hung, two light,
	Demolish and Install	Ea.	278	135.50		413.50	double glazed aluminum window
	Reinstall	Ea.		88.26		88.26	including screen.
	Clean	Ea.	.02	11.35		11.37	
	Minimum Charge	Job		276		276	
3' x 5'							
	Demolish	Ea.		36.50		36.50	Cost includes material and labor to
	Install	Ea.	146	110		256	install 3' x 5' single hung, two light,
	Demolish and Install	Ea.	146	146.50		292.50	double glazed aluminum window
	Reinstall	Ea.		88.26		88.26	including screen.
	Clean	Ea.	.03	13.60		13.63	
	Minimum Charge	Job		276		276	

Exterior Doors and Windows

Aluminum Window	Unit	Material	Labor	Equip.	Total	Specification
3' x 6'						
Demolish	Ea.		36.50		36.50	Cost includes material and labor to
Install	Ea.	380	123		503	install 3' x 6' single hung, two light,
Demolish and Install	Ea.	380	159.50		539.50	double glazed aluminum window
Reinstall	Ea.		98.06		98.06	including screen.
Clean	Ea.	.03	16		16.03	
Minimum Charge	Job		276		276	
Double Hung Two Light						
2' x 2'						
Demolish	Ea.		25.50		25.50	Cost includes material and labor to
Install	Ea.	130	100		230	install 2' x 2' double hung, two light,
Demolish and Install	Ea.	130	125.50		255.50	double glazed aluminum window
Reinstall	Ea.		80.23		80.23	including screen.
Clean	Ea.	.01	8.50		8.51	
Minimum Charge	Job		276		276	
2' x 2' 6"						
Demolish	Ea.		25.50		25.50	Cost includes material and labor to
Install	Ea.	145	100		245	install 2' x 2' 6" double hung, two
Demolish and Install	Ea.	145	125.50		270.50	light, double glazed aluminum
Reinstall	Ea.		80.23		80.23	window including screen.
Clean	Ea.	.01	8.50		8.51	
Minimum Charge	Job		276		276	
3' x 1' 6"						
Demolish	Ea.		25.50		25.50	Cost includes material and labor to
Install	Ea.	298	110		408	install 3' x 1' 6" double hung, two
Demolish and Install	Ea.	298	135.50		433.50	light, double glazed aluminum
Reinstall	Ea.		88.26		88.26	window including screen.
Clean	Ea.	.01	8.50		8.51	
Minimum Charge	Job		276		276	
3' x 2'						
Demolish	Ea.		25.50		25.50	Cost includes material and labor to
Install	Ea.	345	110		455	install 3' x 2' double hung, two light,
Demolish and Install	Ea.	345	135.50		480.50	double glazed aluminum window
Reinstall	Ea.		88.26		88.26	including screen.
Clean	Ea.	.01	8.50		8.51	
Minimum Charge	Job		276		276	
3' x 2' 6"						
Demolish	Ea.		25.50		25.50	Cost includes material and labor to
Install	Ea.	420	110		530	install 3' x 2' 6" double hung, two
Demolish and Install	Ea.	420	135.50		555.50	light, double glazed aluminum
Reinstall	Ea.		88.26		88.26	window including screen.
Clean	Ea.	.02	9.75		9.77	
Minimum Charge	Job		276		276	
3' x 3'						
Demolish	Ea.		25.50		25.50	Cost includes material and labor to
Install	Ea.	470	110		580	install 3' x 3' double hung, two light,
Demolish and Install	Ea.	470	135.50		605.50	double glazed aluminum window
Reinstall	Ea.		88.26		88.26	including screen.
Clean	Ea.	.02	9.75		9.77	
Minimum Charge	Job		276		276	
3' x 3' 6"						
Demolish	Ea.		25.50		25.50	Cost includes material and labor to
Install	Ea.	495	110		605	install 3' x 3' 6" double hung, two
Demolish and Install	Ea.	495	135.50		630.50	light, double glazed aluminum
Reinstall	Ea.		88.26		88.26	window including screen.
Clean	Ea.	.02	11.35		11.37	
Minimum Charge	Job		276		276	

For customer support on your Contractor's Pricing Guide: Residential Repair & Remodeling, call 800.448.8182.

Exterior Doors and Windows

Aluminum Window

	Unit	Material	Labor	Equip.	Total	Specification
3' x 4'						
Demolish	Ea.		25.50		25.50	Cost includes material and labor to
Install	Ea.	535	110		645	install 3' x 4' double hung, two light,
Demolish and Install	Ea.	535	135.50		670.50	double glazed aluminum window
Reinstall	Ea.		88.26		88.26	including screen.
Clean	Ea.	.02	11.35		11.37	
Minimum Charge	Job		276		276	
3' x 5'						
Demolish	Ea.		36.50		36.50	Cost includes material and labor to
Install	Ea.	585	110		695	install 3' x 5' double hung, two light,
Demolish and Install	Ea.	585	146.50		731.50	double glazed aluminum window
Reinstall	Ea.		88.26		88.26	including screen.
Clean	Ea.	.03	13.60		13.63	
Minimum Charge	Job		276		276	
3' x 6'						
Demolish	Ea.		36.50		36.50	Cost includes material and labor to
Install	Ea.	615	123		738	install 3' x 6' double hung, two light,
Demolish and Install	Ea.	615	159.50		774.50	double glazed aluminum window
Reinstall	Ea.		98.06		98.06	including screen.
Clean	Ea.	.03	16		16.03	
Minimum Charge	Job		276		276	

Sliding Sash

	Unit	Material	Labor	Equip.	Total	Specification
1' 6" x 3'						
Demolish	Ea.		25.50		25.50	Cost includes material and labor to
Install	Ea.	248	100		348	install 1' 6" x 3' sliding sash, two
Demolish and Install	Ea.	248	125.50		373.50	light, double glazed aluminum
Reinstall	Ea.		80.23		80.23	window including screen.
Clean	Ea.	.01	8.50		8.51	
Minimum Charge	Job		276		276	
2' x 2'						
Demolish	Ea.		25.50		25.50	Cost includes material and labor to
Install	Ea.	70	110		180	install 2' x 2' sliding sash, two light,
Demolish and Install	Ea.	70	135.50		205.50	double glazed aluminum window
Reinstall	Ea.		88.26		88.26	including screen.
Clean	Ea.	.01	8.50		8.51	
Minimum Charge	Job		276		276	
3' x 2'						
Demolish	Ea.		25.50		25.50	Cost includes material and labor to
Install	Ea.	254	110		364	install 3' x 2' sliding sash, two light,
Demolish and Install	Ea.	254	135.50		389.50	double glazed aluminum window
Reinstall	Ea.		88.26		88.26	including screen.
Clean	Ea.	.01	8.50		8.51	
Minimum Charge	Job		276		276	
3' x 2' 6"						
Demolish	Ea.		25.50		25.50	Cost includes material and labor to
Install	Ea.	93	110		203	install 3' x 2' 6" sliding sash, two
Demolish and Install	Ea.	93	135.50		228.50	light, double glazed aluminum
Reinstall	Ea.		88.26		88.26	window including screen.
Clean	Ea.	.02	9.75		9.77	
Minimum Charge	Job		276		276	
3' x 3'						
Demolish	Ea.		25.50		25.50	Cost includes material and labor to
Install	Ea.	315	110		425	install 3' x 3' sliding sash, two light,
Demolish and Install	Ea.	315	135.50		450.50	double glazed aluminum window
Reinstall	Ea.		88.26		88.26	including screen.
Clean	Ea.	.02	9.75		9.77	
Minimum Charge	Job		276		276	

Aluminum Window	Unit	Material	Labor	Equip.	Total	Specification
3' x 3' 6"						
Demolish	Ea.		25.50		25.50	Cost includes material and labor to
Install	Ea.	112	110		222	install 3' x 3' 6" sliding sash, two
Demolish and Install	Ea.	112	135.50		247.50	light, double glazed aluminum
Reinstall	Ea.		88.26		88.26	window including screen.
Clean	Ea.	.02	11.35		11.37	
Minimum Charge	Job		276		276	
3' x 4'						
Demolish	Ea.		25.50		25.50	Cost includes material and labor to
Install	Ea.	124	123		247	install 3' x 4' sliding sash, two light,
Demolish and Install	Ea.	124	148.50		272.50	double glazed aluminum window
Reinstall	Ea.		98.06		98.06	including screen.
Clean	Ea.	.02	11.35		11.37	
Minimum Charge	Job		276		276	
3' x 5'						
Demolish	Ea.		36.50		36.50	Cost includes material and labor to
Install	Ea.	153	123		276	install 3' x 5' sliding sash, two light,
Demolish and Install	Ea.	153	159.50		312.50	double glazed aluminum window
Reinstall	Ea.		98.06		98.06	including screen.
Clean	Ea.	.03	13.60		13.63	
Minimum Charge	Job		276		276	
4' x 5'						
Demolish	Ea.		36.50		36.50	Cost includes material and labor to
Install	Ea.	155	158		313	install 4' x 5' sliding sash, two light,
Demolish and Install	Ea.	155	194.50		349.50	double glazed aluminum window
Reinstall	Ea.		126.08		126.08	including screen.
Clean	Ea.	.04	19.45		19.49	
Minimum Charge	Job		276		276	
5' x 5'						
Demolish	Ea.		36.50		36.50	Cost includes material and labor to
Install	Ea.	360	158		518	install 5' x 5' sliding sash, two light,
Demolish and Install	Ea.	360	194.50		554.50	double glazed aluminum window
Reinstall	Ea.		126.08		126.08	including screen.
Clean	Ea.	.04	19.45		19.49	
Minimum Charge	Job		276		276	
Awning Sash						
26" x 26", 2 Light						
Demolish	Ea.		25.50		25.50	Cost includes material and labor to
Install	Ea.	350	123		473	install 26" x 26" two light awning,
Demolish and Install	Ea.	350	148.50		498.50	double glazed aluminum window
Reinstall	Ea.		98.06		98.06	including screen.
Clean	Ea.	.02	9.75		9.77	
Minimum Charge	Job		276		276	
26" x 37", 2 Light						
Demolish	Ea.		25.50		25.50	Cost includes material and labor to
Install	Ea.	360	123		483	install 26" x 37" two light awning,
Demolish and Install	Ea.	360	148.50		508.50	double glazed aluminum window
Reinstall	Ea.		98.06		98.06	including screen.
Clean	Ea.	.02	9.75		9.77	
Minimum Charge	Job		276		276	
38" x 26", 3 Light						
Demolish	Ea.		25.50		25.50	Cost includes material and labor to
Install	Ea.	400	123		523	install 38" x 26" three light awning,
Demolish and Install	Ea.	400	148.50		548.50	double glazed aluminum window
Reinstall	Ea.		98.06		98.06	including screen.
Clean	Ea.	.02	11.35		11.37	
Minimum Charge	Job		276		276	

For customer support on your Contractor's Pricing Guide: Residential Repair & Remodeling, call 800.448.8182.

Aluminum Window		Unit	Material	Labor	Equip.	Total	Specification
38" x 37", 3 Light							
	Demolish	Ea.		25.50		25.50	Cost includes material and labor to
	Install	Ea.	405	138		543	install 38" x 37" three light awning,
	Demolish and Install	Ea.	405	163.50		568.50	double glazed aluminum window
	Reinstall	Ea.		110.32		110.32	including screen.
	Clean	Ea.	.02	11.35		11.37	
	Minimum Charge	Job		276		276	
38" x 53", 3 Light							
	Demolish	Ea.		36.50		36.50	Cost includes material and labor to
	Install	Ea.	670	138		808	install 38" x 53" three light awning,
	Demolish and Install	Ea.	670	174.50		844.50	double glazed aluminum window
	Reinstall	Ea.		110.32		110.32	including screen.
	Clean	Ea.	.03	13.60		13.63	
	Minimum Charge	Job		276		276	
50" x 37", 4 Light							
	Demolish	Ea.		36.50		36.50	Cost includes material and labor to
	Install	Ea.	665	138		803	install 50" x 37" four light awning,
	Demolish and Install	Ea.	665	174.50		839.50	double glazed aluminum window
	Reinstall	Ea.		110.32		110.32	including screen.
	Clean	Ea.	.03	13.60		13.63	
	Minimum Charge	Job		276		276	
50" x 53", 4 Light							
	Demolish	Ea.		36.50		36.50	Cost includes material and labor to
	Install	Ea.	1125	158		1283	install 50" x 53" four light awning,
	Demolish and Install	Ea.	1125	194.50		1319.50	double glazed aluminum window
	Reinstall	Ea.		126.08		126.08	including screen.
	Clean	Ea.	.03	16		16.03	
	Minimum Charge	Job		276		276	
63" x 37", 5 Light							
	Demolish	Ea.		36.50		36.50	Cost includes material and labor to
	Install	Ea.	865	158		1023	install 63" x 37" five light awning,
	Demolish and Install	Ea.	865	194.50		1059.50	double glazed aluminum window
	Reinstall	Ea.		126.08		126.08	including screen.
	Clean	Ea.	.03	16		16.03	
	Minimum Charge	Job		276		276	
63" x 53", 5 Light							
	Demolish	Ea.		36.50		36.50	Cost includes material and labor to
	Install	Ea.	1200	158		1358	install 63" x 53" five light awning,
	Demolish and Install	Ea.	1200	194.50		1394.50	double glazed aluminum window
	Reinstall	Ea.		126.08		126.08	including screen.
	Clean	Ea.	.04	19.45		19.49	
	Minimum Charge	Job		276		276	
Mullion Bars							
	Install	Ea.	14.40	44		58.40	Includes labor and material to install
	Minimum Charge	Job		276		276	mullions for sliding aluminum window.
Bay							
4' 8" x 6' 4"							
	Demolish	Ea.		27		27	Cost includes material and labor to
	Install	Ea.	1650	118		1768	install 4' 8" x 6' 4" angle bay double
	Demolish and Install	Ea.	1650	145		1795	hung unit, pine frame with aluminum
	Reinstall	Ea.		94.24		94.24	cladding, picture center sash,
	Clean	Ea.	.07	22.50		22.57	hardware, screens.
	Minimum Charge	Job		276		276	

111

Exterior Doors and Windows

Aluminum Window		Unit	Material	Labor	Equip.	Total	Specification
4′ 8″ x 7′ 2″							
	Demolish	Ea.		27		27	Cost includes material and labor to
	Install	Ea.	1700	118		1818	install 4′ 8″ x 7′ 2″ angle bay double
	Demolish and Install	Ea.	1700	145		1845	hung unit, pine frame with aluminum
	Reinstall	Ea.		94.24		94.24	cladding, picture center sash,
	Clean	Ea.	.07	22.50		22.57	hardware, screens.
	Minimum Charge	Job		276		276	
4′ 8″ x 7′ 10″							
	Demolish	Ea.		27		27	Cost includes material and labor to
	Install	Ea.	1725	118		1843	install 4′ 8″ x 7′ 10″ angle bay
	Demolish and Install	Ea.	1725	145		1870	double hung unit, pine frame with
	Reinstall	Ea.		94.24		94.24	aluminum cladding, picture center
	Clean	Ea.	.07	22.50		22.57	sash, hardware, screens.
	Minimum Charge	Job		276		276	
4′ 8″ x 8′ 6″							
	Demolish	Ea.		31		31	Cost includes material and labor to
	Install	Ea.	1700	135		1835	install 4′ 8″ x 8′ 6″ angle bay double
	Demolish and Install	Ea.	1700	166		1866	hung unit, pine frame with aluminum
	Reinstall	Ea.		107.70		107.70	cladding, picture center sash,
	Clean	Ea.	.01	30.50		30.51	hardware, screens.
	Minimum Charge	Job		276		276	
4′ 8″ x 8′ 10″							
	Demolish	Ea.		31		31	Cost includes material and labor to
	Install	Ea.	1725	135		1860	install 4′ 8″ x 8′ 10″ angle bay
	Demolish and Install	Ea.	1725	166		1891	double hung unit, pine frame with
	Reinstall	Ea.		107.70		107.70	aluminum cladding, picture center
	Clean	Ea.	.01	30.50		30.51	sash, hardware, screens.
	Minimum Charge	Job		276		276	
4′ 8″ x 9′ 2″							
	Demolish	Ea.		31		31	Cost includes material and labor to
	Install	Ea.	1800	236		2036	install 4′ 8″ x 9′ 2″ angle bay double
	Demolish and Install	Ea.	1800	267		2067	hung unit, pine frame with aluminum
	Reinstall	Ea.		188.48		188.48	cladding, picture center sash,
	Clean	Ea.	.01	30.50		30.51	hardware, screens.
	Minimum Charge	Job		276		276	
4′ 8″ x 9′ 10″							
	Demolish	Ea.		31		31	Cost includes material and labor to
	Install	Ea.	1925	236		2161	install 4′ 8″ x 9′ 10″ angle bay
	Demolish and Install	Ea.	1925	267		2192	double hung unit, pine frame with
	Reinstall	Ea.		188.48		188.48	aluminum cladding, picture center
	Clean	Ea.	.01	30.50		30.51	sash, hardware, screens.
	Minimum Charge	Job		276		276	
Casing							
	Demolish	Ea.		10.90		10.90	Cost includes material and labor to
	Install	Opng.	36.50	36.50		73	install 11/16″ x 2-1/2″ pine ranch
	Demolish and Install	Opng.	36.50	47.40		83.90	style window casing.
	Reinstall	Opng.		29		29	
	Clean	Opng.	.21	3.64		3.85	
	Paint	Ea.	5.05	7.80		12.85	
	Minimum Charge	Job		276		276	
Storm Window							
2′ x 3′ 6″							
	Demolish	Ea.		14.95		14.95	Includes material and labor to install
	Install	Ea.	100	31.50		131.50	residential aluminum storm window
	Demolish and Install	Ea.	100	46.45		146.45	with mill finish, 2′ x 3′ 6″ high.
	Reinstall	Ea.		25.13		25.13	
	Clean	Ea.	.01	8.50		8.51	
	Minimum Charge	Job		276		276	

Exterior Doors and Windows

Aluminum Window

	Unit	Material	Labor	Equip.	Total	Specification
2' 6" x 5'						
Demolish	Ea.		14.95		14.95	Includes material and labor to install
Install	Ea.	106	33.50		139.50	residential aluminum storm window
Demolish and Install	Ea.	106	48.45		154.45	with mill finish, 2' 6" x 5' high.
Reinstall	Ea.		26.93		26.93	
Clean	Ea.	.02	11.35		11.37	
Minimum Charge	Job		276		276	
4' x 6'						
Demolish	Ea.		29		29	Includes material and labor to install
Install	Ea.	128	37.50		165.50	residential aluminum storm window, 4'
Demolish and Install	Ea.	128	66.50		194.50	x 6' high.
Reinstall	Ea.		30.16		30.16	
Clean	Ea.	.04	17.05		17.09	
Minimum Charge	Job		276		276	
Hardware						
Weatherstripping						
Install	L.F.	.81	3.93		4.74	Includes labor and material to install
Minimum Charge	Job		157		157	vinyl V strip weatherstripping for double-hung window.
Trim Set						
Demolish	Ea.		10.90		10.90	Cost includes material and labor to
Install	Opng.	36.50	36.50		73	install 11/16" x 2-1/2" pine ranch
Demolish and Install	Opng.	36.50	47.40		83.90	style window casing.
Reinstall	Opng.		29		29	
Clean	Opng.	.21	3.64		3.85	
Paint	Ea.	5.05	7.80		12.85	
Minimum Charge	Job		236		236	

Vinyl Window

	Unit	Material	Labor	Equip.	Total	Specification
Double Hung						
2' x 2' 6"						
Demolish	Ea.		18.35		18.35	Cost includes material and labor to
Install	Ea.	227	63		290	install 2' x 2' 6" double hung, two
Demolish and Install	Ea.	227	81.35		308.35	light, vinyl window with screen.
Reinstall	Ea.		50.26		50.26	
Clean	Ea.	.01	8.50		8.51	
Minimum Charge	Job		157		157	
2' x 3' 6"						
Demolish	Ea.		18.35		18.35	Cost includes material and labor to
Install	Ea.	238	67.50		305.50	install 2' x 3' 6" double hung, two
Demolish and Install	Ea.	238	85.85		323.85	light, vinyl window with screen.
Reinstall	Ea.		53.85		53.85	
Clean	Ea.	.02	9.75		9.77	
Minimum Charge	Job		157		157	
2' 6" x 4' 6"						
Demolish	Ea.		18.35		18.35	Cost includes material and labor to
Install	Ea.	283	72.50		355.50	install 2' 6" x 4' 6" double hung, two
Demolish and Install	Ea.	283	90.85		373.85	light, vinyl window with screen.
Reinstall	Ea.		57.99		57.99	
Clean	Ea.	.02	11.35		11.37	
Minimum Charge	Job		157		157	
3' x 4'						
Demolish	Ea.		18.35		18.35	Includes material and labor to install
Install	Ea.	252	94		346	medium size double hung, 3' x 4',
Demolish and Install	Ea.	252	112.35		364.35	two light, vinyl window with screen.
Reinstall	Ea.		75.39		75.39	
Clean	Ea.	.02	11.35		11.37	
Minimum Charge	Job		157		157	

Exterior Doors and Windows

Vinyl Window		Unit	Material	Labor	Equip.	Total	Specification
3' x 4' 6"							
	Demolish	Ea.		22.50		22.50	Cost includes material and labor to
	Install	Ea.	325	105		430	install 3' x 4' 6" double hung, two
	Demolish and Install	Ea.	325	127.50		452.50	light, vinyl window with screen.
	Reinstall	Ea.		83.77		83.77	
	Clean	Ea.	.03	13.60		13.63	
	Minimum Charge	Job		157		157	
4' x 4' 6"							
	Demolish	Ea.		22.50		22.50	Cost includes material and labor to
	Install	Ea.	340	118		458	install 4' x 4' 6" double hung, two
	Demolish and Install	Ea.	340	140.50		480.50	light, vinyl window with screen.
	Reinstall	Ea.		94.24		94.24	
	Clean	Ea.	.03	16		16.03	
	Minimum Charge	Job		157		157	
4' x 6'							
	Demolish	Ea.		22.50		22.50	Includes material and labor to install
	Install	Ea.	425	135		560	large double hung, 4' x 6', two light,
	Demolish and Install	Ea.	425	157.50		582.50	vinyl window with screen.
	Reinstall	Ea.		107.70		107.70	
	Clean	Ea.	.04	19.45		19.49	
	Minimum Charge	Job		157		157	
Trim Set							
	Demolish	Ea.		10.90		10.90	Cost includes material and labor to
	Install	Opng.	36.50	36.50		73	install 11/16" x 2-1/2" pine ranch
	Demolish and Install	Opng.	36.50	47.40		83.90	style window casing.
	Reinstall	Opng.		29		29	
	Clean	Opng.	.21	3.64		3.85	
	Paint	Ea.	5.05	7.80		12.85	
	Minimum Charge	Job		276		276	

Wood Window		Unit	Material	Labor	Equip.	Total	Specification
Double Hung							
2' 2" x 3' 4"							
	Demolish	Ea.		18.35		18.35	Cost includes material and labor to
	Install	Ea.	240	63		303	install 2' 2" x 3' 4" double hung, two
	Demolish and Install	Ea.	240	81.35		321.35	light, double glazed wood window
	Reinstall	Ea.		50.26		50.26	with screen.
	Clean	Ea.	.02	9.75		9.77	
	Paint	Ea.	3.16	16.30		19.46	
	Minimum Charge	Job		236		236	
2' 2" x 4' 4"							
	Demolish	Ea.		18.35		18.35	Cost includes material and labor to
	Install	Ea.	260	67.50		327.50	install 2' 2" x 4' 4" double hung, two
	Demolish and Install	Ea.	260	85.85		345.85	light, double glazed wood window
	Reinstall	Ea.		53.85		53.85	with screen.
	Clean	Ea.	.02	9.75		9.77	
	Paint	Ea.	3.16	16.30		19.46	
	Minimum Charge	Job		236		236	
2' 6" x 3' 4"							
	Demolish	Ea.		18.35		18.35	Cost includes material and labor to
	Install	Ea.	250	72.50		322.50	install 2' 6" x 3' 4" double hung, two
	Demolish and Install	Ea.	250	90.85		340.85	light, double glazed wood window
	Reinstall	Ea.		57.99		57.99	with screen.
	Clean	Ea.	.02	11.35		11.37	
	Paint	Ea.	4.04	19.55		23.59	
	Minimum Charge	Job		236		236	

Wood Window		Unit	Material	Labor	Equip.	Total	Specification
2' 6" x 4'							
	Demolish	Ea.		18.35		18.35	Cost includes material and labor to
	Install	Ea.	260	78.50		338.50	install 2' 6" x 4' double hung, two
	Demolish and Install	Ea.	260	96.85		356.85	light, double glazed wood window
	Reinstall	Ea.		62.83		62.83	with screen.
	Clean	Ea.	.02	11.35		11.37	
	Paint	Ea.	4.04	19.55		23.59	
	Minimum Charge	Job		236		236	
2' 6" x 4' 8"							
	Demolish	Ea.		18.35		18.35	Cost includes material and labor to
	Install	Ea.	279	78.50		357.50	install 2' 6" x 4' 8" double hung, two
	Demolish and Install	Ea.	279	96.85		375.85	light, double glazed wood window
	Reinstall	Ea.		62.83		62.83	with screen.
	Clean	Ea.	.02	11.35		11.37	
	Paint	Ea.	4.04	19.55		23.59	
	Minimum Charge	Job		236		236	
2' 10" x 3' 4"							
	Demolish	Ea.		18.35		18.35	Cost includes material and labor to
	Install	Ea.	252	94		346	install 2' 10" x 3' 4" double hung, two
	Demolish and Install	Ea.	252	112.35		364.35	light, double glazed wood window
	Reinstall	Ea.		75.39		75.39	with screen.
	Clean	Ea.	.02	11.35		11.37	
	Paint	Ea.	4.04	19.55		23.59	
	Minimum Charge	Job		236		236	
2' 10" x 4'							
	Demolish	Ea.		18.35		18.35	Cost includes material and labor to
	Install	Ea.	279	94		373	install 2' 10" x 4' double hung, two
	Demolish and Install	Ea.	279	112.35		391.35	light, double glazed wood window
	Reinstall	Ea.		75.39		75.39	with screen.
	Clean	Ea.	.02	11.35		11.37	
	Paint	Ea.	4.04	19.55		23.59	
	Minimum Charge	Job		236		236	
3' 7" x 3' 4"							
	Demolish	Ea.		18.35		18.35	Cost includes material and labor to
	Install	Ea.	287	105		392	install 3' 7" x 3' 4" double hung, two
	Demolish and Install	Ea.	287	123.35		410.35	light, double glazed wood window
	Reinstall	Ea.		83.77		83.77	with screen.
	Clean	Ea.	.02	11.35		11.37	
	Paint	Ea.	4.04	19.55		23.59	
	Minimum Charge	Job		236		236	
3' 7" x 5' 4"							
	Demolish	Ea.		22.50		22.50	Cost includes material and labor to
	Install	Ea.	325	105		430	install 3' 7" x 5' 4" double hung, two
	Demolish and Install	Ea.	325	127.50		452.50	light, double glazed wood window
	Reinstall	Ea.		83.77		83.77	with screen.
	Clean	Ea.	.04	19.45		19.49	
	Paint	Ea.	6.70	32.50		39.20	
	Minimum Charge	Job		236		236	
4' 3" x 5' 4"							
	Demolish	Ea.		22.50		22.50	Cost includes material and labor to
	Install	Ea.	570	118		688	install 4' 3" x 5' 4" double hung, two
	Demolish and Install	Ea.	570	140.50		710.50	light, double glazed wood window
	Reinstall	Ea.		94.24		94.24	with screen.
	Clean	Ea.	.04	19.45		19.49	
	Paint	Ea.	4.94	39		43.94	
	Minimum Charge	Job		236		236	

Exterior Doors and Windows

Wood Window	Unit	Material	Labor	Equip.	Total	Specification
Casement						
2' x 3' 4"						
Demolish	Ea.		18.35		18.35	Cost includes material and labor to
Install	Ea.	284	47		331	install 2' x 3' 4 " casement window
Demolish and Install	Ea.	284	65.35		349.35	with 3/4" insulated glass, screens,
Reinstall	Ea.		37.70		37.70	weatherstripping, hardware.
Clean	Ea.	.02	9.75		9.77	
Paint	Ea.	3.16	16.30		19.46	
Minimum Charge	Job		470		470	
2' x 4'						
Demolish	Ea.		18.35		18.35	Cost includes material and labor to
Install	Ea.	300	52.50		352.50	install 2' x 4' casement window with
Demolish and Install	Ea.	300	70.85		370.85	3/4" insulated glass, screens,
Reinstall	Ea.		41.88		41.88	weatherstripping, hardware.
Clean	Ea.	.02	9.75		9.77	
Paint	Ea.	3.16	16.30		19.46	
Minimum Charge	Job		470		470	
2' x 5'						
Demolish	Ea.		18.35		18.35	Cost includes material and labor to
Install	Ea.	350	55.50		405.50	install 2' x 5' casement window with
Demolish and Install	Ea.	350	73.85		423.85	3/4" insulated glass, screens,
Reinstall	Ea.		44.35		44.35	weatherstripping, hardware.
Clean	Ea.	.02	11.35		11.37	
Paint	Ea.	4.04	19.55		23.59	
Minimum Charge	Job		470		470	
2' x 6'						
Demolish	Ea.		18.35		18.35	Cost includes material and labor to
Install	Ea.	345	59		404	install 2' x 6' casement window with
Demolish and Install	Ea.	345	77.35		422.35	3/4" insulated glass, screens,
Reinstall	Ea.		47.12		47.12	weatherstripping, hardware.
Clean	Ea.	.02	11.35		11.37	
Paint	Ea.	4.04	19.55		23.59	
Minimum Charge	Job		470		470	
4' x 3' 4"						
Demolish	Ea.		22.50		22.50	Cost includes material and labor to
Install	Ea.	655	63		718	install 4' x 3' 4 " casement windows
Demolish and Install	Ea.	655	85.50		740.50	with 3/4" insulated glass, screens,
Reinstall	Ea.		50.26		50.26	weatherstripping, hardware.
Clean	Ea.	.02	11.35		11.37	
Paint	Ea.	2.48	19.55		22.03	
Minimum Charge	Job		470		470	
4' x 4'						
Demolish	Ea.		22.50		22.50	Cost includes material and labor to
Install	Ea.	735	63		798	install 4' x 4' casement windows with
Demolish and Install	Ea.	735	85.50		820.50	3/4" insulated glass, screens,
Reinstall	Ea.		50.26		50.26	weatherstripping, hardware.
Clean	Ea.	.03	16		16.03	
Paint	Ea.	2.93	24.50		27.43	
Minimum Charge	Job		470		470	
4' x 5'						
Demolish	Ea.		22.50		22.50	Cost includes material and labor to
Install	Ea.	830	67.50		897.50	install 4' x 5' casement windows with
Demolish and Install	Ea.	830	90		920	3/4" insulated glass, screens,
Reinstall	Ea.		53.85		53.85	weatherstripping, hardware.
Clean	Ea.	.04	19.45		19.49	
Paint	Ea.	3.82	32.50		36.32	
Minimum Charge	Job		470		470	

Exterior Doors and Windows

Wood Window		Unit	Material	Labor	Equip.	Total	Specification
4' x 6'							
	Demolish	Ea.		22.50		22.50	Cost includes material and labor to
	Install	Ea.	935	78.50		1013.50	install 4' x 6' casement windows with
	Demolish and Install	Ea.	935	101		1036	3/4" insulated glass, screens,
	Reinstall	Ea.		62.83		62.83	weatherstripping, hardware.
	Clean	Ea.	.04	19.45		19.49	
	Paint	Ea.	4.49	39		43.49	
	Minimum Charge	Job		470		470	
Casement Bow							
4' 8" x 8' 1"							
	Demolish	Ea.		31		31	Includes material and labor to install
	Install	Ea.	1750	177		1927	casement bow window unit with 4
	Demolish and Install	Ea.	1750	208		1958	lights, pine frame, 1/2" insulated
	Reinstall	Ea.		141.36		141.36	glass, weatherstripping, hardware,
	Clean	Ea.	.07	22.50		22.57	screens.
	Paint	Ea.	8.30	78		86.30	
	Minimum Charge	Job		470		470	
4' 8" x 10'							
	Demolish	Ea.		31		31	Includes material and labor to install
	Install	Ea.	2150	236		2386	casement bow window unit with 5
	Demolish and Install	Ea.	2150	267		2417	lights, pine frame, 1/2" insulated
	Reinstall	Ea.		188.48		188.48	glass, weatherstripping, hardware,
	Clean	Ea.	.01	30.50		30.51	screens.
	Paint	Ea.	10.35	98		108.35	
	Minimum Charge	Job		470		470	
5' 4" x 8' 1"							
	Demolish	Ea.		31		31	Includes material and labor to install
	Install	Ea.	1925	177		2102	casement bow window unit with 4
	Demolish and Install	Ea.	1925	208		2133	lights, pine frame, 1/2" insulated
	Reinstall	Ea.		141.36		141.36	glass, weatherstripping, hardware,
	Clean	Ea.	.01	30.50		30.51	screens.
	Paint	Ea.	9.45	78		87.45	
	Minimum Charge	Job		470		470	
5' 4" x 10'							
	Demolish	Ea.		31		31	Includes material and labor to install
	Install	Ea.	2150	236		2386	casement bow window unit with 5
	Demolish and Install	Ea.	2150	267		2417	lights, pine frame, 1/2" insulated
	Reinstall	Ea.		188.48		188.48	glass, weatherstripping, hardware,
	Clean	Ea.	.01	30.50		30.51	screens.
	Paint	Ea.	11.90	98		109.90	
	Minimum Charge	Job		470		470	
6' x 8' 1"							
	Demolish	Ea.		31		31	Includes material and labor to install
	Install	Ea.	2150	236		2386	casement bow window unit with 5
	Demolish and Install	Ea.	2150	267		2417	lights, pine frame, 1/2" insulated
	Reinstall	Ea.		188.48		188.48	glass, weatherstripping, hardware,
	Clean	Ea.	.01	30.50		30.51	screens.
	Paint	Ea.	10.80	98		108.80	
	Minimum Charge	Job		470		470	
6' x 10'							
	Demolish	Ea.		31		31	Includes material and labor to install
	Install	Ea.	2250	236		2486	casement bow window unit with 5
	Demolish and Install	Ea.	2250	267		2517	lights, pine frame, 1/2" insulated
	Reinstall	Ea.		188.48		188.48	glass, weatherstripping, hardware,
	Clean	Ea.	.01	30.50		30.51	screens.
	Paint	Ea.	13.50	130		143.50	
	Minimum Charge	Job		470		470	

Exterior Doors and Windows

Wood Window	Unit	Material	Labor	Equip.	Total	Specification
Casing						
Demolish	Ea.		10.90		10.90	Cost includes material and labor to
Install	Opng.	36.50	36.50		73	install 11/16" x 2-1/2" pine ranch
Demolish and Install	Opng.	36.50	47.40		83.90	style window casing.
Reinstall	Opng.		29		29	
Clean	Opng.	.21	3.64		3.85	
Paint	Ea.	5.05	7.80		12.85	
Minimum Charge	Job		276		276	
Trim Set						
Demolish	Ea.		8.30		8.30	Cost includes material and labor to
Install	Opng.	36.50	36.50		73	install 11/16" x 2-1/2" pine ranch
Demolish and Install	Opng.	36.50	44.80		81.30	style window casing.
Reinstall	Opng.		29		29	
Clean	Opng.	.21	3.64		3.85	
Paint	Ea.	5.05	7.80		12.85	
Minimum Charge	Job		236		236	
Wood Picture Window						
3' 6" x 4'						
Demolish	Ea.		22.50		22.50	Includes material and labor to install a
Install	Ea.	490	78.50		568.50	wood framed picture window
Demolish and Install	Ea.	490	101		591	including exterior trim.
Clean	Ea.	.04	19.45		19.49	
Paint	Ea.	3.58	30		33.58	
Minimum Charge	Job		470		470	
4' x 4' 6"						
Demolish	Ea.		22.50		22.50	Includes material and labor to install a
Install	Ea.	590	85.50		675.50	wood framed picture window
Demolish and Install	Ea.	590	108		698	including exterior trim.
Clean	Ea.	.04	19.45		19.49	
Paint	Ea.	3.58	30		33.58	
Minimum Charge	Job		470		470	
5' x 4'						
Demolish	Ea.		22.50		22.50	Includes material and labor to install a
Install	Ea.	670	85.50		755.50	wood framed picture window
Demolish and Install	Ea.	670	108		778	including exterior trim.
Clean	Ea.	.04	19.45		19.49	
Paint	Ea.	7.65	49		56.65	
Minimum Charge	Job		470		470	
6' x 4' 6"						
Demolish	Ea.		22.50		22.50	Includes material and labor to install a
Install	Ea.	710	94		804	wood framed picture window
Demolish and Install	Ea.	710	116.50		826.50	including exterior trim.
Clean	Ea.	.04	19.45		19.49	
Paint	Ea.	7.65	49		56.65	
Minimum Charge	Job		470		470	

Window Screens	Unit	Material	Labor	Equip.	Total	Specification
Small						
Demolish	Ea.		4.21		4.21	Includes material and labor to install
Install	Ea.	22	14.30		36.30	small window screen, 4 S.F.
Demolish and Install	Ea.	22	18.51		40.51	
Reinstall	Ea.		11.42		11.42	
Clean	Ea.	.15	7.60		7.75	
Minimum Charge	Job		118		118	

For customer support on your Contractor's Pricing Guide: Residential Repair & Remodeling, call 800.448.8182.

Exterior Doors and Windows

Window Screens		Unit	Material	Labor	Equip.	Total	Specification
Standard Size							
	Demolish	Ea.		4.21		4.21	Includes material and labor to install
	Install	Ea.	53	14.30		67.30	medium window screen, 12 S.F.
	Demolish and Install	Ea.	53	18.51		71.51	
	Reinstall	Ea.		11.42		11.42	
	Clean	Ea.	.15	7.60		7.75	
	Minimum Charge	Job		118		118	
Large							
	Demolish	Ea.		4.21		4.21	Includes material and labor to install
	Install	Ea.	70.50	14.30		84.80	large window screen, 16 S.F.
	Demolish and Install	Ea.	70.50	18.51		89.01	
	Reinstall	Ea.		11.42		11.42	
	Clean	Ea.	.15	7.60		7.75	
	Minimum Charge	Job		118		118	

Window Glass		Unit	Material	Labor	Equip.	Total	Specification
Plate, Clear, 1/4"							
	Demolish	S.F.		2.02		2.02	Includes material and labor to install
	Install	S.F.	7.80	7.65		15.45	clear glass.
	Demolish and Install	S.F.	7.80	9.67		17.47	
	Reinstall	S.F.		6.10		6.10	
	Clean	S.F.	.01	.21		.22	
	Minimum Charge	Job		229		229	
Insulated							
	Demolish	S.F.		2.02		2.02	Includes material and labor to install
	Install	S.F.	16.45	13.05		29.50	insulated glass.
	Demolish and Install	S.F.	16.45	15.07		31.52	
	Reinstall	S.F.		10.46		10.46	
	Clean	S.F.	.01	.21		.22	
	Minimum Charge	Job		229		229	
Tinted							
	Demolish	S.F.		2.02		2.02	Includes material and labor to install
	Install	S.F.	7.70	7.05		14.75	tinted glass.
	Demolish and Install	S.F.	7.70	9.07		16.77	
	Reinstall	S.F.		5.63		5.63	
	Clean	S.F.	.01	.21		.22	
	Minimum Charge	Job		229		229	
Obscure / Florentine							
	Demolish	S.F.		2.02		2.02	Includes material and labor to install
	Install	S.F.	13.80	6.55		20.35	obscure glass.
	Demolish and Install	S.F.	13.80	8.57		22.37	
	Reinstall	S.F.		5.23		5.23	
	Clean	S.F.	.01	.21		.22	
	Minimum Charge	Job		229		229	
Laminated Safety							
	Demolish	S.F.		2.02		2.02	Includes material and labor to install
	Install	S.F.	14.35	10.15		24.50	clear laminated glass.
	Demolish and Install	S.F.	14.35	12.17		26.52	
	Reinstall	S.F.		8.14		8.14	
	Clean	S.F.	.01	.21		.22	
	Minimum Charge	Job		229		229	

Window Glass

Window Glass	Unit	Material	Labor	Equip.	Total	Specification
Wired						
Demolish	S.F.		2.02		2.02	Includes material and labor to install
Install	S.F.	29.50	6.80		36.30	rough obscure wire glass.
Demolish and Install	S.F.	29.50	8.82		38.32	
Reinstall	S.F.		5.42		5.42	
Clean	S.F.	.01	.21		.22	
Minimum Charge	Job		229		229	
Tempered						
Demolish	S.F.		2.02		2.02	Includes material and labor to install
Install	S.F.	7.80	7.65		15.45	clear glass.
Demolish and Install	S.F.	7.80	9.67		17.47	
Reinstall	S.F.		6.10		6.10	
Clean	S.F.	.01	.21		.22	
Minimum Charge	Job		229		229	
Stained						
Demolish	S.F.		2.02		2.02	Includes material and labor to install
Install	S.F.	70	7.65		77.65	stained glass.
Demolish and Install	S.F.	70	9.67		79.67	
Reinstall	S.F.		6.10		6.10	
Clean	S.F.	.01	.21		.22	
Minimum Charge	Job		229		229	
Etched						
Demolish	S.F.		2.02		2.02	Includes material and labor to install
Install	S.F.	58	7.65		65.65	etched glass.
Demolish and Install	S.F.	58	9.67		67.67	
Reinstall	S.F.		6.10		6.10	
Clean	S.F.	.01	.21		.22	
Minimum Charge	Job		229		229	
Wall Mirror						
Demolish	S.F.		2.02		2.02	Includes material and labor to install
Install	S.F.	10.85	7.30		18.15	distortion-free float glass mirror, 1/4"
Demolish and Install	S.F.	10.85	9.32		20.17	thick, cut and polished edges.
Reinstall	S.F.		5.86		5.86	
Clean	S.F.	.01	.21		.22	
Minimum Charge	Job		229		229	
Bullet Resistant						
Demolish	S.F.		2.02		2.02	Includes material and labor to install
Install	S.F.	143	91.50		234.50	bullet resistant clear laminated glass.
Demolish and Install	S.F.	143	93.52		236.52	
Reinstall	S.F.		73.22		73.22	
Clean	S.F.	.01	.21		.22	
Minimum Charge	Job		229		229	

Storefront System

Storefront System	Unit	Material	Labor	Equip.	Total	Specification
Clear Glass						
Stub Wall to 8'						
Demolish	S.F.		3.16		3.16	Includes material and labor to install
Install	S.F.	27	12.80		39.80	commercial grade 4-1/2" section,
Demolish and Install	S.F.	27	15.96		42.96	anodized aluminum with clear glass in
Reinstall	S.F.		10.23		10.23	upper section and safety glass in
Clean	S.F.		.45		.45	lower section.
Minimum Charge	Job		276		276	

Exterior Doors and Windows

Storefront System	Unit	Material	Labor	Equip.	Total	Specification
Floor to 10′						
Demolish	S.F.		3.16		3.16	Includes material and labor to install
Install	S.F.	27.50	12.80		40.30	commercial grade 4-1/2″ section,
Demolish and Install	S.F.	27.50	15.96		43.46	anodized aluminum with clear glass in
Reinstall	S.F.		10.23		10.23	upper section and safety glass in
Clean	S.F.		.45		.45	lower section.
Minimum Charge	Job		276		276	
Polished Plate Glass						
Stub Wall to 9′						
Demolish	S.F.		3.16		3.16	Cost includes material and labor to
Install	S.F.	17.90	12.80		30.70	install 4′ x 9′ polished plate glass
Demolish and Install	S.F.	17.90	15.96		33.86	window built on 8″ stub wall.
Reinstall	S.F.		10.23		10.23	
Clean	S.F.		.45		.45	
Minimum Charge	Job		276		276	
Stub Wall to 13′						
Demolish	S.F.		3.16		3.16	Cost includes material and labor to
Install	S.F.	19.30	12.80		32.10	install 4′ x 13′ polished plate glass
Demolish and Install	S.F.	19.30	15.96		35.26	window built on 8″ stub wall.
Reinstall	S.F.		10.23		10.23	
Clean	S.F.		.45		.45	
Minimum Charge	Job		276		276	
Floor to 9′						
Demolish	S.F.		3.16		3.16	Cost includes material and labor to
Install	S.F.	19.90	12.80		32.70	install 4′ x 9′ polished plate glass
Demolish and Install	S.F.	19.90	15.96		35.86	window built on floor.
Reinstall	S.F.		10.23		10.23	
Clean	S.F.		.45		.45	
Minimum Charge	Job		276		276	
Floor to 13′						
Demolish	S.F.		3.16		3.16	Cost includes material and labor to
Install	S.F.	21.50	12.80		34.30	install 4′ x 13′ polished plate glass
Demolish and Install	S.F.	21.50	15.96		37.46	window built on floor.
Reinstall	S.F.		10.23		10.23	
Clean	S.F.		.45		.45	
Minimum Charge	Job		276		276	
Clear Plate Glass						
Floor to 9′						
Demolish	S.F.		3.16		3.16	Cost includes material and labor to
Install	S.F.	28.50	12.80		41.30	install 4′ x 9′ clear plate glass window
Demolish and Install	S.F.	28.50	15.96		44.46	built on floor.
Reinstall	S.F.		10.23		10.23	
Clean	S.F.		.45		.45	
Minimum Charge	Job		276		276	
Floor to 11′						
Demolish	S.F.		3.16		3.16	Cost includes material and labor to
Install	S.F.	28	12.80		40.80	install 4′ x 11′ clear plate glass
Demolish and Install	S.F.	28	15.96		43.96	window built on floor.
Reinstall	S.F.		10.23		10.23	
Clean	S.F.		.45		.45	
Minimum Charge	Job		276		276	
Floor to 13′						
Demolish	S.F.		3.16		3.16	Cost includes material and labor to
Install	S.F.	27.50	12.80		40.30	install 4′ x 13′ clear plate glass
Demolish and Install	S.F.	27.50	15.96		43.46	window built on floor.
Reinstall	S.F.		10.23		10.23	
Clean	S.F.		.45		.45	
Minimum Charge	Job		276		276	

For customer support on your Contractor's Pricing Guide: Residential Repair & Remodeling, call 800.448.8182.

Exterior Doors and Windows

Storefront System		Unit	Material	Labor	Equip.	Total	Specification
Entrance							
Narrow Stile 3' x 7'							
	Demolish	Ea.		26		26	Cost includes material and labor to
	Install	Ea.	1100	315		1415	install 3' x 7' narrow stile door, center
	Demolish and Install	Ea.	1100	341		1441	pivot concealed closer.
	Reinstall	Ea.		251.31		251.31	
	Clean	Ea.	1.87	45.50		47.37	
	Minimum Charge	Job		276		276	
Tempered Glass							
	Demolish	Ea.		26		26	Cost includes material and labor to
	Install	Ea.	3375	595		3970	install 3' x 7' tempered glass door,
	Demolish and Install	Ea.	3375	621		3996	center pivot concealed closer.
	Reinstall	Ea.		477.06		477.06	
	Clean	Ea.	1.87	45.50		47.37	
	Minimum Charge	Job		276		276	
Heavy Section							
	Demolish	Ea.		26		26	Includes material and labor to install
	Install	Ea.	1100	550		1650	wide stile door with 3' 6" x 7'
	Demolish and Install	Ea.	1100	576		1676	opening.
	Reinstall	Ea.		441.28		441.28	
	Clean	Ea.	.37	15.15		15.52	
	Minimum Charge	Job		915		915	

Curtain Wall System		Unit	Material	Labor	Equip.	Total	Specification
Regular Weight							
Bronze Anodized							
	Demolish	S.F.		3.16		3.16	Includes material and labor to install
	Install	S.F.	44	12.80		56.80	structural curtain wall system with
	Demolish and Install	S.F.	44	15.96		59.96	bronze anodized frame of regular
	Clean	S.F.		.45		.45	weight, float glass.
	Minimum Charge	Job		236		236	
Black Anodized							
	Demolish	S.F.		3.16		3.16	Includes material and labor to install
	Install	S.F.	48	12.80		60.80	structural curtain wall system with
	Demolish and Install	S.F.	48	15.96		63.96	black anodized frame of regular
	Clean	S.F.		.45		.45	weight, float glass.
	Minimum Charge	Job		236		236	
Heavy Weight							
Bronze Anodized							
	Demolish	S.F.		3.16		3.16	Includes material and labor to install
	Install	S.F.	44	12.80		56.80	structural curtain wall system with
	Demolish and Install	S.F.	44	15.96		59.96	bronze anodized frame of regular
	Clean	S.F.		.45		.45	weight, float glass.
	Minimum Charge	Job		236		236	
Black Anodized							
	Demolish	S.F.		3.16		3.16	Includes material and labor to install
	Install	S.F.	45.50	12.80		58.30	structural curtain wall system with
	Demolish and Install	S.F.	45.50	15.96		61.46	black anodized frame of heavy
	Clean	S.F.		.45		.45	weight, float glass.
	Minimum Charge	Job		236		236	
Glass							
1" Insulated							
	Install	S.F.	19.45	12.20		31.65	Cost includes labor and material to
	Minimum Charge	Job		229		229	install 1" insulated glass.

Skylight	Unit	Material	Labor	Equip.	Total	Specification
Dome						
22″ x 22″						
Demolish	Ea.		93		93	Cost includes material and labor to
Install	Ea.	238	147		385	install 22″ x 22″ fixed dome skylight.
Demolish and Install	Ea.	238	240		478	
Reinstall	Ea.		117.50		117.50	
Clean	Ea.	.17	13		13.17	
Minimum Charge	Job		880		880	
22″ x 46″						
Demolish	Ea.		93		93	Cost includes material and labor to
Install	Ea.	298	176		474	install 22″ x 46″ fixed dome skylight.
Demolish and Install	Ea.	298	269		567	
Reinstall	Ea.		141		141	
Clean	Ea.		20		20	
Minimum Charge	Job		880		880	
30″ x 30″						
Demolish	Ea.		93		93	Cost includes material and labor to
Install	Ea.	315	147		462	install 30″ x 30″ fixed dome skylight.
Demolish and Install	Ea.	315	240		555	
Reinstall	Ea.		117.50		117.50	
Clean	Ea.		20		20	
Minimum Charge	Job		880		880	
30″ x 46″						
Demolish	Ea.		93		93	Cost includes material and labor to
Install	Ea.	415	176		591	install 30″ x 46″ fixed dome skylight.
Demolish and Install	Ea.	415	269		684	
Reinstall	Ea.		141		141	
Clean	Ea.		20		20	
Minimum Charge	Job		880		880	
Fixed						
22″ x 27″						
Demolish	Ea.		93		93	Includes material and labor to install
Install	Ea.	305	147		452	double glazed skylight, 22″ x 27″,
Demolish and Install	Ea.	305	240		545	fixed.
Reinstall	Ea.		117.50		117.50	
Clean	Ea.	.17	13		13.17	
Minimum Charge	Job		880		880	
22″ x 46″						
Demolish	Ea.		93		93	Includes material and labor to install
Install	Ea.	320	176		496	double glazed skylight, 22″ x 46″,
Demolish and Install	Ea.	320	269		589	fixed.
Reinstall	Ea.		141		141	
Clean	Ea.		20		20	
Minimum Charge	Job		880		880	
44″ x 46″						
Demolish	Ea.		93		93	Includes material and labor to install
Install	Ea.	475	176		651	double glazed skylight, 44″ x 46″,
Demolish and Install	Ea.	475	269		744	fixed.
Reinstall	Ea.		141		141	
Clean	Ea.		22.50		22.50	
Minimum Charge	Job		880		880	
Operable						
22″ x 27″						
Demolish	Ea.		93		93	Includes material and labor to install
Install	Ea.	445	147		592	double glazed skylight, 22″ x 27″,
Demolish and Install	Ea.	445	240		685	operable.
Reinstall	Ea.		117.50		117.50	
Clean	Ea.	.17	13		13.17	
Minimum Charge	Job		880		880	

123

Exterior Doors and Windows

Skylight

Skylight	Unit	Material	Labor	Equip.	Total	Specification
22" x 46"						
Demolish	Ea.		93		93	Includes material and labor to install
Install	Ea.	535	176		711	double glazed skylight, 22" x 46",
Demolish and Install	Ea.	535	269		804	operable.
Reinstall	Ea.		141		141	
Clean	Ea.		20		20	
Minimum Charge	Job		880		880	
44" x 46"						
Demolish	Ea.		93		93	Includes material and labor to install
Install	Ea.	1025	176		1201	double glazed skylight, 44" x 46",
Demolish and Install	Ea.	1025	269		1294	operable.
Reinstall	Ea.		141		141	
Clean	Ea.		22.50		22.50	
Minimum Charge	Job		880		880	

Industrial Steel Window

Industrial Steel Window	Unit	Material	Labor	Equip.	Total	Specification
100% Fixed						
Demolish	S.F.		2.59		2.59	Includes material and labor to install
Install	S.F.	32.50	5.50		38	commercial grade steel windows,
Demolish and Install	S.F.	32.50	8.09		40.59	including glazing.
Reinstall	S.F.		4.41		4.41	
Clean	Ea.	.07	1.42		1.49	
Paint	S.F.	.18	.87		1.05	
Minimum Charge	Job		276		276	
50% Vented						
Demolish	S.F.		2.59		2.59	Includes material and labor to install
Install	S.F.	61	5.50		66.50	commercial grade steel windows,
Demolish and Install	S.F.	61	8.09		69.09	including glazing.
Reinstall	S.F.		4.41		4.41	
Clean	Ea.	.07	1.42		1.49	
Paint	S.F.	.18	.87		1.05	
Minimum Charge	Job		276		276	

Brick Veneer

Concrete Masonry Units

Masonry Wall	Unit	Material	Labor	Equip.	Total	Specification
Unreinforced CMU						
4" x 8" x 16"						
Demolish	S.F.		.14	.30	.44	Includes material and labor to install
Install	S.F.	2.11	4.78		6.89	normal weight 4" x 8" x 16" block.
Demolish and Install	S.F.	2.11	4.92	.30	7.33	
Clean	S.F.	.10	.56		.66	
Paint	S.F.	.15	.89		1.04	
Minimum Charge	Job		235		235	
6" x 8" x 16"						
Demolish	S.F.		.18	.41	.59	Includes material and labor to install
Install	S.F.	2.83	4.99		7.82	normal weight 6" x 8" x 16" block.
Demolish and Install	S.F.	2.83	5.17	.41	8.41	
Clean	S.F.	.10	.56		.66	
Paint	S.F.	.15	.89		1.04	
Minimum Charge	Job		235		235	
8" x 8" x 16"						
Demolish	S.F.		.37	.81	1.18	Includes material and labor to install
Install	S.F.	2.96	5.50		8.46	normal weight 8" x 8" x 16" block.
Demolish and Install	S.F.	2.96	5.87	.81	9.64	
Clean	S.F.	.10	.56		.66	
Paint	S.F.	.15	.89		1.04	
Minimum Charge	Job		235		235	
12" x 8" x 16"						
Demolish	S.F.		.37	.81	1.18	Includes material and labor to install
Install	S.F.	4.73	8.35		13.08	normal weight 12" x 8" x 16" block.
Demolish and Install	S.F.	4.73	8.72	.81	14.26	
Clean	S.F.	.10	.56		.66	
Paint	S.F.	.15	.89		1.04	
Minimum Charge	Job		235		235	
Reinforced CMU						
4" x 8" x 16"						
Demolish	S.F.		2.24	.75	2.99	Includes material and labor to install
Install	S.F.	2.31	4.88		7.19	normal weight 4" x 8" x 16" block,
Demolish and Install	S.F.	2.31	7.12	.75	10.18	mortar and 1/2" re-bar installed
Clean	S.F.	.10	.56		.66	vertically every 24" O.C.
Paint	S.F.	.15	.89		1.04	
Minimum Charge	Job		430		430	
6" x 8" x 16"						
Demolish	S.F.		2.36	.52	2.88	Includes material and labor to install
Install	S.F.	3.04	5.10		8.14	normal weight 6" x 8" x 16" block,
Demolish and Install	S.F.	3.04	7.46	.52	11.02	mortar and 1/2" rebar installed
Clean	S.F.	.10	.56		.66	vertically every 24" O.C.
Paint	S.F.	.15	.89		1.04	
Minimum Charge	Job		430		430	

Masonry

Masonry Wall		Unit	Material	Labor	Equip.	Total	Specification
8" x 8" x 16"							
	Demolish	S.F.		2.49	.55	3.04	Includes material and labor to install
	Install	S.F.	3.18	5.55		8.73	normal weight 8" x 8" x 16" block,
	Demolish and Install	S.F.	3.18	8.04	.55	11.77	mortar and 1/2" rebar installed
	Clean	S.F.	.10	.56		.66	vertically every 24" O.C.
	Paint	S.F.	.15	.89		1.04	
	Minimum Charge	Job		430		430	
12" x 8" x 16"							
	Demolish	S.F.		2.57	.57	3.14	Includes material and labor to install
	Install	S.F.	4.94	8.65		13.59	normal weight 12" x 8" x 16" block,
	Demolish and Install	S.F.	4.94	11.22	.57	16.73	mortar and 1/2" rebar installed
	Clean	S.F.	.10	.56		.66	vertically every 24" O.C.
	Paint	S.F.	.15	.89		1.04	
	Minimum Charge	Job		430		430	
Structural Brick							
4" Single Row							
	Demolish	S.F.		3.23		3.23	Cost includes material and labor to
	Install	S.F.	4.48	10.20		14.68	install 4" red brick wall, running bond
	Demolish and Install	S.F.	4.48	13.43		17.91	with 3/8" concave joints, mortar and
	Clean	S.F.	.10	.48		.58	ties.
	Paint	S.F.	.15	.89		1.04	
	Minimum Charge	Job		430		430	
8" Double Row							
	Demolish	S.F.		3.23		3.23	Cost includes material and labor to
	Install	S.F.	10.10	16.30		26.40	install 8" red brick wall, running bond,
	Demolish and Install	S.F.	10.10	19.53		29.63	with 3/8" concave joints, mortar and
	Clean	S.F.	.10	.48		.58	wall ties.
	Paint	S.F.	.15	.89		1.04	
	Minimum Charge	Job		430		430	
12" Triple Row							
	Demolish	S.F.		4.49		4.49	Cost includes material and labor to
	Install	S.F.	15.25	23		38.25	install 12" red brick wall, running
	Demolish and Install	S.F.	15.25	27.49		42.74	bond, with 3/8" concave joints,
	Clean	S.F.	.10	.48		.58	mortar and wall ties.
	Paint	S.F.	.15	.89		1.04	
	Minimum Charge	Job		430		430	
Brick Veneer							
4" Red Common							
	Demolish	S.F.		3.23		3.23	Cost includes material and labor to
	Install	S.F.	4.81	9.55		14.36	install 4" common red face brick,
	Demolish and Install	S.F.	4.81	12.78		17.59	running bond with 3/8" concave
	Clean	S.F.	.10	.48		.58	joints, mortar and wall ties.
	Paint	S.F.	.15	.89		1.04	
	Minimum Charge	Job		235		235	
4" Used							
	Demolish	S.F.		3.23		3.23	Cost includes material and labor to
	Install	S.F.	4.48	10.20		14.68	install 4" red brick wall, running bond
	Demolish and Install	S.F.	4.48	13.43		17.91	with 3/8" concave joints, mortar and
	Clean	S.F.	.10	.48		.58	ties.
	Paint	S.F.	.15	.89		1.04	
	Minimum Charge	Job		235		235	

Masonry Wall		Unit	Material	Labor	Equip.	Total	Specification
6" Jumbo							
	Demolish	S.F.		3.23		3.23	Cost includes material and labor to
	Install	S.F.	5.90	5.05		10.95	install 6" red jumbo brick veneer,
	Demolish and Install	S.F.	5.90	8.28		14.18	running bond with 3/8" concave
	Clean	S.F.	.10	.48		.58	joints, mortar and ties.
	Paint	S.F.	.15	.89		1.04	
	Minimum Charge	Job		235		235	
Norman (11-1/2")							
	Demolish	S.F.		3.23		3.23	Cost includes material and labor to
	Install	S.F.	7.25	6.85		14.10	install 4" norman red brick wall,
	Demolish and Install	S.F.	7.25	10.08		17.33	running bond with 3/8" concave
	Clean	S.F.	.10	.48		.58	joints, mortar and ties.
	Paint	S.F.	.15	.89		1.04	
	Minimum Charge	Job		235		235	
Glazed							
	Demolish	S.F.		3.23		3.23	Includes material, labor and
	Install	S.F.	21.50	10.45		31.95	equipment to install glazed brick.
	Demolish and Install	S.F.	21.50	13.68		35.18	
	Clean	S.F.	.10	.48		.58	
	Paint	S.F.	.15	.89		1.04	
	Minimum Charge	Job		235		235	
Slumpstone							
6" x 4"							
	Demolish	S.F.		.14	.30	.44	Cost includes material and labor to
	Install	S.F.	6.90	5.40		12.30	install 6" x 4" x 16" concrete slump
	Demolish and Install	S.F.	6.90	5.54	.30	12.74	block.
	Clean	S.F.	.10	.56		.66	
	Paint	S.F.	.15	.89		1.04	
	Minimum Charge	Job		235		235	
6" x 6"							
	Demolish	S.F.		.18	.41	.59	Cost includes material and labor to
	Install	S.F.	6.75	5.55		12.30	install 6" x 6" x 16" concrete slump
	Demolish and Install	S.F.	6.75	5.73	.41	12.89	block.
	Clean	S.F.	.10	.56		.66	
	Paint	S.F.	.15	.89		1.04	
	Minimum Charge	Job		235		235	
8" x 6"							
	Demolish	S.F.		.18	.41	.59	Cost includes material and labor to
	Install	S.F.	9.95	5.75		15.70	install 8" x 6" x 16" concrete slump
	Demolish and Install	S.F.	9.95	5.93	.41	16.29	block.
	Clean	S.F.	.10	.56		.66	
	Paint	S.F.	.15	.89		1.04	
	Minimum Charge	Job		235		235	
12" x 4"							
	Demolish	S.F.		.14	.30	.44	Cost includes material and labor to
	Install	S.F.	14.35	6.65		21	install 12" x 4" x 16" concrete slump
	Demolish and Install	S.F.	14.35	6.79	.30	21.44	block.
	Clean	S.F.	.10	.56		.66	
	Paint	S.F.	.15	.89		1.04	
	Minimum Charge	Job		235		235	
12" x 6"							
	Demolish	S.F.		.18	.41	.59	Cost includes material and labor to
	Install	S.F.	15.80	7.20		23	install 12" x 6" x 16" concrete slump
	Demolish and Install	S.F.	15.80	7.38	.41	23.59	block.
	Clean	S.F.	.10	.56		.66	
	Paint	S.F.	.15	.89		1.04	
	Minimum Charge	Job		235		235	

Masonry

Masonry Wall	Unit	Material	Labor	Equip.	Total	Specification
Stucco						
Three Coat						
Demolish	S.Y.		2.79		2.79	Includes material and labor to install
Install	S.Y.	5.10	10.75	.74	16.59	stucco on exterior masonry walls.
Demolish and Install	S.Y.	5.10	13.54	.74	19.38	
Clean	S.Y.		5.05		5.05	
Paint	S.F.	.25	.65		.90	
Minimum Charge	Job		228		228	
Three Coat w / Wire Mesh						
Demolish	S.Y.		5.05		5.05	Includes material and labor to install
Install	S.Y.	9.30	41.50	2.36	53.16	stucco on wire mesh over wood
Demolish and Install	S.Y.	9.30	46.55	2.36	58.21	framing.
Clean	S.Y.		5.05		5.05	
Paint	S.F.	.25	.65		.90	
Minimum Charge	Job		228		228	
Fill Cracks						
Install	S.F.	.46	3.65		4.11	Includes labor and material to fill in
Minimum Charge	Job		228		228	hairline cracks up to 1/8" with filler, including sand and fill per S.F.
Furring						
1" x 2"						
Demolish	L.F.		.16		.16	Cost includes material and labor to
Install	L.F.	.32	.95		1.27	install 1" x 2" furring on masonry wall.
Demolish and Install	L.F.	.32	1.11		1.43	
Minimum Charge	Job		236		236	
2" x 2"						
Demolish	L.F.		.18		.18	Cost includes material and labor to
Install	L.F.	.24	1.05		1.29	install 2" x 2" furring on masonry wall.
Demolish and Install	L.F.	.24	1.23		1.47	
Minimum Charge	Job		236		236	
2" x 4"						
Demolish	L.F.		.19		.19	Cost includes material and labor to
Install	L.F.	.57	1.19		1.76	install 2" x 4" furring on masonry wall.
Demolish and Install	L.F.	.57	1.38		1.95	
Reinstall	L.F.		.95		.95	
Minimum Charge	Job		236		236	
Maintenance / Cleaning						
Repoint						
Install	S.F.	.62	5.90		6.52	Includes labor and material to cut and
Minimum Charge	Hr.		59		59	repoint brick, hard mortar, running bond.
Acid Etch						
Install	S.F.	.06	.84		.90	Includes labor and material to acid
Minimum Charge	Job		182	40	222	wash smooth brick.
Pressure Wash						
Clean	S.F.		.97	.18	1.15	Includes labor and material to high
Minimum Charge	Job		182	40	222	pressure water wash masonry, average.
Waterproof						
Install	S.F.	.44	.89		1.33	Includes labor and material to install
Minimum Charge	Job		221		221	masonry waterproofing.

For customer support on your Contractor's Pricing Guide: Residential Repair & Remodeling, call 800.448.8182.

Masonry

Chimney

Chimney		Unit	Material	Labor	Equip.	Total	Specification
Demo & Haul Exterior							
	Demolish	C.F.		6.60		6.60	Includes minimum labor and
	Minimum Charge	Job		202		202	equipment to remove masonry and haul debris to truck or dumpster.
Demo & Haul Interior							
	Demolish	C.F.		9.10		9.10	Includes minimum labor and
	Minimum Charge	Job		202		202	equipment to remove masonry and haul debris to truck or dumpster.
Install New (w / flue)							
Brick							
	Install	V.L.F.	44	54		98	Includes material and labor to install
	Minimum Charge	Job		235		235	brick chimney with flue. Foundation not included.
Natural Stone							
	Install	L.F.	75.50	86.50		162	Includes labor and material to install
	Minimum Charge	Job		235		235	natural stone chimney.
Install From Roof Up							
Brick							
	Install	V.L.F.	44	54		98	Includes material and labor to install
	Minimum Charge	Job		235		235	brick chimney with flue. Foundation not included.
Natural Stone							
	Install	L.F.	75.50	72		147.50	Includes labor and material to install
	Minimum Charge	Job		235		235	natural stone chimney.
Install Base Pad							
	Install	Ea.	310	261	1.21	572.21	Includes labor and material to replace
	Minimum Charge	Job		1150	29.50	1179.50	spread footing, 5' square x 16" deep.
Metal Cap							
	Install	Ea.	60	29.50		89.50	Includes labor and materials to install
	Minimum Charge	Job		236		236	bird and squirrel screens in a chimney flue.

Fireplace

Fireplace		Unit	Material	Labor	Equip.	Total	Specification
Common Brick							
30"							
	Demolish	Ea.		1450		1450	Includes material and labor to install a
	Install	Ea.	2175	4325		6500	complete unit which includes common
	Demolish and Install	Ea.	2175	5775		7950	brick, foundation, damper, flue lining
	Minimum Charge	Job		235		235	stack to 15' high and a 30" fire box.
36"							
	Demolish	Ea.		1450		1450	Includes material and labor to install a
	Install	Ea.	2300	4800		7100	complete unit which includes common
	Demolish and Install	Ea.	2300	6250		8550	brick, foundation, damper, flue lining
	Minimum Charge	Job		235		235	stack to 15' high and a 36" fire box.
42"							
	Demolish	Ea.		1450		1450	Includes material and labor to install a
	Install	Ea.	2500	5400		7900	complete unit which includes common
	Demolish and Install	Ea.	2500	6850		9350	brick, foundation, damper, flue lining
	Minimum Charge	Job		235		235	stack to 15' high and a 42" fire box.

Masonry

Fireplace		Unit	Material	Labor	Equip.	Total	Specification
48″							Includes material and labor to install a
	Demolish	Ea.		1450		1450	complete unit which includes common
	Install	Ea.	2625	6150		8775	brick, foundation, damper, flue lining
	Demolish and Install	Ea.	2625	7600		10225	stack to 15′ high and a 48″ fire box.
	Minimum Charge	Job		235		235	
Raised Hearth							
Brick Hearth							Includes material and labor to install a
	Demolish	S.F.		6.30		6.30	brick fireplace hearth based on 30″ x
	Install	S.F.	19.30	36		55.30	29″ opening.
	Demolish and Install	S.F.	19.30	42.30		61.60	
	Clean	S.F.	.10	.48		.58	
	Minimum Charge	Job		235		235	
Marble Hearth / Facing							Includes material and labor to install
	Demolish	Ea.		40.50		40.50	marble fireplace hearth.
	Install	Ea.	375	145		520	
	Demolish and Install	Ea.	375	185.50		560.50	
	Reinstall	Ea.		115.75		115.75	
	Clean	Ea.	48.50	7.25		55.75	
	Minimum Charge	Job		235		235	
Breast (face) Brick							Includes material and labor to install
	Demolish	S.F.		6.30		6.30	standard red brick, running bond.
	Install	S.F.	4.57	10		14.57	
	Demolish and Install	S.F.	4.57	16.30		20.87	
	Clean	S.F.		.97	.18	1.15	
	Minimum Charge	Hr.		59		59	
Fire Box							Includes material and labor to install
	Demolish	Ea.		325		325	fireplace box only (110 brick).
	Install	Ea.	178	430		608	
	Demolish and Install	Ea.	178	755		933	
	Clean	Ea.	.06	17.05		17.11	
	Minimum Charge	Hr.		59		59	
Natural Stone							Includes material and labor to install
	Demolish	S.F.		6.30		6.30	high priced stone, split or rock face.
	Install	S.F.	19.35	18.30		37.65	
	Demolish and Install	S.F.	19.35	24.60		43.95	
	Clean	S.F.	.10	1.08		1.18	
	Minimum Charge	Job		430		430	
Slumpstone Veneer							Cost includes material and labor to
	Demolish	S.F.		.18	.41	.59	install 8″ x 6″ x 16″ concrete slump
	Install	S.F.	9.95	5.75		15.70	block.
	Demolish and Install	S.F.	9.95	5.93	.41	16.29	
	Clean	S.F.	.10	.56		.66	
	Paint	S.F.	.15	.89		1.04	
	Minimum Charge	Job		235		235	
Prefabricated							
Firebox							Includes material and labor to install
	Demolish	Ea.		57.50		57.50	up to 43″ radiant heat, zero
	Install	Ea.	2625	430		3055	clearance, prefabricated fireplace.
	Demolish and Install	Ea.	2625	487.50		3112.50	Flues, doors or blowers not included.
	Reinstall	Ea.		342.69		342.69	
	Clean	Ea.	.75	22.50		23.25	
	Minimum Charge	Job		236		236	

Fireplace	Unit	Material	Labor	Equip.	Total	Specification
Flue						
Demolish	V.L.F.		6.75		6.75	Includes material and labor to install
Install	V.L.F.	25.50	17.90		43.40	prefabricated chimney vent, 8"
Demolish and Install	V.L.F.	25.50	24.65		50.15	exposed metal flue.
Reinstall	V.L.F.		14.33		14.33	
Clean	L.F.	.30	4.55		4.85	
Minimum Charge	Job		340		340	
Blower						
Demolish	Ea.		5.40		5.40	Includes material and labor to install
Install	Ea.	157	32.50		189.50	blower for metal fireplace flue.
Demolish and Install	Ea.	157	37.90		194.90	
Reinstall	Ea.		25.88		25.88	
Clean	Ea.	1.50	45.50		47	
Minimum Charge	Job		340		340	
Accessories						
Wood Mantle						
Demolish	L.F.		6.75		6.75	Includes material and labor to install
Install	L.F.	12.10	13.45		25.55	wood fireplace mantel.
Demolish and Install	L.F.	12.10	20.20		32.30	
Reinstall	L.F.		13.46		13.46	
Clean	S.F.		.45		.45	
Paint	L.F.	.13	10.65		10.78	
Minimum Charge	Job		236		236	
Colonial Style						
Demolish	Ea.		58		58	Includes material and labor to install
Install	Opng.	2100	236		2336	wood fireplace mantel.
Demolish and Install	Opng.	2100	294		2394	
Reinstall	Opng.		235.60		235.60	
Clean	Ea.		.76		.76	
Paint	Ea.	21.50	147		168.50	
Minimum Charge	Job		236		236	
Modern Design						
Demolish	Ea.		58		58	Includes material and labor to install
Install	Opng.	700	94		794	wood fireplace mantel.
Demolish and Install	Opng.	700	152		852	
Reinstall	Opng.		94.24		94.24	
Clean	Ea.		.76		.76	
Paint	Ea.	21.50	147		168.50	
Minimum Charge	Job		236		236	
Glass Door Unit						
Demolish	Ea.		8.40		8.40	Includes material and labor to install a
Install	Ea.	330	9.80		339.80	fireplace glass door unit.
Demolish and Install	Ea.	330	18.20		348.20	
Reinstall	Ea.		9.79		9.79	
Clean	Ea.	.18	14		14.18	
Minimum Charge	Job		236		236	
Fire Screen						
Demolish	Ea.		4.21		4.21	Includes material and labor to install a
Install	Ea.	133	4.90		137.90	fireplace fire screen.
Demolish and Install	Ea.	133	9.11		142.11	
Reinstall	Ea.		4.90		4.90	
Clean	Ea.	.09	12.10		12.19	
Minimum Charge	Job		236		236	

For customer support on your Contractor's Pricing Guide: Residential Repair & Remodeling, call 800.448.8182.

Siding

Shingles

Board & Batten

Texture 1-11

Exterior Siding		Unit	Material	Labor	Equip.	Total	Specification
Trim							
1″ x 2″							
	Demolish	L.F.		.34		.34	Cost includes material and labor to
	Install	L.F.	.32	1.43		1.75	install 1″ x 2″ pine trim.
	Demolish and Install	L.F.	.32	1.77		2.09	
	Clean	L.F.		.24		.24	
	Paint	L.F.	.02	.61		.63	
	Minimum Charge	Job		236		236	
1″ x 3″							
	Demolish	L.F.		.34		.34	Cost includes material and labor to
	Install	L.F.	.34	1.63		1.97	install 1″ x 3″ pine trim.
	Demolish and Install	L.F.	.34	1.97		2.31	
	Clean	L.F.		.24		.24	
	Paint	L.F.	.02	.61		.63	
	Minimum Charge	Job		236		236	
1″ x 4″							
	Demolish	L.F.		.34		.34	Cost includes material and labor to
	Install	L.F.	.62	1.88		2.50	install 1″ x 4″ pine trim.
	Demolish and Install	L.F.	.62	2.22		2.84	
	Clean	L.F.		.24		.24	
	Paint	L.F.	.02	.61		.63	
	Minimum Charge	Job		236		236	
1″ x 6″							
	Demolish	L.F.		.34		.34	Cost includes material and labor to
	Install	L.F.	.90	1.88		2.78	install 1″ x 6″ pine trim.
	Demolish and Install	L.F.	.90	2.22		3.12	
	Clean	L.F.	.01	.24		.25	
	Paint	L.F.	.09	.61		.70	
	Minimum Charge	Job		236		236	
1″ x 8″							
	Demolish	L.F.		.37		.37	Cost includes material and labor to
	Install	L.F.	1.52	2.36		3.88	install 1″ x 8″ pine trim.
	Demolish and Install	L.F.	1.52	2.73		4.25	
	Clean	L.F.	.01	.26		.27	
	Paint	L.F.	.09	.61		.70	
	Minimum Charge	Job		236		236	
Aluminum							
8″ Non-Insulated							
	Demolish	S.F.		.91		.91	Includes material and labor to install
	Install	S.F.	2.77	1.91		4.68	horizontal, colored clapboard, 8″ to
	Demolish and Install	S.F.	2.77	2.82		5.59	10″ wide, plain.
	Reinstall	S.F.		1.91		1.91	
	Clean	S.F.		.36		.36	
	Minimum Charge	Job		236		236	

Siding

Exterior Siding	Unit	Material	Labor	Equip.	Total	Specification
8″ Insulated						
Demolish	S.F.		.91		.91	Includes material and labor to install
Install	S.F.	3.18	1.91		5.09	horizontal, colored insulated
Demolish and Install	S.F.	3.18	2.82		6	clapboard, 8″ wide, plain.
Reinstall	S.F.		1.91		1.91	
Clean	S.F.		.36		.36	
Minimum Charge	Job		236		236	
12″ Non-Insulated						
Demolish	S.F.		.91		.91	Includes material and labor to install
Install	S.F.	2.99	1.64		4.63	horizontal, colored clapboard, plain.
Demolish and Install	S.F.	2.99	2.55		5.54	
Reinstall	S.F.		1.64		1.64	
Clean	S.F.		.36		.36	
Minimum Charge	Job		236		236	
12″ Insulated						
Demolish	S.F.		.91		.91	Includes material and labor to install
Install	S.F.	3.27	1.64		4.91	horizontal, colored insulated
Demolish and Install	S.F.	3.27	2.55		5.82	clapboard, plain.
Reinstall	S.F.		1.64		1.64	
Clean	S.F.		.36		.36	
Minimum Charge	Job		236		236	
Clapboard 8″ to 10″						
Demolish	S.F.		.91		.91	Includes material and labor to install
Install	S.F.	2.77	1.91		4.68	horizontal, colored clapboard, 8″ to
Demolish and Install	S.F.	2.77	2.82		5.59	10″ wide, plain.
Reinstall	S.F.		1.91		1.91	
Clean	S.F.		.36		.36	
Minimum Charge	Job		236		236	
Corner Strips						
Demolish	L.F.		.48		.48	Includes material and labor to install
Install	V.L.F.	4.07	1.91		5.98	corners for horizontal lapped
Demolish and Install	V.L.F.	4.07	2.39		6.46	aluminum siding.
Reinstall	V.L.F.		1.91		1.91	
Clean	L.F.		.24		.24	
Minimum Charge	Job		236		236	
Vinyl						
Non-Insulated						
Demolish	S.F.		.71		.71	Cost includes material and labor to
Install	S.F.	1.84	1.71		3.55	install 10″ non-insulated solid vinyl
Demolish and Install	S.F.	1.84	2.42		4.26	siding.
Reinstall	S.F.		1.37		1.37	
Clean	S.F.		.36		.36	
Minimum Charge	Job		236		236	
Insulated						
Demolish	S.F.		.71		.71	Cost includes material and labor to
Install	S.F.	3.82	2.36		6.18	install insulated solid vinyl siding
Demolish and Install	S.F.	3.82	3.07		6.89	panels.
Reinstall	S.F.		1.88		1.88	
Clean	S.F.		.36		.36	
Minimum Charge	Job		236		236	
Corner Strips						
Demolish	L.F.		.45		.45	Cost includes material and labor to
Install	L.F.	2.37	1.35		3.72	install corner posts for vinyl siding
Demolish and Install	L.F.	2.37	1.80		4.17	panels.
Reinstall	L.F.		1.08		1.08	
Clean	L.F.		.24		.24	
Minimum Charge	Job		236		236	

For customer support on your Contractor's Pricing Guide: Residential Repair & Remodeling, call 800.448.8182.

Siding

Exterior Siding	Unit	Material	Labor	Equip.	Total	Specification
Fiber Cement Siding						
4′ x 8′ Panel						
Demolish	S.F.		.71		.71	Cost includes material and labor to
Install	S.F.	1.09	1.03		2.12	install 4′ x 8′ fiber cement 7/16″
Demolish and Install	S.F.	1.09	1.74		2.83	thick.
Clean	S.F.		.36		.36	
Paint	S.F.	.34	1.33		1.67	
Minimum Charge	Job		236		236	
Lap						
Demolish	S.F.		.70		.70	Cost includes material and labor to
Install	S.F.	1.17	1.86		3.03	install 8″ or 12″ fiber cement lap
Demolish and Install	S.F.	1.17	2.56		3.73	siding 7/16″ thick with wood grain
Clean	S.F.		.36		.36	finish.
Paint	S.F.	.34	1.33		1.67	
Minimum Charge	Job		236		236	
Plywood Panel						
Douglas Fir (T-1-11)						
Demolish	S.F.		.56		.56	Includes material and labor to install
Install	S.F.	1.50	1.46		2.96	Douglas fir 4′ x 8′ plywood exterior
Demolish and Install	S.F.	1.50	2.02		3.52	siding plain or patterns.
Reinstall	S.F.		1.46		1.46	
Clean	S.F.		.36		.36	
Paint	S.F.	.34	1.33		1.67	
Minimum Charge	Job		236		236	
Redwood						
Demolish	S.F.		.56		.56	Includes material and labor to install
Install	S.F.	2.23	1.46		3.69	redwood 4′ x 8′ plywood exterior
Demolish and Install	S.F.	2.23	2.02		4.25	siding plain or patterns.
Reinstall	S.F.		1.46		1.46	
Clean	S.F.		.36		.36	
Paint	S.F.	.34	1.33		1.67	
Minimum Charge	Job		236		236	
Cedar						
Demolish	S.F.		.56		.56	Cost includes material and labor to
Install	S.F.	2.88	1.46		4.34	install 4′ x 8′ cedar plywood exterior
Demolish and Install	S.F.	2.88	2.02		4.90	siding plain or patterns.
Reinstall	S.F.		1.46		1.46	
Clean	S.F.		.36		.36	
Paint	S.F.	.34	1.33		1.67	
Minimum Charge	Job		236		236	
Tongue and Groove						
Fir, Tongue and Groove						
Demolish	S.F.		1.01		1.01	Includes material and labor to install
Install	S.F.	6.40	1.43		7.83	fir tongue and groove exterior siding
Demolish and Install	S.F.	6.40	2.44		8.84	1″ x 8″.
Reinstall	S.F.		1.14		1.14	
Clean	S.F.		.36		.36	
Paint	S.F.	.34	1.33		1.67	
Minimum Charge	Job		236		236	
Pine, Tongue and Groove						
Demolish	S.F.		1.01		1.01	Includes labor and material to install
Install	S.F.	2.79	1.43		4.22	Pine tongue and groove exterior
Demolish and Install	S.F.	2.79	2.44		5.23	siding 1″ x 8″.
Reinstall	S.F.		1.43		1.43	
Clean	S.F.		.36		.36	
Paint	S.F.	.34	1.33		1.67	
Minimum Charge	Job		236		236	

Exterior Siding		Unit	Material	Labor	Equip.	Total	Specification
Cedar, Tongue and Groove							
	Demolish	S.F.		1.01		1.01	Includes labor and material to install
	Install	S.F.	8.25	2.14		10.39	cedar tongue and groove exterior
	Demolish and Install	S.F.	8.25	3.15		11.40	siding 1" x 8".
	Reinstall	S.F.		1.71		1.71	
	Clean	S.F.		.36		.36	
	Paint	S.F.	.34	1.33		1.67	
	Minimum Charge	Job		236		236	
Redwood, Tongue and Groove							
	Demolish	S.F.		1.01		1.01	Includes labor and material to install
	Install	S.F.	7.55	2.14		9.69	redwood tongue and groove exterior
	Demolish and Install	S.F.	7.55	3.15		10.70	siding 1" x 4".
	Reinstall	S.F.		2.14		2.14	
	Clean	S.F.		.36		.36	
	Paint	S.F.	.34	1.33		1.67	
	Minimum Charge	Job		236		236	
Board and Batten							
Fir							
	Demolish	S.F.		1.01		1.01	Cost includes material and labor to
	Install	S.F.	6.50	1.43		7.93	install 1" x 8" fir boards with 1" x 3"
	Demolish and Install	S.F.	6.50	2.44		8.94	battens.
	Reinstall	S.F.		1.14		1.14	
	Clean	S.F.		.36		.36	
	Paint	S.F.	.34	1.33		1.67	
	Minimum Charge	Job		236		236	
Pine							
	Demolish	S.F.		1.01		1.01	Cost includes material and labor to
	Install	S.F.	4.04	1.26		5.30	install 1" x 10" pine boards with 1" x
	Demolish and Install	S.F.	4.04	2.27		6.31	3" battens.
	Reinstall	S.F.		1.01		1.01	
	Clean	S.F.		.36		.36	
	Paint	S.F.	.34	1.33		1.67	
	Minimum Charge	Job		236		236	
Cedar							
	Demolish	S.F.		1.01		1.01	Cost includes material and labor to
	Install	S.F.	7.30	1.26		8.56	install 1" x 10" cedar boards with 1" x
	Demolish and Install	S.F.	7.30	2.27		9.57	3" battens.
	Reinstall	S.F.		1.01		1.01	
	Clean	S.F.		.36		.36	
	Paint	S.F.	.34	1.33		1.67	
	Minimum Charge	Job		236		236	
Redwood							
	Demolish	S.F.		1.01		1.01	Cost includes material and labor to
	Install	S.F.	5.95	1.26		7.21	install 1" x 10" redwood boards with
	Demolish and Install	S.F.	5.95	2.27		8.22	1" x 3" battens.
	Reinstall	S.F.		1.01		1.01	
	Clean	S.F.		.36		.36	
	Paint	S.F.	.34	1.33		1.67	
	Minimum Charge	Job		236		236	
Wood Shingle							
	Demolish	Sq.		93		93	Includes material and labor to install
	Install	Sq.	206	209		415	wood shingles/shakes.
	Demolish and Install	Sq.	206	302		508	
	Clean	Sq.		38.50		38.50	
	Paint	Sq.	22.50	119		141.50	
	Minimum Charge	Job		236		236	

Siding

Exterior Siding	Unit	Material	Labor	Equip.	Total	Specification
Synthetic Stucco System						
Demolish	S.F.		3.37		3.37	Cost includes material and labor to
Install	S.F.	2.85	8.05	.56	11.46	install 2 coats of adhesive mixture,
Demolish and Install	S.F.	2.85	11.42	.56	14.83	glass fiber mesh, 1" insulation board,
Clean	S.F.		.33		.33	and textured finish coat.
Paint	S.F.	.25	.65		.90	
Minimum Charge	Job		1075	74.50	1149.50	

Front-end Loader   **Bulldozer**

Grading		Unit	Material	Labor	Equip.	Total	Specification
Hand (fine)							
	Install	S.F.		.73		.73	Includes labor and material to install
	Minimum Charge	Job		182		182	fine grade for slab on grade, hand grade.
Machine (fine)							
	Install	S.Y.		.63	.53	1.16	Includes labor and material to install
	Minimum Charge	Job		430	360	790	fine grade for roadway, base or leveling course.

Asphalt Paving		Unit	Material	Labor	Equip.	Total	Specification
Seal Coat							
	Install	S.F.	.10	.13		.23	Includes labor and material to install
	Minimum Charge	Job		182		182	sealcoat for a small area.
Wearing Course							
	Install	S.F.	.18	.39	.27	.84	Cost includes labor and material to
	Minimum Charge	Job		182		182	install 1″ thick wearing course.

Masonry Pavers		Unit	Material	Labor	Equip.	Total	Specification
Natural Concrete							
Sand Base							
	Demolish	S.F.		4.04		4.04	Includes material and labor to install
	Install	S.F.	4.03	7.85		11.88	concrete pavers on sand base.
	Demolish and Install	S.F.	4.03	11.89		15.92	
	Reinstall	S.F.		6.27		6.27	
	Clean	S.F.		1.45		1.45	
	Minimum Charge	Job		235		235	
Mortar Base							
	Demolish	S.F.		6.65	1.96	8.61	Includes material and labor to install
	Install	S.F.	3.67	6.55		10.22	concrete pavers on mortar base.
	Demolish and Install	S.F.	3.67	13.20	1.96	18.83	
	Reinstall	S.F.		5.22		5.22	
	Clean	S.F.		1.32		1.32	
	Minimum Charge	Job		235		235	

Site Work

Masonry Pavers

	Unit	Material	Labor	Equip.	Total	Specification
Adobe						
Sand Base						
Demolish	S.F.		4.04		4.04	Includes material and labor to install
Install	S.F.	2.05	5.20		7.25	adobe pavers on sand base.
Demolish and Install	S.F.	2.05	9.24		11.29	
Reinstall	S.F.		4.18		4.18	
Clean	S.F.		1.45		1.45	
Minimum Charge	Job		235		235	
Mortar Base						
Demolish	S.F.		6.65	1.96	8.61	Includes material and labor to install
Install	S.F.	2.93	6.55		9.48	adobe pavers on mortar base.
Demolish and Install	S.F.	2.93	13.20	1.96	18.09	
Reinstall	S.F.		5.22		5.22	
Clean	S.F.		1.32		1.32	
Minimum Charge	Job		235		235	
Standard Brick						
Sand Base						
Demolish	S.F.		4.04		4.04	Includes material and labor to install
Install	S.F.	3.20	8.65		11.85	standard brick pavers on sand base.
Demolish and Install	S.F.	3.20	12.69		15.89	
Reinstall	S.F.		6.91		6.91	
Clean	S.F.		1.45		1.45	
Minimum Charge	Job		235		235	
Mortar Base						
Demolish	S.F.		6.65	1.96	8.61	Includes material and labor to install
Install	S.F.	3.18	3.13		6.31	standard brick pavers on mortar base.
Demolish and Install	S.F.	3.18	9.78	1.96	14.92	
Reinstall	S.F.		2.51		2.51	
Clean	S.F.		1.32		1.32	
Minimum Charge	Job		235		235	
Deluxe Brick						
Sand Base						
Demolish	S.F.		4.04		4.04	Includes material and labor to install
Install	S.F.	3.73	6.55		10.28	deluxe brick pavers on sand base.
Demolish and Install	S.F.	3.73	10.59		14.32	
Reinstall	S.F.		5.22		5.22	
Clean	S.F.		1.45		1.45	
Minimum Charge	Job		235		235	
Mortar Base						
Demolish	S.F.		6.65	1.96	8.61	Includes material and labor to install
Install	S.F.	4.36	11.75		16.11	deluxe brick pavers on mortar base.
Demolish and Install	S.F.	4.36	18.40	1.96	24.72	
Reinstall	S.F.		9.40		9.40	
Clean	S.F.		1.32		1.32	
Minimum Charge	Job		235		235	

Fencing

	Unit	Material	Labor	Equip.	Total	Specification
Block w / Footing						
4"						
Demolish	S.F.		.14	.30	.44	Cost includes material and labor to
Install	S.F.	2	5.10		7.10	install 4" non-reinforced concrete
Demolish and Install	S.F.	2	5.24	.30	7.54	block.
Clean	S.F.	.29	1.21	.23	1.73	
Paint	S.F.	.34	.75		1.09	
Minimum Charge	Job		235		235	

For customer support on your Contractor's Pricing Guide: Residential Repair & Remodeling, call 800.448.8182.

Fencing	Unit	Material	Labor	Equip.	Total	Specification
4" Reinforced						
Demolish	S.F.		2.24	.75	2.99	Cost includes material and labor to
Install	S.F.	2.21	5.15		7.36	install 4" reinforced concrete block.
Demolish and Install	S.F.	2.21	7.39	.75	10.35	
Clean	S.F.	.29	1.21	.23	1.73	
Paint	S.F.	.34	.75		1.09	
Minimum Charge	Job		235		235	
6"						
Demolish	S.F.		.18	.41	.59	Cost includes material and labor to
Install	S.F.	2.73	5.50		8.23	install 6" non-reinforced concrete
Demolish and Install	S.F.	2.73	5.68	.41	8.82	block.
Clean	S.F.	.29	1.21	.23	1.73	
Paint	S.F.	.34	.75		1.09	
Minimum Charge	Job		235		235	
6" Reinforced						
Demolish	S.F.		2.36	.52	2.88	Cost includes material and labor to
Install	S.F.	2.94	5.55		8.49	install 6" reinforced concrete block.
Demolish and Install	S.F.	2.94	7.91	.52	11.37	
Clean	S.F.	.29	1.21	.23	1.73	
Paint	S.F.	.34	.75		1.09	
Minimum Charge	Job		235		235	
Cap for Block Wall						
Demolish	L.F.		2.53		2.53	Includes labor and equipment to install
Install	L.F.	26.50	11.50		38	precast concrete coping.
Demolish and Install	L.F.	26.50	14.03		40.53	
Clean	L.F.	.02	3.79		3.81	
Paint	S.F.	.34	.75		1.09	
Minimum Charge	Job		235		235	
Block Pilaster						
Demolish	Ea.		173		173	Includes material and labor to install
Install	Ea.	54.50	173		227.50	block pilaster for fence.
Demolish and Install	Ea.	54.50	346		400.50	
Reinstall	Ea.		172.74		172.74	
Paint	Ea.	4.79	17.40		22.19	
Minimum Charge	Job		235		235	
Block w / o Footing						
4"						
Demolish	S.F.		.14	.30	.44	Cost includes material and labor to
Install	S.F.	2.15	3.93		6.08	install 4" non-reinforced hollow
Demolish and Install	S.F.	2.15	4.07	.30	6.52	lightweight block.
Clean	S.F.	.29	1.21	.23	1.73	
Paint	S.F.	.34	.75		1.09	
Minimum Charge	Job		235		235	
4" Reinforced						
Demolish	S.F.		2.24	.75	2.99	Cost includes material and labor to
Install	S.F.	2.33	5.05		7.38	install 4" reinforced hollow lightweight
Demolish and Install	S.F.	2.33	7.29	.75	10.37	block.
Clean	S.F.	.29	1.21	.23	1.73	
Paint	S.F.	.34	.75		1.09	
Minimum Charge	Job		235		235	
6"						
Demolish	S.F.		.18	.41	.59	Cost includes material and labor to
Install	S.F.	3.07	4.32		7.39	install 6" non-reinforced hollow
Demolish and Install	S.F.	3.07	4.50	.41	7.98	lightweight block.
Clean	S.F.	.29	1.21	.23	1.73	
Paint	S.F.	.34	.75		1.09	
Minimum Charge	Job		235		235	

Fencing		Unit	Material	Labor	Equip.	Total	Specification
6" Reinforced							
	Demolish	S.F.		2.36	.52	2.88	Cost includes material and labor to
	Install	S.F.	3.25	5.45		8.70	install 6" reinforced hollow lightweight
	Demolish and Install	S.F.	3.25	7.81	.52	11.58	block.
	Clean	S.F.	.29	1.21	.23	1.73	
	Paint	S.F.	.34	.75		1.09	
	Minimum Charge	Job		235		235	
Cap for Block Wall							
	Demolish	L.F.		2.53		2.53	Includes labor and equipment to install
	Install	L.F.	26.50	11.50		38	precast concrete coping.
	Demolish and Install	L.F.	26.50	14.03		40.53	
	Clean	L.F.	.02	3.79		3.81	
	Paint	S.F.	.34	.75		1.09	
	Minimum Charge	Job		235		235	
Block Pilaster							
	Demolish	Ea.		173		173	Includes material and labor to install
	Install	Ea.	54.50	173		227.50	block pilaster for fence.
	Demolish and Install	Ea.	54.50	346		400.50	
	Reinstall	Ea.		172.74		172.74	
	Paint	Ea.	4.79	17.40		22.19	
	Minimum Charge	Job		235		235	
Chain Link							
3' High w / o Top Rail							
	Demolish	L.F.		3.12	.92	4.04	Includes material and labor to install
	Install	L.F.	5.95	2.20	.42	8.57	residential galvanized fence 3' high
	Demolish and Install	L.F.	5.95	5.32	1.34	12.61	with 1-3/8" steel post 10' O.C. set
	Reinstall	L.F.		1.76	.34	2.10	without concrete.
	Clean	L.F.	.04	1.32		1.36	
	Minimum Charge	Job		560		560	
4' High w / o Top Rail							
	Demolish	L.F.		3.12	.92	4.04	Includes material and labor to install
	Install	L.F.	6.20	2.71	.52	9.43	residential galvanized fence 4' high
	Demolish and Install	L.F.	6.20	5.83	1.44	13.47	with 1-3/8" steel post 10' O.C. set
	Reinstall	L.F.		2.17	.42	2.59	without concrete.
	Clean	L.F.	.06	1.45		1.51	
	Minimum Charge	Job		560		560	
5' High w / o Top Rail							
	Demolish	L.F.		3.12	.92	4.04	Includes material and labor to install
	Install	L.F.	7.90	3.55	.69	12.14	residential galvanized fence 5' high
	Demolish and Install	L.F.	7.90	6.67	1.61	16.18	with 1-3/8" steel post 10' O.C. set
	Reinstall	L.F.		2.84	.55	3.39	without concrete.
	Clean	L.F.	.08	1.62		1.70	
	Minimum Charge	Job		560		560	
3' High w / Top Rail							
	Demolish	L.F.		3.32	.98	4.30	Includes material and labor to install
	Install	L.F.	2.23	2.31	.45	4.99	residential galvanized fence 3' high
	Demolish and Install	L.F.	2.23	5.63	1.43	9.29	with 1-3/8" steel post 10' O.C. set
	Reinstall	L.F.		1.84	.36	2.20	without concrete.
	Clean	L.F.	.04	1.35		1.39	
	Minimum Charge	Job		560		560	
4' High w / Top Rail							
	Demolish	L.F.		3.32	.98	4.30	Includes material and labor to install
	Install	L.F.	7.85	2.88	.56	11.29	residential galvanized fence 4' high
	Demolish and Install	L.F.	7.85	6.20	1.54	15.59	with 1-5/8" steel post 10' O.C. set
	Reinstall	L.F.		2.30	.45	2.75	without concrete and 1-5/8" top rail
	Clean	L.F.	.06	1.48		1.54	with sleeves.
	Minimum Charge	Job		560		560	

For customer support on your Contractor's Pricing Guide: Residential Repair & Remodeling, call 800.448.8182.

Fencing	Unit	Material	Labor	Equip.	Total	Specification
5' High w / Top Rail						
Demolish	L.F.		3.32	.98	4.30	Includes material and labor to install
Install	L.F.	8.40	3.84	.74	12.98	residential galvanized fence 5' high
Demolish and Install	L.F.	8.40	7.16	1.72	17.28	with 1-5/8" steel post 10' O.C. set
Reinstall	L.F.		3.07	.59	3.66	without concrete and 1-5/8" top rail
Clean	L.F.	.08	1.65		1.73	with sleeves.
Minimum Charge	Job		560		560	
6' High w / Top Rail						
Demolish	L.F.		3.32	.98	4.30	Includes material and labor to install
Install	L.F.	10	5.75	1.11	16.86	residential galvanized fence 6' high
Demolish and Install	L.F.	10	9.07	2.09	21.16	with 2" steel post 10' O.C. set without
Reinstall	L.F.		4.61	.89	5.50	concrete and 1-3/8" top rail with
Clean	L.F.	.09	1.86		1.95	sleeves.
Minimum Charge	Job		560		560	
Barbed Wire						
Demolish	L.F.		1.88		1.88	Cost includes material and labor to
Install	L.F.	.43	.73	.18	1.34	install 3 strand galvanized barbed
Demolish and Install	L.F.	.43	2.61	.18	3.22	wire fence.
Reinstall	L.F.		.58	.14	.72	
Minimum Charge	Job		560	350	910	
6' Redwood						
Demolish	L.F.		12.40		12.40	Cost includes material and labor to
Install	L.F.	31	9.20	1.78	41.98	install 6' high redwood fencing.
Demolish and Install	L.F.	31	21.60	1.78	54.38	
Reinstall	L.F.		7.38	1.43	8.81	
Clean	L.F.	.09	1.69		1.78	
Paint	L.F.	.92	4.12		5.04	
Minimum Charge	Job		560		560	
6' Red Cedar						
Demolish	L.F.		12.40		12.40	Cost includes material and labor to
Install	L.F.	23.50	9.20	1.78	34.48	install 6' cedar fence.
Demolish and Install	L.F.	23.50	21.60	1.78	46.88	
Reinstall	L.F.		7.38	1.43	8.81	
Clean	L.F.	.09	1.69		1.78	
Paint	L.F.	.92	4.12		5.04	
Minimum Charge	Job		560		560	
6' Shadowbox						
Demolish	L.F.		12.40		12.40	Cost includes material and labor to
Install	L.F.	27.50	7.70	1.49	36.69	install 6' pressure treated pine fence.
Demolish and Install	L.F.	27.50	20.10	1.49	49.09	
Reinstall	L.F.		6.15	1.19	7.34	
Clean	L.F.	.09	1.69		1.78	
Paint	L.F.	.92	4.12		5.04	
Minimum Charge	Job		560		560	
3' Picket						
Demolish	L.F.		12.40		12.40	Includes material, labor and
Install	L.F.	8.70	8.25	1.59	18.54	equipment to install pressure treated
Demolish and Install	L.F.	8.70	20.65	1.59	30.94	pine picket fence.
Reinstall	L.F.		6.59	1.27	7.86	
Clean	L.F.	.04	1.45		1.49	
Paint	L.F.	.46	2.24		2.70	
Minimum Charge	Job		560		560	

143

Fencing	Unit	Material	Labor	Equip.	Total	Specification
Wrought Iron						
Demolish	L.F.		27		27	Cost includes material and labor to
Install	L.F.	40.50	14.90		55.40	install 5' high wrought iron fence with
Demolish and Install	L.F.	40.50	41.90		82.40	pickets at 4-1/2" O.C.
Reinstall	L.F.		11.92		11.92	
Clean	L.F.		1.45		1.45	
Paint	L.F.	4.19	4.60		8.79	
Minimum Charge	Job		560		560	
Post						
Dirt Based Wood						
Demolish	Ea.		31		31	Includes material and labor to install
Install	Ea.	32.50	140		172.50	wood post set in earth.
Demolish and Install	Ea.	32.50	171		203.50	
Reinstall	Ea.		111.72		111.72	
Clean	Ea.		4.55		4.55	
Paint	Ea.	1.01	1.56		2.57	
Minimum Charge	Job		560		560	
Concrete Based Wood						
Demolish	Ea.		44.50	13.05	57.55	Includes material and labor to install
Install	Ea.	20.50	186		206.50	wood post set in concrete.
Demolish and Install	Ea.	20.50	230.50	13.05	264.05	
Clean	Ea.		4.55		4.55	
Paint	Ea.	1.01	1.56		2.57	
Minimum Charge	Job		560		560	
Block Pilaster						
Demolish	Ea.		173		173	Includes material and labor to install
Install	Ea.	54.50	173		227.50	block pilaster for fence.
Demolish and Install	Ea.	54.50	346		400.50	
Reinstall	Ea.		172.74		172.74	
Paint	Ea.	4.79	17.40		22.19	
Minimum Charge	Job		235		235	
Gate						
Chain Link						
Demolish	Ea.		41.50		41.50	Cost includes material and labor to
Install	Ea.	114	62		176	install 3' wide chain link pass gate
Demolish and Install	Ea.	114	103.50		217.50	and hardware.
Reinstall	Ea.		49.65		49.65	
Clean	Ea.		5.05		5.05	
Minimum Charge	Job		560		560	
Redwood						
Demolish	Ea.		35.50		35.50	Includes material, labor and
Install	Ea.	305	62		367	equipment to install redwood pass
Demolish and Install	Ea.	305	97.50		402.50	gate and hardware.
Reinstall	Ea.		49.65		49.65	
Clean	Ea.		5.05		5.05	
Paint	Ea.	1.52	7.80		9.32	
Minimum Charge	Job		560		560	
Red Cedar						
Demolish	Ea.		35.50		35.50	Cost includes material and labor to
Install	Ea.	102	62		164	install 3' wide red cedar pass gate
Demolish and Install	Ea.	102	97.50		199.50	and hardware.
Reinstall	Ea.		49.65		49.65	
Clean	Ea.		5.05		5.05	
Paint	Ea.	1.52	7.80		9.32	
Minimum Charge	Job		560		560	

Fencing	Unit	Material	Labor	Equip.	Total	Specification
Wrought Iron						
Demolish	Ea.		27		27	Includes material, labor and
Install	Ea.	287	140		427	equipment to install wrought iron gate
Demolish and Install	Ea.	287	167		454	with hardware.
Reinstall	Ea.		111.72		111.72	
Clean	Ea.		5.05		5.05	
Paint	Ea.	8.35	7.80		16.15	
Minimum Charge	Job		560		560	

For customer support on your Contractor's Pricing Guide: Residential Repair & Remodeling, call 800.448.8182.

Fire-extinguishing System

Space Heater

Hot-air Furnace

Plumbing	Unit	Material	Labor	Equip.	Total	Specification
Components						
3/4" Supply						
Demolish	L.F.		2.95		2.95	Includes material and labor to install
Install	L.F.	4.83	7		11.83	type L copper tubing including
Demolish and Install	L.F.	4.83	9.95		14.78	couplings and hangers at 10' O.C.
Reinstall	L.F.		6.98		6.98	
Minimum Charge	Job		265		265	
1/2" Fixture Supply						
Install	Ea.	50.50	177		227.50	Includes labor and materials to install
						supply pipe only to a plumbing fixture.
2" Single Vent Stack						
Demolish	Ea.		49		49	Includes material and labor to install
Install	Ea.	27.50	265		292.50	PVC vent stack pipe with couplings at
Demolish and Install	Ea.	27.50	314		341.50	10' O.C.
Reinstall	Ea.		265.20		265.20	
Minimum Charge	Job		265		265	
4" Drain						
Demolish	L.F.		3.93		3.93	Includes material and labor to install
Install	L.F.	6.60	8.70		15.30	DVW type PVC pipe with couplings at
Demolish and Install	L.F.	6.60	12.63		19.23	10' O.C. and 3 strap hangers per 10'
Reinstall	L.F.		8.68		8.68	run.
Minimum Charge	Job		265		265	
1-1/2" Drain						
Demolish	L.F.		2.95		2.95	Includes material and labor to install
Install	L.F.	1.66	5.65		7.31	DVW type PVC pipe with couplings at
Demolish and Install	L.F.	1.66	8.60		10.26	10' O.C. and 3 strap hangers per 10'
Reinstall	L.F.		5.64		5.64	run.
Minimum Charge	Job		265		265	
Maintenance						
Flush Out / Unclog						
Clean	Job		265		265	Includes 4 hour minimum labor charge
Minimum Charge	Job		265		265	for a plumber.
Vent Cap						
Large						
Demolish	Ea.		13.40		13.40	Includes labor and material to install a
Install	Ea.	50.50	24		74.50	PVC plumbing pipe vent cap.
Demolish and Install	Ea.	50.50	37.40		87.90	
Clean	Ea.	.21	8.25		8.46	
Paint	Ea.	.20	8.90		9.10	
Minimum Charge	Job		133		133	

Plumbing

	Unit	Material	Labor	Equip.	Total	Specification
Medium						
Demolish	Ea.		13.40		13.40	Includes labor and material to install a
Install	Ea.	43.50	23		66.50	PVC plumbing pipe vent cap.
Demolish and Install	Ea.	43.50	36.40		79.90	
Clean	Ea.	.21	8.25		8.46	
Paint	Ea.	.20	8.90		9.10	
Minimum Charge	Job		133		133	
Small						
Demolish	Ea.		13.40		13.40	Includes material and labor to install a
Install	Ea.	15.15	22		37.15	PVC plumbing pipe vent cap.
Demolish and Install	Ea.	15.15	35.40		50.55	
Clean	Ea.	.21	8.25		8.46	
Paint	Ea.	.20	8.90		9.10	
Minimum Charge	Job		133		133	

Plumbing

	Unit	Material	Labor	Equip.	Total	Specification
Sump Pump						
1/4 HP						
Demolish	Ea.		53		53	Includes material and labor to install a
Install	Ea.	179	83		262	submersible sump pump.
Demolish and Install	Ea.	179	136		315	
Reinstall	Ea.		66.30		66.30	
Clean	Ea.	.53	22.50		23.03	
Minimum Charge	Job		265		265	
1/3 HP						
Demolish	Ea.		53		53	Includes material and labor to install a
Install	Ea.	243	88.50		331.50	submersible sump pump.
Demolish and Install	Ea.	243	141.50		384.50	
Reinstall	Ea.		70.72		70.72	
Clean	Ea.	.53	22.50		23.03	
Minimum Charge	Job		265		265	
1/2 HP						
Demolish	Ea.		53		53	Includes material and labor to install a
Install	Ea.	281	98		379	submersible sump pump.
Demolish and Install	Ea.	281	151		432	
Reinstall	Ea.		78.58		78.58	
Clean	Ea.	.53	22.50		23.03	
Minimum Charge	Job		265		265	
Clean and Service						
Clean	Ea.		33		33	Includes labor to clean and service a
Minimum Charge	Job		265		265	sump pump.

Water Heater

	Unit	Material	Labor	Equip.	Total	Specification
Electric						
30 Gallon						
Demolish	Ea.		166		166	Includes material and labor to install a
Install	Ea.	1100	241		1341	30 gallon electric water heater.
Demolish and Install	Ea.	1100	407		1507	
Reinstall	Ea.		192.87		192.87	
Clean	Ea.	.53	22.50		23.03	
Minimum Charge	Job		265		265	

Water Heater		Unit	Material	Labor	Equip.	Total	Specification
40 Gallon							
	Demolish	Ea.		166		166	Includes material and labor to install a
	Install	Ea.	1175	265		1440	40 gallon electric water heater.
	Demolish and Install	Ea.	1175	431		1606	
	Reinstall	Ea.		212.16		212.16	
	Clean	Ea.	.53	22.50		23.03	
	Minimum Charge	Job		265		265	
52 Gallon							
	Demolish	Ea.		166		166	Includes material and labor to install a
	Install	Ea.	1325	265		1590	52 gallon electric water heater.
	Demolish and Install	Ea.	1325	431		1756	
	Reinstall	Ea.		212.16		212.16	
	Clean	Ea.	.53	22.50		23.03	
	Minimum Charge	Job		265		265	
82 Gallon							
	Demolish	Ea.		166		166	Includes material and labor to install
	Install	Ea.	2025	330		2355	an 82 gallon electric water heater.
	Demolish and Install	Ea.	2025	496		2521	
	Reinstall	Ea.		265.20		265.20	
	Clean	Ea.	.53	22.50		23.03	
	Minimum Charge	Job		265		265	
Gas-fired							
30 Gallon Gas-fired							
	Demolish	Ea.		204		204	Includes material and labor to install a
	Install	Ea.	1950	265		2215	30 gallon gas water heater.
	Demolish and Install	Ea.	1950	469		2419	
	Reinstall	Ea.		212.16		212.16	
	Clean	Ea.	.53	22.50		23.03	
	Minimum Charge	Job		265		265	
30 Gallon Gas-fired, Flue							
	Install	Ea.	66	76		142	Includes labor and materials to install
	Minimum Charge	Job		265		265	a flue for a water heater (up to 50 gallon size).
40 Gallon Gas-fired							
	Demolish	Ea.		204		204	Includes material and labor to install a
	Install	Ea.	1950	279		2229	40 gallon gas water heater.
	Demolish and Install	Ea.	1950	483		2433	
	Reinstall	Ea.		223.33		223.33	
	Clean	Ea.	.53	22.50		23.03	
	Minimum Charge	Job		265		265	
40 Gallon Gas-fired, Flue							
	Install	Ea.	66	76		142	Includes labor and materials to install
	Minimum Charge	Job		265		265	a flue for a water heater (up to 50 gallon size).
50 Gallon Gas-fired							
	Demolish	Ea.		204		204	Includes material and labor to install a
	Install	Ea.	2125	295		2420	50 gallon gas water heater.
	Demolish and Install	Ea.	2125	499		2624	
	Reinstall	Ea.		235.73		235.73	
	Clean	Ea.	.53	22.50		23.03	
	Minimum Charge	Job		265		265	
50 Gallon Gas-fired, Flue							
	Install	Ea.	66	76		142	Includes labor and materials to install
	Minimum Charge	Job		265		265	a flue for a water heater (up to 50 gallon size).

149

For customer support on your **Contractor's Pricing Guide: Residential Repair & Remodeling**, call 800.448.8182.

Rough Mechanical

Water Heater

Water Heater	Unit	Material	Labor	Equip.	Total	Specification
Solar Hot Water System						
Water Heater						
Demolish	Ea.		133		133	Includes material and labor to install a
Install	Ea.	1800	355		2155	solar hot water heat exchanger
Demolish and Install	Ea.	1800	488		2288	storage tank, 100 gallon.
Reinstall	Ea.		282.88		282.88	
Minimum Charge	Job		273		273	
Solar Panel						
Demolish	S.F.		1.94		1.94	Includes material and labor to install
Install	S.F.	35	3.49		38.49	solar panel.
Demolish and Install	S.F.	35	5.43		40.43	
Reinstall	S.F.		2.79		2.79	
Minimum Charge	Job		273		273	
Repair Plumbing						
Install	Job		273		273	Includes 4 hour minimum labor charge
Minimum Charge	Job		273		273	for a steamfitter or pipefitter.
Insulating Wrap						
Demolish	Ea.		6.15		6.15	Includes material and labor to install
Install	Ea.	20.50	16.60		37.10	insulating wrap around a 100 gallon
Demolish and Install	Ea.	20.50	22.75		43.25	solar hot water heat exchanger
Reinstall	Ea.		13.26		13.26	storage tank.
Minimum Charge	Job		265		265	
Water Softener						
Demolish	Ea.		98		98	Includes material and labor to install a
Install	Ea.	2650	239		2889	14 GPM water softener.
Demolish and Install	Ea.	2650	337		2987	
Reinstall	Ea.		190.88		190.88	
Clean	Ea.	.53	22.50		23.03	
Minimum Charge	Ea.		265		265	

Ductwork

Ductwork	Unit	Material	Labor	Equip.	Total	Specification
Galvanized, Square or Rectangular						
6"						
Demolish	L.F.		2.40		2.40	Includes material and labor to install
Install	L.F.	1.45	11.50		12.95	rectangular ductwork with a greatest
Demolish and Install	L.F.	1.45	13.90		15.35	dimension of 6".
Reinstall	L.F.		9.20		9.20	
Clean	L.F.		2.76		2.76	
Minimum Charge	Job		259		259	
8"						
Demolish	L.F.		2.40		2.40	Includes material and labor to install
Install	L.F.	1.93	14.80		16.73	rectangular ductwork with a greatest
Demolish and Install	L.F.	1.93	17.20		19.13	dimension of 8".
Reinstall	L.F.		11.83		11.83	
Clean	L.F.		3.68		3.68	
Minimum Charge	Job		259		259	
10"						
Demolish	L.F.		2.40		2.40	Includes material and labor to install
Install	L.F.	2.41	19.15		21.56	rectangular ductwork with a greatest
Demolish and Install	L.F.	2.41	21.55		23.96	dimension of 10".
Reinstall	L.F.		15.34		15.34	
Clean	L.F.		4.60		4.60	
Minimum Charge	Job		259		259	

Ductwork

Ductwork	Unit	Material	Labor	Equip.	Total	Specification
Galvanized, Spiral						
4"						
Demolish	L.F.		2.88		2.88	Cost includes material and labor to
Install	L.F.	2.08	2.59		4.67	install 4" diameter spiral pre-formed
Demolish and Install	L.F.	2.08	5.47		7.55	steel galvanized ductwork.
Reinstall	L.F.		2.07		2.07	
Clean	L.F.		1.45		1.45	
Minimum Charge	Job		259		259	
6"						
Demolish	L.F.		2.88		2.88	Cost includes material and labor to
Install	L.F.	2.23	3.33		5.56	install 6" diameter spiral pre-formed
Demolish and Install	L.F.	2.23	6.21		8.44	steel galvanized ductwork.
Reinstall	L.F.		2.66		2.66	
Clean	L.F.		2.17		2.17	
Minimum Charge	Job		259		259	
8"						
Demolish	L.F.		2.88		2.88	Cost includes material and labor to
Install	L.F.	2.99	4.66		7.65	install 8" diameter spiral pre-formed
Demolish and Install	L.F.	2.99	7.54		10.53	steel galvanized ductwork.
Reinstall	L.F.		3.73		3.73	
Clean	L.F.		2.89		2.89	
Minimum Charge	Job		259		259	
10"						
Demolish	L.F.		3.59		3.59	Cost includes material and labor to
Install	L.F.	3.72	5.80		9.52	install 10" diameter spiral pre-formed
Demolish and Install	L.F.	3.72	9.39		13.11	steel galvanized ductwork.
Reinstall	L.F.		4.66		4.66	
Clean	L.F.		3.61		3.61	
Minimum Charge	Job		259		259	
12"						
Demolish	L.F.		4.79		4.79	Cost includes material and labor to
Install	L.F.	4.37	7.75		12.12	install 12" diameter spiral pre-formed
Demolish and Install	L.F.	4.37	12.54		16.91	steel galvanized ductwork.
Reinstall	L.F.		6.21		6.21	
Clean	L.F.		4.34		4.34	
Minimum Charge	Job		259		259	
16"						
Demolish	L.F.		9.60		9.60	Cost includes material and labor to
Install	L.F.	7.05	15.55		22.60	install 16" diameter spiral pre-formed
Demolish and Install	L.F.	7.05	25.15		32.20	steel galvanized ductwork.
Reinstall	L.F.		12.42		12.42	
Clean	L.F.		5.80		5.80	
Minimum Charge	Job		259		259	
Flexible Fiberglass						
6" Non-insulated						
Demolish	L.F.		2.21		2.21	Cost includes material and labor to
Install	L.F.	1.77	3.33		5.10	install 6" diameter non-insulated
Demolish and Install	L.F.	1.77	5.54		7.31	flexible fiberglass fabric on corrosion
Reinstall	L.F.		2.66		2.66	resistant metal ductwork.
Clean	L.F.		2.17		2.17	
Minimum Charge	Job		259		259	
8" Non-insulated						
Demolish	L.F.		3.19		3.19	Cost includes material and labor to
Install	L.F.	2.21	4.66		6.87	install 8" diameter non-insulated
Demolish and Install	L.F.	2.21	7.85		10.06	flexible fiberglass fabric on corrosion
Reinstall	L.F.		3.73		3.73	resistant metal ductwork.
Clean	L.F.		2.89		2.89	
Minimum Charge	Job		259		259	

Ductwork	Unit	Material	Labor	Equip.	Total	Specification
10″ Non-insulated						
Demolish	L.F.		4.11		4.11	Cost includes material and labor to
Install	L.F.	2.86	5.80		8.66	install 10″ diameter non-insulated
Demolish and Install	L.F.	2.86	9.91		12.77	flexible fiberglass fabric on corrosion
Reinstall	L.F.		4.66		4.66	resistant metal ductwork.
Clean	L.F.		3.61		3.61	
Minimum Charge	Job		259		259	
12″ Non-insulated						
Demolish	L.F.		5.75		5.75	Cost includes material and labor to
Install	L.F.	3.40	7.75		11.15	install 12″ diameter non-insulated
Demolish and Install	L.F.	3.40	13.50		16.90	flexible fiberglass fabric on corrosion
Reinstall	L.F.		6.21		6.21	resistant metal ductwork.
Clean	L.F.		4.34		4.34	
Minimum Charge	Job		259		259	
4″ Insulated						
Demolish	L.F.		1.69		1.69	Cost includes material and labor to
Install	L.F.	3.11	2.74		5.85	install 4″ diameter insulated flexible
Demolish and Install	L.F.	3.11	4.43		7.54	fiberglass fabric on corrosion resistant
Reinstall	L.F.		2.19		2.19	metal ductwork.
Clean	L.F.		1.45		1.45	
Minimum Charge	Job		259		259	
6″ Insulated						
Demolish	L.F.		2.21		2.21	Cost includes material and labor to
Install	L.F.	3.45	3.58		7.03	install 6″ diameter insulated flexible
Demolish and Install	L.F.	3.45	5.79		9.24	fiberglass fabric on corrosion resistant
Reinstall	L.F.		2.87		2.87	metal ductwork.
Clean	L.F.		2.17		2.17	
Minimum Charge	Job		259		259	
8″ Insulated						
Demolish	L.F.		3.19		3.19	Cost includes material and labor to
Install	L.F.	4.36	5.20		9.56	install 8″ diameter insulated flexible
Demolish and Install	L.F.	4.36	8.39		12.75	fiberglass fabric on corrosion resistant
Reinstall	L.F.		4.14		4.14	metal ductwork.
Clean	L.F.		2.89		2.89	
Minimum Charge	Job		259		259	
10″ Insulated						
Demolish	L.F.		4.11		4.11	Cost includes material and labor to
Install	L.F.	5.05	6.65		11.70	install 10″ diameter insulated flexible
Demolish and Install	L.F.	5.05	10.76		15.81	fiberglass fabric on corrosion resistant
Reinstall	L.F.		5.32		5.32	metal ductwork.
Clean	L.F.		3.61		3.61	
Minimum Charge	Job		259		259	
12″ Insulated						
Demolish	L.F.		5.75		5.75	Cost includes material and labor to
Install	L.F.	6.10	9.30		15.40	install 12″ diameter insulated flexible
Demolish and Install	L.F.	6.10	15.05		21.15	fiberglass fabric on corrosion resistant
Reinstall	L.F.		7.45		7.45	metal ductwork.
Clean	L.F.		4.34		4.34	
Minimum Charge	Job		259		259	

Heating

	Unit	Material	Labor	Equip.	Total	Specification
Hot Water Baseboard						
Baseboard						
Demolish	L.F.		9.30		9.30	Includes material and labor to install
Install	L.F.	12.95	16.35		29.30	baseboard radiation.
Demolish and Install	L.F.	12.95	25.65		38.60	
Reinstall	L.F.		13.09		13.09	
Clean	L.F.	.53	5.70		6.23	
Minimum Charge	Job		265		265	
3'						
Demolish	Ea.		29.50		29.50	Includes material and labor to install
Install	Ea.	27	54.50		81.50	fin tube baseboard radiation in 3'
Demolish and Install	Ea.	27	84		111	sections.
Reinstall	Ea.		43.63		43.63	
Clean	Ea.	.15	17.30		17.45	
Minimum Charge	Job		265		265	
4'						
Demolish	Ea.		39.50		39.50	Includes material and labor to install
Install	Ea.	36	61.50		97.50	fin tube baseboard radiation in 4'
Demolish and Install	Ea.	36	101		137	sections.
Reinstall	Ea.		49.08		49.08	
Clean	Ea.	.30	22.50		22.80	
Minimum Charge	Job		265		265	
5'						
Demolish	Ea.		49		49	Includes material and labor to install
Install	Ea.	44.50	70		114.50	fin tube baseboard radiation in 5'
Demolish and Install	Ea.	44.50	119		163.50	sections.
Reinstall	Ea.		56.09		56.09	
Clean	Ea.	.30	28		28.30	
Minimum Charge	Job		265		265	
6'						
Demolish	Ea.		59		59	Includes material and labor to install
Install	Ea.	53.50	82		135.50	fin tube baseboard radiation in 6'
Demolish and Install	Ea.	53.50	141		194.50	sections.
Reinstall	Ea.		65.44		65.44	
Clean	Ea.	.30	33		33.30	
Minimum Charge	Job		265		265	
7'						
Demolish	Ea.		65.50		65.50	Includes material and labor to install
Install	Ea.	62	106		168	fin tube baseboard radiation in 7'
Demolish and Install	Ea.	62	171.50		233.50	sections.
Reinstall	Ea.		84.86		84.86	
Clean	Ea.	.37	40.50		40.87	
Minimum Charge	Job		265		265	
9'						
Demolish	Ea.		98		98	Includes material and labor to install
Install	Ea.	80.50	152		232.50	fin tube baseboard radiation in 9'
Demolish and Install	Ea.	80.50	250		330.50	sections.
Reinstall	Ea.		121.23		121.23	
Clean	Ea.	.59	52		52.59	
Minimum Charge	Job		265		265	

Heating

Heating	Unit	Material	Labor	Equip.	Total	Specification
Cabinet Unit Heater						
50 MBH						
Demolish	Ea.		49		49	Includes material and labor to install a
Install	Ea.	1975	140		2115	cabinet space heater with an output of
Demolish and Install	Ea.	1975	189		2164	50 MBH.
Reinstall	Ea.		112.18		112.18	
Clean	Ea.	.37	45.50		45.87	
Minimum Charge	Job		273		273	
75 MBH						
Demolish	Ea.		49		49	Includes material and labor to install a
Install	Ea.	2200	164		2364	cabinet space heater with an output of
Demolish and Install	Ea.	2200	213		2413	75 MBH.
Reinstall	Ea.		130.88		130.88	
Clean	Ea.	.37	45.50		45.87	
Minimum Charge	Job		273		273	
125 MBH						
Demolish	Ea.		59		59	Includes material and labor to install a
Install	Ea.	2575	196		2771	cabinet space heater with an output of
Demolish and Install	Ea.	2575	255		2830	125 MBH.
Reinstall	Ea.		157.06		157.06	
Clean	Ea.	.37	45.50		45.87	
Minimum Charge	Job		273		273	
175 MBH						
Demolish	Ea.		73.50		73.50	Includes material and labor to install a
Install	Ea.	3500	280		3780	cabinet space heater with an output of
Demolish and Install	Ea.	3500	353.50		3853.50	175 MBH.
Reinstall	Ea.		224.37		224.37	
Clean	Ea.	.37	45.50		45.87	
Minimum Charge	Job		273		273	

Furnace Unit

Furnace Unit	Unit	Material	Labor	Equip.	Total	Specification
Wall (gas-fired)						
25 MBH w / Thermostat						
Demolish	Ea.		116		116	Includes material and labor to install a
Install	Ea.	925	186		1111	gas fired hot air furnace rated at 25
Demolish and Install	Ea.	925	302		1227	MBH.
Reinstall	Ea.		149.07		149.07	
Clean	Ea.	.75	101		101.75	
Minimum Charge	Job		273		273	
35 MBH w / Thermostat						
Demolish	Ea.		116		116	Includes material and labor to install a
Install	Ea.	810	207		1017	gas fired hot air furnace rated at 35
Demolish and Install	Ea.	810	323		1133	MBH.
Reinstall	Ea.		165.63		165.63	
Clean	Ea.	.75	101		101.75	
Minimum Charge	Job		273		273	
50 MBH w / Thermostat						
Demolish	Ea.		116		116	Includes material and labor to install a
Install	Ea.	1650	233		1883	gas fired hot air furnace rated at 50
Demolish and Install	Ea.	1650	349		1999	MBH.
Reinstall	Ea.		186.34		186.34	
Clean	Ea.	.75	101		101.75	
Minimum Charge	Job		273		273	

Furnace Unit	Unit	Material	Labor	Equip.	Total	Specification
60 MBH w / Thermostat						
Demolish	Ea.		116		116	Includes material and labor to install a
Install	Ea.	1475	259		1734	gas fired hot air furnace rated at 60
Demolish and Install	Ea.	1475	375		1850	MBH.
Reinstall	Ea.		207.04		207.04	
Clean	Ea.	.75	101		101.75	
Minimum Charge	Job		273		273	
Floor (gas-fired)						
30 MBH w / Thermostat						
Demolish	Ea.		116		116	Includes material and labor to install a
Install	Ea.	1400	186		1586	gas fired hot air floor furnace rated at
Demolish and Install	Ea.	1400	302		1702	30 MBH.
Reinstall	Ea.		149.07		149.07	
Clean	Ea.	.75	101		101.75	
Minimum Charge	Job		273		273	
50 MBH w / Thermostat						
Demolish	Ea.		116		116	Includes material and labor to install a
Install	Ea.	1600	233		1833	gas fired hot air floor furnace rated at
Demolish and Install	Ea.	1600	349		1949	50 MBH.
Reinstall	Ea.		186.34		186.34	
Clean	Ea.	.75	101		101.75	
Minimum Charge	Job		273		273	
Gas Forced Air						
50 MBH						
Demolish	Ea.		116		116	Includes material and labor to install
Install	Ea.	1275	233		1508	an up flow gas fired hot air furnace
Demolish and Install	Ea.	1275	349		1624	rated at 50 MBH.
Reinstall	Ea.		186.34		186.34	
Clean	Ea.	.75	101		101.75	
Minimum Charge	Job		273		273	
80 MBH						
Demolish	Ea.		116		116	Includes material and labor to install
Install	Ea.	1450	259		1709	an up flow gas fired hot air furnace
Demolish and Install	Ea.	1450	375		1825	rated at 80 MBH.
Reinstall	Ea.		207.04		207.04	
Clean	Ea.	.75	101		101.75	
Minimum Charge	Job		273		273	
100 MBH						
Demolish	Ea.		116		116	Includes material and labor to install
Install	Ea.	1625	291		1916	an up flow gas fired hot air furnace
Demolish and Install	Ea.	1625	407		2032	rated at 100 MBH.
Reinstall	Ea.		232.92		232.92	
Clean	Ea.	.75	101		101.75	
Minimum Charge	Job		273		273	
120 MBH						
Demolish	Ea.		116		116	Includes material and labor to install
Install	Ea.	1575	310		1885	an up flow gas fired hot air furnace
Demolish and Install	Ea.	1575	426		2001	rated at 120 MBH.
Reinstall	Ea.		248.45		248.45	
Clean	Ea.	.75	101		101.75	
Minimum Charge	Job		273		273	

Furnace Unit		Unit	Material	Labor	Equip.	Total	Specification
Electric Forced Air							
31.4 MBH							
	Demolish	Ea.		116		116	Includes material and labor to install
	Install	Ea.	2725	267		2992	an electric hot air furnace rated at
	Demolish and Install	Ea.	2725	383		3108	31.4 MBH.
	Reinstall	Ea.		213.65		213.65	
	Clean	Ea.	.75	101		101.75	
	Minimum Charge	Job		273		273	
47.1 MBH							
	Demolish	Ea.		116		116	Includes material and labor to install
	Install	Ea.	3200	279		3479	an electric hot air furnace rated at
	Demolish and Install	Ea.	3200	395		3595	47.1 MBH.
	Reinstall	Ea.		223.59		223.59	
	Clean	Ea.	.75	101		101.75	
	Minimum Charge	Job		273		273	
62.9 MBH							
	Demolish	Ea.		116		116	Includes material and labor to install
	Install	Ea.	3375	300		3675	an electric hot air furnace rated at
	Demolish and Install	Ea.	3375	416		3791	62.9 MBH.
	Reinstall	Ea.		240.36		240.36	
	Clean	Ea.	.75	101		101.75	
	Minimum Charge	Job		273		273	
78.5 MBH							
	Demolish	Ea.		116		116	Includes material and labor to install
	Install	Ea.	4100	310		4410	an electric hot air furnace rated at
	Demolish and Install	Ea.	4100	426		4526	78.5 MBH.
	Reinstall	Ea.		246.52		246.52	
	Clean	Ea.	.75	101		101.75	
	Minimum Charge	Job		273		273	
110 MBH							
	Demolish	Ea.		116		116	Includes material and labor to install
	Install	Ea.	5200	325		5525	an electric hot air furnace rated at
	Demolish and Install	Ea.	5200	441		5641	110 MBH.
	Reinstall	Ea.		259.85		259.85	
	Clean	Ea.	.75	101		101.75	
	Minimum Charge	Job		273		273	
A/C Junction Box							
	Install	Ea.	58.50	135		193.50	Includes labor and materials to install
	Minimum Charge	Job		273		273	a heating system emergency shut-off switch. Cost includes up to 40 feet of non-metallic cable.

Boiler		Unit	Material	Labor	Equip.	Total	Specification
Gas-Fired							
60 MBH							
	Demolish	Ea.		990		990	Includes material and labor to install a
	Install	Ea.	2825	1775		4600	gas fired cast iron boiler rated at 60
	Demolish and Install	Ea.	2825	2765		5590	MBH.
	Reinstall	Ea.		1427.78		1427.78	
	Clean	Ea.	1.50	60.50		62	
	Minimum Charge	Job		273		273	

Boiler

Boiler	Unit	Material	Labor	Equip.	Total	Specification
Gas-Fired						
125 MBH						
Demolish	Ea.		2725		2725	Includes material and labor to install a
Install	Ea.	4400	2175		6575	gas fired cast iron boiler rated at 125
Demolish and Install	Ea.	4400	4900		9300	MBH.
Reinstall	Ea.		1745.07		1745.07	
Clean	Ea.	1.50	60.50		62	
Minimum Charge	Job		273		273	
Gas-Fired						
400 MBH						
Demolish	Ea.		2725		2725	Includes material and labor to install a
Install	Ea.	5675	3075		8750	gas fired cast iron boiler rated at 400
Demolish and Install	Ea.	5675	5800		11475	MBH.
Reinstall	Ea.		3067.50		3067.50	
Clean	Ea.	1.50	60.50		62	
Minimum Charge	Job		273		273	

Heat Pump

Heat Pump	Unit	Material	Labor	Equip.	Total	Specification
2 Ton						
Demolish	Ea.		273		273	Includes material and labor to install a
Install	Ea.	1950	490		2440	2 ton heat pump.
Demolish and Install	Ea.	1950	763		2713	
Reinstall	Ea.		392.64		392.64	
Clean	Ea.	1.50	45.50		47	
Minimum Charge	Job		273		273	
3 Ton						
Demolish	Ea.		455		455	Includes material and labor to install a
Install	Ea.	2375	820		3195	3 ton heat pump.
Demolish and Install	Ea.	2375	1275		3650	
Reinstall	Ea.		654.40		654.40	
Clean	Ea.	1.50	45.50		47	
Minimum Charge	Job		273		273	
4 Ton						
Demolish	Ea.		680		680	Includes material and labor to install a
Install	Ea.	2725	1225		3950	4 ton heat pump.
Demolish and Install	Ea.	2725	1905		4630	
Reinstall	Ea.		981.60		981.60	
Clean	Ea.	1.50	45.50		47	
Minimum Charge	Job		273		273	
5 Ton						
Demolish	Ea.		1100		1100	Includes material and labor to install a
Install	Ea.	2975	1975		4950	5 ton heat pump.
Demolish and Install	Ea.	2975	3075		6050	
Reinstall	Ea.		1570.56		1570.56	
Clean	Ea.	1.50	45.50		47	
Minimum Charge	Job		273		273	
Thermostat						
Demolish	Ea.		15.15		15.15	Includes material and labor to install
Install	Ea.	246	68		314	an electric thermostat, 2 set back.
Demolish and Install	Ea.	246	83.15		329.15	
Reinstall	Ea.		54.52		54.52	
Clean	Ea.	.03	11.35		11.38	
Minimum Charge	Job		273		273	

Heat Pump

A/C Junction Box	Unit	Material	Labor	Equip.	Total	Specification
Install	Ea.	555	415		970	Includes labor and materials to install
Minimum Charge	Job		270		270	a heat pump circuit including up to 40 feet of non-metallic cable.

Heating

Electric Baseboard	Unit	Material	Labor	Equip.	Total	Specification
Demolish	L.F.		7.50		7.50	Includes material and labor to install
Install	L.F.	10.20	18		28.20	electric baseboard heating elements.
Demolish and Install	L.F.	10.20	25.50		35.70	
Reinstall	L.F.		14.41		14.41	
Clean	L.F.	.53	5.70		6.23	
Minimum Charge	Job		270		270	
2' 6"						
Demolish	Ea.		25		25	Includes material and labor to install a
Install	Ea.	30.50	67.50		98	2' 6" long section of electric
Demolish and Install	Ea.	30.50	92.50		123	baseboard heating elements.
Reinstall	Ea.		54.04		54.04	
Clean	Ea.	.15	14		14.15	
Minimum Charge	Job		270		270	
3'						
Demolish	Ea.		27.50		27.50	Includes material and labor to install a
Install	Ea.	35	67.50		102.50	3' long section of electric baseboard
Demolish and Install	Ea.	35	95		130	heating elements.
Reinstall	Ea.		54.04		54.04	
Clean	Ea.	.15	17.30		17.45	
Minimum Charge	Job		270		270	
4'						
Demolish	Ea.		30		30	Includes material and labor to install a
Install	Ea.	43	80.50		123.50	4' long section of electric baseboard
Demolish and Install	Ea.	43	110.50		153.50	heating elements.
Reinstall	Ea.		64.53		64.53	
Clean	Ea.	.06	3.79		3.85	
Minimum Charge	Job		270		270	
5'						
Demolish	Ea.		33.50		33.50	Includes material and labor to install a
Install	Ea.	51	95		146	5' long section of electric baseboard
Demolish and Install	Ea.	51	128.50		179.50	heating elements.
Reinstall	Ea.		75.85		75.85	
Clean	Ea.	.30	28		28.30	
Minimum Charge	Job		270		270	
6'						
Demolish	Ea.		37.50		37.50	Includes material and labor to install a
Install	Ea.	56.50	108		164.50	6' long section of electric baseboard
Demolish and Install	Ea.	56.50	145.50		202	heating elements.
Reinstall	Ea.		86.46		86.46	
Clean	Ea.	.06	4.55		4.61	
Minimum Charge	Job		270		270	
8'						
Demolish	Ea.		50		50	Includes material and labor to install
Install	Ea.	72	135		207	an 8' long section of electric
Demolish and Install	Ea.	72	185		257	baseboard heating elements.
Reinstall	Ea.		108.08		108.08	
Clean	Ea.	.06	5.70		5.76	
Minimum Charge	Job		270		270	

158

For customer support on your Contractor's Pricing Guide: Residential Repair & Remodeling, call 800.448.8182.

Rough Mechanical

Heating

	Unit	Material	Labor	Equip.	Total	Specification
10'						
Demolish	Ea.		60		60	Includes material and labor to install a
Install	Ea.	169	164		333	10' long section of electric baseboard
Demolish and Install	Ea.	169	224		393	heating elements.
Reinstall	Ea.		131.01		131.01	
Clean	Ea.	.59	57		57.59	
Minimum Charge	Job		270		270	
Electric Radiant Wall Panel						
Demolish	Ea.		40		40	Includes material and labor to install
Install	Ea.	565	108		673	an 18" x 23" electric radiant heat wall
Demolish and Install	Ea.	565	148		713	panel.
Reinstall	Ea.		86.46		86.46	
Clean	Ea.	.08	11.35		11.43	
Minimum Charge	Job		270		270	

Heating Control

	Unit	Material	Labor	Equip.	Total	Specification
Thermostat						
Demolish	Ea.		15.15		15.15	Includes material and labor to install
Install	Ea.	246	68		314	an electric thermostat, 2 set back.
Demolish and Install	Ea.	246	83.15		329.15	
Reinstall	Ea.		54.52		54.52	
Clean	Ea.	.03	11.35		11.38	
Minimum Charge	Job		273		273	

Air Conditioning

	Unit	Material	Labor	Equip.	Total	Specification
Self-contained						
2 Ton						
Demolish	Ea.		490		490	Includes material and labor to install a
Install	Ea.	4375	845		5220	rooftop air conditioning unit, 2 ton.
Demolish and Install	Ea.	4375	1335		5710	No hoisting is included.
Reinstall	Ea.		846.94		846.94	
Clean	Ea.	1.50	45.50		47	
Minimum Charge	Job		273		273	
3 Ton						
Demolish	Ea.		490		490	Includes material and labor to install a
Install	Ea.	4450	1125		5575	rooftop air conditioning unit, 3 ton.
Demolish and Install	Ea.	4450	1615		6065	No hoisting is included.
Reinstall	Ea.		1116.72		1116.72	
Clean	Ea.	1.50	45.50		47	
Minimum Charge	Job		273		273	
4 Ton						
Demolish	Ea.		490		490	Includes material and labor to install a
Install	Ea.	5375	1300		6675	rooftop air conditioning unit, 4 ton.
Demolish and Install	Ea.	5375	1790		7165	No hoisting is included.
Reinstall	Ea.		1294.99		1294.99	
Clean	Ea.	1.50	45.50		47	
Minimum Charge	Job		273		273	
5 Ton						
Demolish	Ea.		490		490	Includes material and labor to install a
Install	Ea.	5475	1400		6875	rooftop air conditioning unit, 5 ton.
Demolish and Install	Ea.	5475	1890		7365	No hoisting is included.
Reinstall	Ea.		1398.29		1398.29	
Clean	Ea.	1.50	45.50		47	
Minimum Charge	Job		273		273	

Air Conditioning	Unit	Material	Labor	Equip.	Total	Specification
Compressor						
2 Ton						
Demolish	Ea.		455		455	Includes material and labor to install a
Install	Ea.	315	273		588	refrigeration compressor.
Demolish and Install	Ea.	315	728		1043	
Reinstall	Ea.		218.08		218.08	
Clean	Ea.	1.50	45.50		47	
Minimum Charge	Job		273		273	
2.5 Ton						
Demolish	Ea.		455		455	Includes material and labor to install a
Install	Ea.	370	295		665	refrigeration compressor.
Demolish and Install	Ea.	370	750		1120	
Reinstall	Ea.		236.27		236.27	
Clean	Ea.	1.50	45.50		47	
Minimum Charge	Job		273		273	
3 Ton						
Demolish	Ea.		455		455	Includes material and labor to install a
Install	Ea.	405	310		715	refrigeration compressor.
Demolish and Install	Ea.	405	765		1170	
Reinstall	Ea.		246.56		246.56	
Clean	Ea.	1.50	45.50		47	
Minimum Charge	Job		273		273	
4 Ton						
Demolish	Ea.		455		455	Includes material and labor to install a
Install	Ea.	515	248		763	refrigeration compressor.
Demolish and Install	Ea.	515	703		1218	
Reinstall	Ea.		198.25		198.25	
Clean	Ea.	1.50	45.50		47	
Minimum Charge	Job		273		273	
5 Ton						
Demolish	Ea.		455		455	Includes material and labor to install a
Install	Ea.	570	340		910	refrigeration compressor.
Demolish and Install	Ea.	570	795		1365	
Reinstall	Ea.		270.07		270.07	
Clean	Ea.	1.50	45.50		47	
Minimum Charge	Job		273		273	
Condensing Unit						
2 Ton Air-cooled						
Demolish	Ea.		130		130	Includes material and labor to install a
Install	Ea.	1150	465		1615	2 ton condensing unit.
Demolish and Install	Ea.	1150	595		1745	
Reinstall	Ea.		373.94		373.94	
Clean	Ea.	1.50	45.50		47	
Minimum Charge	Job		273		273	
2.5 Ton Air-cooled						
Demolish	Ea.		151		151	Includes material and labor to install a
Install	Ea.	1375	575		1950	2.5 ton condensing unit.
Demolish and Install	Ea.	1375	726		2101	
Reinstall	Ea.		461.93		461.93	
Clean	Ea.	1.50	45.50		47	
Minimum Charge	Job		273		273	
3 Ton Air-cooled						
Demolish	Ea.		182		182	Includes material and labor to install a
Install	Ea.	1450	755		2205	3 ton condensing unit.
Demolish and Install	Ea.	1450	937		2387	
Reinstall	Ea.		604.06		604.06	
Clean	Ea.	1.50	45.50		47	
Minimum Charge	Job		273		273	

Rough Mechanical

Air Conditioning

	Unit	Material	Labor	Equip.	Total	Specification
4 Ton Air-cooled						
Demolish	Ea.		273		273	Includes material and labor to install a
Install	Ea.	1950	1100		3050	4 ton condensing unit.
Demolish and Install	Ea.	1950	1373		3323	
Reinstall	Ea.		872.53		872.53	
Clean	Ea.	1.50	45.50		47	
Minimum Charge	Job		273		273	
5 Ton Air-cooled						
Demolish	Ea.		410		410	Includes material and labor to install a
Install	Ea.	2350	1625		3975	5 ton condensing unit.
Demolish and Install	Ea.	2350	2035		4385	
Reinstall	Ea.		1308.80		1308.80	
Clean	Ea.	1.50	45.50		47	
Minimum Charge	Job		273		273	
Air Handler						
1,200 CFM						
Demolish	Ea.		330		330	Includes material and labor to install a
Install	Ea.	4900	755		5655	1200 CFM capacity air handling unit.
Demolish and Install	Ea.	4900	1085		5985	
Reinstall	Ea.		604.06		604.06	
Clean	Ea.	1.50	45.50		47	
Minimum Charge	Job		273		273	
2,000 CFM						
Demolish	Ea.		330		330	Includes material and labor to install a
Install	Ea.	7350	890		8240	2000 CFM capacity air handling unit.
Demolish and Install	Ea.	7350	1220		8570	
Reinstall	Ea.		713.89		713.89	
Clean	Ea.	1.50	45.50		47	
Minimum Charge	Job		273		273	
3,000 CFM						
Demolish	Ea.		365		365	Includes material and labor to install a
Install	Ea.	10800	980		11780	3000 CFM capacity air handling unit.
Demolish and Install	Ea.	10800	1345		12145	
Reinstall	Ea.		785.28		785.28	
Clean	Ea.	1.50	45.50		47	
Minimum Charge	Job		273		273	
Evaporative Cooler						
4,300 CFM, 1/3 HP						
Demolish	Ea.		131		131	Includes material and labor to install
Install	Ea.	815	291		1106	an evaporative cooler.
Demolish and Install	Ea.	815	422		1237	
Reinstall	Ea.		232.92		232.92	
Clean	Ea.	.59	45.50		46.09	
Minimum Charge	Job		273		273	
4,700 CFM, 1/2 HP						
Demolish	Ea.		131		131	Includes material and labor to install
Install	Ea.	815	300		1115	an evaporative cooler.
Demolish and Install	Ea.	815	431		1246	
Reinstall	Ea.		240.43		240.43	
Clean	Ea.	.59	45.50		46.09	
Minimum Charge	Job		273		273	
5,600 CFM, 3/4 HP						
Demolish	Ea.		131		131	Includes material and labor to install
Install	Ea.	1275	310		1585	an evaporative cooler.
Demolish and Install	Ea.	1275	441		1716	
Reinstall	Ea.		248.45		248.45	
Clean	Ea.	.59	45.50		46.09	
Minimum Charge	Job		273		273	

Air Conditioning	Unit	Material	Labor	Equip.	Total	Specification
Master Cool Brand						
Demolish	Ea.		131		131	Includes material and labor to install
Install	Ea.	815	320		1135	an evaporative cooler.
Demolish and Install	Ea.	815	451		1266	
Reinstall	Ea.		257.02		257.02	
Clean	Ea.	.59	45.50		46.09	
Minimum Charge	Job		273		273	
Motor						
Demolish	Ea.		73.50		73.50	Includes material and labor to install
Install	Ea.	223	226		449	an evaporative cooler motor.
Demolish and Install	Ea.	223	299.50		522.50	
Reinstall	Ea.		181.14		181.14	
Clean	Ea.	.59	45.50		46.09	
Minimum Charge	Job		273		273	
Water Pump						
Demolish	Ea.		73.50		73.50	Includes material and labor to install
Install	Ea.	62.50	129		191.50	an evaporative cooler water pump.
Demolish and Install	Ea.	62.50	202.50		265	
Reinstall	Ea.		103.52		103.52	
Clean	Ea.	.59	45.50		46.09	
Minimum Charge	Job		273		273	
Fan Belt						
Demolish	Ea.		14.75		14.75	Includes material and labor to install
Install	Ea.	45	26		71	an evaporative cooler fan belt.
Demolish and Install	Ea.	45	40.75		85.75	
Reinstall	Ea.		20.70		20.70	
Minimum Charge	Job		273		273	
Pads (4 ea.)						
Demolish	Ea.		14.75		14.75	Includes material and labor to install
Install	Ea.	22	64.50		86.50	pre-fabricated roof pads for an
Demolish and Install	Ea.	22	79.25		101.25	evaporative cooler.
Reinstall	Ea.		51.76		51.76	
Clean	Ea.	.59	45.50		46.09	
Minimum Charge	Job		273		273	
Service call						
Install	Job		259		259	Includes 4 hour minimum labor charge for a sheet metal worker.
Window Mounted						
Low Output BTU						
Install	Ea.	277	59		336	Includes material and labor to install a
Reinstall	Ea.		47.12		47.12	window air conditioning unit.
Clean	Ea.	.22	11.35		11.57	
Minimum Charge	Job		273		273	
Mid-range Output BTU						
Demolish	Ea.		41.50		41.50	Includes material and labor to install a
Install	Ea.	390	78.50		468.50	window air conditioning unit.
Demolish and Install	Ea.	390	120		510	
Reinstall	Ea.		62.83		62.83	
Clean	Ea.	.22	11.35		11.57	
Minimum Charge	Job		273		273	
High Output BTU						
Demolish	Ea.		52.50		52.50	Includes material and labor to install a
Install	Ea.	660	78.50		738.50	window air conditioning unit.
Demolish and Install	Ea.	660	131		791	
Reinstall	Ea.		62.83		62.83	
Clean	Ea.	.22	11.35		11.57	
Minimum Charge	Job		273		273	

Air Conditioning

	Unit	Material	Labor	Equip.	Total	Specification
Hotel-type Hot / Cold						
Demolish	Ea.		87.50		87.50	Includes material and labor to install
Install	Ea.	1600	325		1925	packaged air conditioning unit,
Demolish and Install	Ea.	1600	412.50		2012.50	15000 BTU cooling, 13900 BTU
Reinstall	Ea.		261.76		261.76	heating.
Clean	Ea.	.30	22.50		22.80	
Minimum Charge	Job		273		273	
Thermostat						
Demolish	Ea.		15.15		15.15	Includes material and labor to install
Install	Ea.	246	68		314	an electric thermostat, 2 set back.
Demolish and Install	Ea.	246	83.15		329.15	
Reinstall	Ea.		54.52		54.52	
Clean	Ea.	.03	11.35		11.38	
Minimum Charge	Job		273		273	
Concrete Pad						
Demolish	S.F.		2.53		2.53	Includes material and labor to install
Install	S.F.	3.04	5.75		8.79	air conditioner pad.
Demolish and Install	S.F.	3.04	8.28		11.32	
Reinstall	S.F.		4.62		4.62	
Clean	S.F.	.03	.73		.76	
Paint	S.F.	.17	.40		.57	
Minimum Charge	Job		273		273	
A/C Junction Box						
Install	Ea.	187	154		341	Includes labor and materials to install
Minimum Charge	Job		270		270	an air conditioning circuit including up to 40 feet of non-metallic cable.

Sprinkler System

	Unit	Material	Labor	Equip.	Total	Specification
Exposed						
Wet Type						
Demolish	S.F. Flr.		.24		.24	Includes material and labor to install
Install	S.F. Flr.	2.93	2.31		5.24	an exposed wet pipe fire sprinkler
Demolish and Install	S.F. Flr.	2.93	2.55		5.48	system.
Reinstall	S.F. Flr.		2.31		2.31	
Minimum Charge	Job		265		265	
Dry Type						
Demolish	S.F. Flr.		.24		.24	Includes material and labor to install
Install	S.F. Flr.	3.37	2.31		5.68	an exposed dry pipe fire sprinkler
Demolish and Install	S.F. Flr.	3.37	2.55		5.92	system.
Reinstall	S.F. Flr.		2.31		2.31	
Minimum Charge	Job		265		265	
Concealed						
Wet Type						
Demolish	S.F. Flr.		.47		.47	Includes material and labor to install a
Install	S.F. Flr.	2.93	3.15		6.08	concealed wet pipe fire sprinkler
Demolish and Install	S.F. Flr.	2.93	3.62		6.55	system.
Reinstall	S.F. Flr.		3.15		3.15	
Minimum Charge	Job		265		265	
Dry Type						
Demolish	S.F. Flr.		.47		.47	Includes material and labor to install a
Install	S.F. Flr.	3.37	3.15		6.52	concealed dry pipe fire sprinkler
Demolish and Install	S.F. Flr.	3.37	3.62		6.99	system.
Reinstall	S.F. Flr.		3.15		3.15	
Minimum Charge	Job		265		265	

Junction Boxes	Cable	Switch Box

Electrical Per S.F.

	Unit	Material	Labor	Equip.	Total	Specification
Residential						
Wiring, Lighting and Fixtures						
Install	S.F.	1.53	2.25		3.78	Includes labor and material to replace
Minimum Charge	Job		270		270	house wiring, lighting fixtures, and
						built-in appliances. Apply S.F. cost to
						floor area including garages but not
						basements.
Wiring Only						
Install	S.F.	.15	.89		1.04	Includes labor and material to replace
						house wiring only. Apply S.F. cost to
						floor area including garages but not
						basements.

Main Service

	Unit	Material	Labor	Equip.	Total	Specification
Complete Panel						
100 Amp.						
Demolish	Ea.		222		222	Cost includes material and labor to
Install	Ea.	360	455		815	install 100 Amp. residential service
Demolish and Install	Ea.	360	677		1037	panel including 12 breaker interior
Reinstall	Ea.		363.29		363.29	panel, 10 single pole breakers and
Clean	Ea.	.15	91		91.15	hook-up. Wiring is not included.
Minimum Charge	Job		270		270	
150 Amp.						
Demolish	Ea.		222		222	Cost includes material and labor to
Install	Ea.	930	525		1455	install 150 Amp. residential service,
Demolish and Install	Ea.	930	747		1677	incl. 20 breaker exterior panel, main
Reinstall	Ea.		419.73		419.73	switch, 1 GFI and 14 single pole
Clean	Ea.	.15	91		91.15	breakers, meter socket including
Minimum Charge	Job		270		270	hookup. Wiring is not included.
200 Amp.						
Demolish	Ea.		222		222	Cost includes material and labor to
Install	Ea.	1100	600		1700	install 200 Amp. residential service,
Demolish and Install	Ea.	1100	822		1922	incl. 20 breaker exterior panel, main
Reinstall	Ea.		480.36		480.36	switch, 1 GFI and 18 single pole
Clean	Ea.	.15	91		91.15	breakers, meter socket including
Minimum Charge	Job		270		270	hookup. Wiring is not included.

For customer support on your Contractor's Pricing Guide: Residential Repair & Remodeling, call 800.448.8182.

Rough Electrical

Main Service	Unit	Material	Labor	Equip.	Total	Specification
Masthead						
Demolish	Ea.		37.50		37.50	Includes material and labor to install
Install	Ea.	51	225		276	galvanized steel conduit including
Demolish and Install	Ea.	51	262.50		313.50	masthead.
Reinstall	Ea.		180.13		180.13	
Minimum Charge	Job		270		270	
Allowance to Repair						
Install	Job		270		270	Minimum labor charge for residential
Minimum Charge	Job		270		270	wiring.

For customer support on your Contractor's Pricing Guide: Residential Repair & Remodeling, call 800.448.8182.

Wall Insulation

Floor Insulation	Unit	Material	Labor	Equip.	Total	Specification
Kraft Faced Batt						
3-1/2″ (R-13)						
Demolish	S.F.		.17		.17	Cost includes material and labor to
Install	S.F.	.45	.67		1.12	install 3-1/2″ (R-13) kraft paper or foil
Demolish and Install	S.F.	.45	.84		1.29	faced roll and batt fiberglass
Minimum Charge	Job		236		236	insulation, joists 16″ O.C.
6″ (R-19)						
Demolish	S.F.		.17		.17	Cost includes material and labor to
Install	S.F.	.55	.79		1.34	install 6″ (R-19) kraft paper or foil
Demolish and Install	S.F.	.55	.96		1.51	faced roll and batt fiberglass
Minimum Charge	Job		236		236	insulation, joists 16″ O.C.
10″ (R-30)						
Demolish	S.F.		.17		.17	Cost includes material and labor to
Install	S.F.	.80	.94		1.74	install 10″ (R-30) kraft paper-faced roll
Demolish and Install	S.F.	.80	1.11		1.91	and batt fiberglass insulation, joists
Minimum Charge	Job		236		236	16″ O.C.
12″ (R-38)						
Demolish	S.F.		.17		.17	Cost includes material and labor to
Install	S.F.	1.17	.99		2.16	install 12″ (R-38) kraft paper-faced roll
Demolish and Install	S.F.	1.17	1.16		2.33	and batt fiberglass insulation, joists
Minimum Charge	Job		236		236	16″ O.C.
Foil Faced Batt						
3-1/2″ (R-13)						
Demolish	S.F.		.17		.17	Cost includes material and labor to
Install	S.F.	.45	.67		1.12	install 3-1/2″ (R-13) kraft paper or foil
Demolish and Install	S.F.	.45	.84		1.29	faced roll and batt fiberglass
Minimum Charge	Job		236		236	insulation, joists 16″ O.C.
6″ (R-19)						
Demolish	S.F.		.17		.17	Cost includes material and labor to
Install	S.F.	.55	.79		1.34	install 6″ (R-19) kraft paper or foil
Demolish and Install	S.F.	.55	.96		1.51	faced roll and batt fiberglass
Minimum Charge	Job		236		236	insulation, joists 16″ O.C.
Unfaced Batt						
3-1/2″ (R-13)						
Demolish	S.F.		.17		.17	Cost includes material and labor to
Install	S.F.	.36	.79		1.15	install 3-1/2″ (R-13) unfaced roll and
Demolish and Install	S.F.	.36	.96		1.32	batt fiberglass insulation, joists 16″
Minimum Charge	Job		236		236	O.C.

Insulation

Floor Insulation

Floor Insulation		Unit	Material	Labor	Equip.	Total	Specification
6" (R-19)							
	Demolish	S.F.		.17		.17	Cost includes material and labor to
	Install	S.F.	.43	.94		1.37	install 6" (R-19) unfaced roll and batt
	Demolish and Install	S.F.	.43	1.11		1.54	fiberglass insulation, joists 16" O.C.
	Minimum Charge	Job		236		236	
10" (R-30)							
	Demolish	S.F.		.17		.17	Cost includes material and labor to
	Install	S.F.	.67	1.05		1.72	install 10" (R-30) unfaced roll and batt
	Demolish and Install	S.F.	.67	1.22		1.89	fiberglass insulation, joists 16" O.C.
	Minimum Charge	Job		236		236	
12" (R-38)							
	Demolish	S.F.		.17		.17	Cost includes material and labor to
	Install	S.F.	.81	1.11		1.92	install 12" (R-38) unfaced roll and batt
	Demolish and Install	S.F.	.81	1.28		2.09	fiberglass insulation, joists 16" O.C.
	Minimum Charge	Job		236		236	

Rigid Board

Rigid Board		Unit	Material	Labor	Equip.	Total	Specification
1/2"							
	Demolish	S.F.		.17		.17	Includes material and labor to install
	Install	S.F.	.34	.59		.93	rigid foam insulation board foil faced
	Demolish and Install	S.F.	.34	.76		1.10	two sides.
	Minimum Charge	Job		236		236	
3/4"							
	Demolish	S.F.		.17		.17	Includes material and labor to install
	Install	S.F.	.51	.59		1.10	rigid foam insulation board foil faced
	Demolish and Install	S.F.	.51	.76		1.27	two sides.
	Minimum Charge	Job		236		236	
1-1/2"							
	Demolish	S.F.		.17		.17	Includes material and labor to install
	Install	S.F.	.74	.63		1.37	rigid foam insulation board foil faced
	Demolish and Install	S.F.	.74	.80		1.54	two sides.
	Minimum Charge	Job		236		236	

Support Wire

Support Wire		Unit	Material	Labor	Equip.	Total	Specification
	Demolish	S.F.		.07		.07	Includes material and labor to install
	Install	S.F.	.02	.13		.15	friction fit support wire.
	Demolish and Install	S.F.	.02	.20		.22	
	Minimum Charge	Job		236		236	

Vapor Barrier

Vapor Barrier		Unit	Material	Labor	Equip.	Total	Specification
	Demolish	S.F.		.09		.09	Cost includes material and labor to
	Install	S.F.	.03	.13		.16	install 4 mil polyethylene vapor barrier
	Demolish and Install	S.F.	.03	.22		.25	10' wide sheets with 6" overlaps.
	Minimum Charge	Job		236		236	

Wall Insulation

Wall Insulation		Unit	Material	Labor	Equip.	Total	Specification
Kraft Faced Batt							
3-1/2" (R-13)							
	Demolish	S.F.		.16		.16	Cost includes material and labor to
	Install	S.F.	.37	.35		.72	install 3-1/2" (R-13) kraft paper-faced
	Demolish and Install	S.F.	.37	.51		.88	roll and batt fiberglass insulation, joist
	Minimum Charge	Job		236		236	or studs 16" O.C.
6" (R-19)							
	Demolish	S.F.		.16		.16	Cost includes material and labor to
	Install	S.F.	.47	.35		.82	install 6" (R-19) kraft paper-faced roll
	Demolish and Install	S.F.	.47	.51		.98	and batt fiberglass insulation, joist or
	Minimum Charge	Job		236		236	studs 16" O.C.

For customer support on your Contractor's Pricing Guide: Residential Repair & Remodeling, call 800.448.8182.

Insulation

Wall Insulation		Unit	Material	Labor	Equip.	Total	Specification
10″ (R-30)							
	Demolish	S.F.		.16		.16	Cost includes material and labor to
	Install	S.F.	.80	.41		1.21	install 10″ (R-30) kraft paper-faced roll
	Demolish and Install	S.F.	.80	.57		1.37	and batt fiberglass insulation, joist or
	Minimum Charge	Job		236		236	studs 16″ O.C.
12″ (R-38)							
	Demolish	S.F.		.16		.16	Cost includes material and labor to
	Install	S.F.	1.17	.41		1.58	install 12″ (R-38) kraft paper-faced roll
	Demolish and Install	S.F.	1.17	.57		1.74	and batt fiberglass insulation, joist or
	Minimum Charge	Job		236		236	studs 16″ O.C.
Foil Faced Batt							
3-1/2″ (R-13)							
	Demolish	S.F.		.16		.16	Cost includes material and labor to
	Install	S.F.	.52	.35		.87	install 3-1/2″ (R-13) foil faced roll and
	Demolish and Install	S.F.	.52	.51		1.03	batt fiberglass insulation, joist or studs
	Minimum Charge	Job		236		236	16″ O.C.
6″ (R-19)							
	Demolish	S.F.		.16		.16	Cost includes material and labor to
	Install	S.F.	.67	.35		1.02	install 6″ (R-19) foil faced roll and batt
	Demolish and Install	S.F.	.67	.51		1.18	fiberglass insulation, joist or studs 16″
	Minimum Charge	Job		236		236	O.C.
Unfaced Batt							
3-1/2″ (R-13)							
	Demolish	S.F.		.16		.16	Cost includes material and labor to
	Install	S.F.	.36	.35		.71	install 3-1/2″ (R-13) unfaced roll and
	Demolish and Install	S.F.	.36	.51		.87	batt fiberglass insulation, joist or studs
	Minimum Charge	Job		236		236	16″ O.C.
6″ (R-19)							
	Demolish	S.F.		.16		.16	Cost includes material and labor to
	Install	S.F.	.43	.41		.84	install 6″ (R-19) unfaced roll and batt
	Demolish and Install	S.F.	.43	.57		1	fiberglass insulation, joist or studs 16″
	Minimum Charge	Job		236		236	O.C.
10″ (R-30)							
	Demolish	S.F.		.16		.16	Cost includes material and labor to
	Install	S.F.	.67	.41		1.08	install 10″ (R-30) unfaced roll and batt
	Demolish and Install	S.F.	.67	.57		1.24	fiberglass insulation, joist or studs 16″
	Minimum Charge	Job		236		236	O.C.
12″ (R-38)							
	Demolish	S.F.		.16		.16	Cost includes material and labor to
	Install	S.F.	.81	.47		1.28	install 12″ (R-38) unfaced roll and batt
	Demolish and Install	S.F.	.81	.63		1.44	fiberglass insulation, joist or studs 16″
	Minimum Charge	Job		236		236	O.C.
Blown							
R-13							
	Demolish	S.F.		.16		.16	Includes material and labor to install
	Install	S.F.	.34	1.62	.50	2.46	fiberglass blown-in insulation in wall
	Demolish and Install	S.F.	.34	1.78	.50	2.62	with wood siding.
	Minimum Charge	Job		560	171	731	
R-19							
	Demolish	S.F.		.16		.16	Includes material and labor to install
	Install	S.F.	.54	1.86	.57	2.97	fiberglass blown-in insulation in wall
	Demolish and Install	S.F.	.54	2.02	.57	3.13	with wood siding.
	Minimum Charge	Job		560	171	731	

Insulation

Wall Insulation

Wall Insulation		Unit	Material	Labor	Equip.	Total	Specification
Rigid Board							
1/2″	Demolish	S.F.		.20		.20	Includes material and labor to install
	Install	S.F.	.34	.59		.93	rigid foam insulation board, foil faced,
	Demolish and Install	S.F.	.34	.79		1.13	two sides.
	Minimum Charge	Job		236		236	
3/4″	Demolish	S.F.		.20		.20	Includes material and labor to install
	Install	S.F.	.51	.59		1.10	rigid foam insulation board, foil faced,
	Demolish and Install	S.F.	.51	.79		1.30	two sides.
	Minimum Charge	Job		236		236	
1-1/2″	Demolish	S.F.		.20		.20	Includes material and labor to install
	Install	S.F.	.74	.65		1.39	rigid foam insulation board, foil faced,
	Demolish and Install	S.F.	.74	.85		1.59	two sides.
	Minimum Charge	Job		236		236	
Vapor Barrier	Demolish	S.F.		.05		.05	Cost includes material and labor to
	Install	S.F.	.03	.13		.16	install 4 mil polyethylene vapor barrier
	Demolish and Install	S.F.	.03	.18		.21	10′ wide sheets with 6″ overlaps.
	Minimum Charge	Job		236		236	

Ceiling Insulation

Ceiling Insulation		Unit	Material	Labor	Equip.	Total	Specification
Kraft Faced Batt							
3-1/2″ (R-13)	Demolish	S.F.		.17		.17	Cost includes material and labor to
	Install	S.F.	.37	.35		.72	install 3-1/2″ (R-13) kraft paper-faced
	Demolish and Install	S.F.	.37	.52		.89	roll and batt fiberglass insulation, joist
	Minimum Charge	Job		236		236	or studs 16″ O.C.
6″ (R-19)	Demolish	S.F.		.17		.17	Cost includes material and labor to
	Install	S.F.	.47	.35		.82	install 6″ (R-19) kraft paper-faced roll
	Demolish and Install	S.F.	.47	.52		.99	and batt fiberglass insulation, joist or
	Minimum Charge	Job		236		236	studs 16″ O.C.
10″ (R-30)	Demolish	S.F.		.17		.17	Cost includes material and labor to
	Install	S.F.	.80	.41		1.21	install 10″ (R-30) kraft paper-faced roll
	Demolish and Install	S.F.	.80	.58		1.38	and batt fiberglass insulation, joist or
	Minimum Charge	Job		236		236	studs 16″ O.C.
12″ (R-38)	Demolish	S.F.		.17		.17	Cost includes material and labor to
	Install	S.F.	1.17	.41		1.58	install 12″ (R-38) kraft paper-faced roll
	Demolish and Install	S.F.	1.17	.58		1.75	and batt fiberglass insulation, joist or
	Minimum Charge	Job		236		236	studs 16″ O.C.
Foil Faced Batt							
3-1/2″ (R-13)	Demolish	S.F.		.17		.17	Cost includes material and labor to
	Install	S.F.	.52	.35		.87	install 3-1/2″ (R-13) foil faced roll and
	Demolish and Install	S.F.	.52	.52		1.04	batt fiberglass insulation, joist or studs
	Minimum Charge	Job		236		236	16″ O.C.

Insulation

Ceiling Insulation		Unit	Material	Labor	Equip.	Total	Specification
6" (R-19)							
	Demolish	S.F.		.17		.17	Cost includes material and labor to
	Install	S.F.	.67	.35		1.02	install 6" (R-19) foil faced roll and batt
	Demolish and Install	S.F.	.67	.52		1.19	fiberglass insulation, joist or studs 16"
	Minimum Charge	Job		236		236	O.C.
Unfaced Batt							
3-1/2" (R-13)							
	Demolish	S.F.		.17		.17	Cost includes material and labor to
	Install	S.F.	.36	.35		.71	install 3-1/2" (R-13) unfaced roll and
	Demolish and Install	S.F.	.36	.52		.88	batt fiberglass insulation, joist or studs
	Minimum Charge	Job		236		236	16" O.C.
6" (R-19)							
	Demolish	S.F.		.17		.17	Cost includes material and labor to
	Install	S.F.	.43	.41		.84	install 6" (R-19) unfaced roll and batt
	Demolish and Install	S.F.	.43	.58		1.01	fiberglass insulation, joist or studs 16"
	Minimum Charge	Job		236		236	O.C.
10" (R-30)							
	Demolish	S.F.		.17		.17	Cost includes material and labor to
	Install	S.F.	.67	.41		1.08	install 10" (R-30) unfaced roll and batt
	Demolish and Install	S.F.	.67	.58		1.25	fiberglass insulation, joist or studs 16"
	Minimum Charge	Job		236		236	O.C.
12" (R-38)							
	Demolish	S.F.		.17		.17	Cost includes material and labor to
	Install	S.F.	.81	.47		1.28	install 12" (R-38) unfaced roll and batt
	Demolish and Install	S.F.	.81	.64		1.45	fiberglass insulation, joist or studs 16"
	Minimum Charge	Job		236		236	O.C.
Blown							
5"							
	Demolish	S.F.		.50		.50	Includes material and labor to install
	Install	S.F.	.26	.29	.09	.64	fiberglass blown-in insulation.
	Demolish and Install	S.F.	.26	.79	.09	1.14	
	Minimum Charge	Job		560	171	731	
6"							
	Demolish	S.F.		.50		.50	Includes material and labor to install
	Install	S.F.	.37	.37	.11	.85	fiberglass blown-in insulation.
	Demolish and Install	S.F.	.37	.87	.11	1.35	
	Minimum Charge	Job		560	171	731	
9"							
	Demolish	S.F.		.50		.50	Includes material and labor to install
	Install	S.F.	.46	.51	.16	1.13	fiberglass blown-in insulation.
	Demolish and Install	S.F.	.46	1.01	.16	1.63	
	Minimum Charge	Job		560	171	731	
12"							
	Demolish	S.F.		.50		.50	Includes material and labor to install
	Install	S.F.	.64	.74	.23	1.61	fiberglass blown-in insulation.
	Demolish and Install	S.F.	.64	1.24	.23	2.11	
	Minimum Charge	Job		560	171	731	
Rigid Board							
1/2"							
	Demolish	S.F.		.28		.28	Includes material and labor to install
	Install	S.F.	.34	.59		.93	rigid foam insulation board, foil faced,
	Demolish and Install	S.F.	.34	.87		1.21	two sides.
	Minimum Charge	Job		236		236	

Insulation

Ceiling Insulation

	Unit	Material	Labor	Equip.	Total	Specification
3/4"						
Demolish	S.F.		.28		.28	Includes material and labor to install
Install	S.F.	.51	.59		1.10	rigid foam insulation board, foil faced,
Demolish and Install	S.F.	.51	.87		1.38	two sides.
Minimum Charge	Job		236		236	
1-1/2"						
Demolish	S.F.		.28		.28	Includes material and labor to install
Install	S.F.	.74	.65		1.39	rigid foam insulation board, foil faced,
Demolish and Install	S.F.	.74	.93		1.67	two sides.
Minimum Charge	Job		236		236	

Exterior Insulation

	Unit	Material	Labor	Equip.	Total	Specification
Rigid Board						
1/2"						
Demolish	S.F.		.16		.16	Includes material and labor to install
Install	S.F.	.34	.59		.93	rigid foam insulation board foil faced
Demolish and Install	S.F.	.34	.75		1.09	two sides.
Minimum Charge	Job		236		236	
3/4"						
Demolish	S.F.		.16		.16	Includes material and labor to install
Install	S.F.	.51	.59		1.10	rigid foam insulation board foil faced
Demolish and Install	S.F.	.51	.75		1.26	two sides.
Minimum Charge	Job		236		236	
1-1/2"						
Demolish	S.F.		.16		.16	Includes material and labor to install
Install	S.F.	.74	.63		1.37	rigid foam insulation board foil faced
Demolish and Install	S.F.	.74	.79		1.53	two sides.
Minimum Charge	Job		236		236	
Vapor Barrier						
Demolish	S.F.		.09		.09	Cost includes material and labor to
Install	S.F.	.03	.13		.16	install 4 mil polyethylene vapor barrier
Demolish and Install	S.F.	.03	.22		.25	10' wide sheets with 6" overlaps.
Minimum Charge	Job		236		236	

Roof Insulation Board

	Unit	Material	Labor	Equip.	Total	Specification
Fiberglass						
3/4" Fiberglass						
Demolish	Sq.		52.50		52.50	Includes material and labor to install
Install	Sq.	68	27		95	fiberglass board insulation 3/4" thick,
Demolish and Install	Sq.	68	79.50		147.50	R-2.80 and C-0.36 values.
Minimum Charge	Job		236		236	
1" Fiberglass						
Demolish	Sq.		52.50		52.50	Includes material and labor to install
Install	Sq.	96.50	30		126.50	fiberglass board insulation 1" thick,
Demolish and Install	Sq.	96.50	82.50		179	R-4.20 and C-0.24 values.
Minimum Charge	Job		236		236	
1-3/8" Fiberglass						
Demolish	Sq.		52.50		52.50	Includes material and labor to install
Install	Sq.	157	33		190	fiberglass board insulation 1-3/8"
Demolish and Install	Sq.	157	85.50		242.50	thick, R-5.30 and C-0.19 values.
Minimum Charge	Job		236		236	

Insulation

Roof Insulation Board

	Unit	Material	Labor	Equip.	Total	Specification
1-5/8″ Fiberglass						
Demolish	Sq.		52.50		52.50	Includes material and labor to install
Install	Sq.	158	34.50		192.50	fiberglass board insulation 1-5/8″
Demolish and Install	Sq.	158	87		245	thick, R-6.70 and C-0.15 values.
Minimum Charge	Job		236		236	
2-1/4″ Fiberglass						
Demolish	Sq.		52.50		52.50	Includes material and labor to install
Install	Sq.	190	38		228	fiberglass board insulation 2-1/4″
Demolish and Install	Sq.	190	90.50		280.50	thick, R-8.30 and C-0.12 values.
Minimum Charge	Job		236		236	
Perlite						
1″ Perlite						
Demolish	Sq.		52.50		52.50	Includes material and labor to install
Install	Sq.	60.50	30		90.50	perlite board insulation 1″ thick,
Demolish and Install	Sq.	60.50	82.50		143	R-2.80 and C-0.36 values.
Minimum Charge	Job		236		236	
4″ Perlite						
Demolish	Sq.		52.50		52.50	Includes material and labor to install
Install	Sq.	242	58		300	perlite board insulation 4″ thick,
Demolish and Install	Sq.	242	110.50		352.50	R-10.0 and C-0.10 values.
Minimum Charge	Job		236		236	
1-1/2″ Perlite						
Demolish	Sq.		52.50		52.50	Includes material and labor to install
Install	Sq.	89	33		122	perlite board insulation 1-1/2″ thick,
Demolish and Install	Sq.	89	85.50		174.50	R-4.20 and C-0.24 values.
Minimum Charge	Job		236		236	
2″ Perlite						
Demolish	Sq.		52.50		52.50	Includes material and labor to install
Install	Sq.	121	40.50		161.50	perlite board insulation 2″ thick,
Demolish and Install	Sq.	121	93		214	R-5.30 and C-0.19 values.
Minimum Charge	Job		236		236	
2-1/2″ Perlite						
Demolish	Sq.		52.50		52.50	Includes material and labor to install
Install	Sq.	160	45.50		205.50	perlite board insulation 2-1/2″ thick,
Demolish and Install	Sq.	160	98		258	R-6.70 and C-0.15 values.
Minimum Charge	Job		236		236	
3″ Perlite						
Demolish	Sq.		52.50		52.50	Includes material and labor to install
Install	Sq.	182	48.50		230.50	perlite board insulation 3″ thick,
Demolish and Install	Sq.	182	101		283	R-8.30 and C-0.12 values.
Minimum Charge	Job		236		236	
Polystyrene						
2″ Polystyrene						
Demolish	Sq.		52.50		52.50	Includes material and labor to install
Install	Sq.	59.50	38		97.50	extruded polystyrene board, R5.0/ 2″
Demolish and Install	Sq.	59.50	90.50		150	in thickness @25PSI compressive
Minimum Charge	Job		236		236	strength.
1″ Polystyrene						
Demolish	Sq.		52.50		52.50	Includes material and labor to install
Install	Sq.	29.50	26		55.50	extruded polystyrene board, R5.0/ 1″
Demolish and Install	Sq.	29.50	78.50		108	in thickness @25PSI compressive
Minimum Charge	Job		236		236	strength.
1-1/2″ Polystyrene						
Demolish	Sq.		52.50		52.50	Includes material and labor to install
Install	Sq.	44.50	30		74.50	extruded polystyrene board, R5.0/
Demolish and Install	Sq.	44.50	82.50		127	1-1/2″ in thickness @25PSI
Minimum Charge	Job		236		236	compressive strength.

173

Insulation

Roof Insulation Board	Unit	Material	Labor	Equip.	Total	Specification
Urethane						
3/4" Urethane						
Demolish	Sq.		52.50		52.50	Includes material and labor to install
Install	Sq.	53	22.50		75.50	urethane board insulation 3/4" thick,
Demolish and Install	Sq.	53	75		128	R-5.30 and C-0.19 values.
Minimum Charge	Job		236		236	
1" Urethane						
Demolish	Sq.		52.50		52.50	Includes material and labor to install
Install	Sq.	94.50	24		118.50	urethane board insulation 1" thick,
Demolish and Install	Sq.	94.50	76.50		171	R-6.70 and C-0.15 values.
Minimum Charge	Job		236		236	
1-1/2" Urethane						
Demolish	Sq.		52.50		52.50	Includes material and labor to install
Install	Sq.	77	28		105	urethane board insulation 1-1/2"
Demolish and Install	Sq.	77	80.50		157.50	thick, R-10.1 and C-0.09 values.
Minimum Charge	Job		236		236	
2" Urethane						
Demolish	Sq.		52.50		52.50	Includes material and labor to install
Install	Sq.	94	33		127	urethane board insulation 2" thick,
Demolish and Install	Sq.	94	85.50		179.50	R-13.4 and C-0.06 values.
Minimum Charge	Job		236		236	

For customer support on your Contractor's Pricing Guide: Residential Repair & Remodeling, call 800.448.8182.

Gypsum Wallboard **Tile Ceiling** **Gypsum Wallboard Ceiling**

Wall Plaster	Unit	Material	Labor	Equip.	Total	Specification
Interior						
Cement						
Demolish	S.Y.		6.75		6.75	Includes material and labor to install
Install	S.Y.	15.25	29	2.01	46.26	cement plaster on interior wood
Demolish and Install	S.Y.	15.25	35.75	2.01	53.01	framing with metal lath.
Clean	S.Y.		3.31		3.31	
Paint	S.Y.	1.91	5.20		7.11	
Minimum Charge	Job		228		228	
Acoustic						
Demolish	S.Y.		6.75		6.75	Includes material and labor to install
Install	S.Y.	11.65	35	2.01	48.66	acoustical cement plaster on interior
Demolish and Install	S.Y.	11.65	41.75	2.01	55.41	wood framing with metal lath.
Clean	S.Y.		3.31		3.31	
Paint	S.Y.	1.91	5.20		7.11	
Minimum Charge	Job		228		228	
Gypsum						
Demolish	S.Y.		6.75		6.75	Includes material and labor to install
Install	S.Y.	10.45	30	1.71	42.16	gypsum wall or ceiling plaster on
Demolish and Install	S.Y.	10.45	36.75	1.71	48.91	interior wood framing with metal lath.
Clean	S.Y.		3.31		3.31	
Paint	S.Y.	1.91	5.20		7.11	
Minimum Charge	Job		228		228	
Stucco						
Demolish	S.Y.		16.15		16.15	Includes material and labor to install
Install	S.Y.	9.30	41.50	2.36	53.16	stucco on wire mesh over wood
Demolish and Install	S.Y.	9.30	57.65	2.36	69.31	framing.
Clean	S.Y.		3.31		3.31	
Paint	S.Y.	1.91	5.20		7.11	
Minimum Charge	Job		228		228	
Thin Coat						
Veneer						
Demolish	S.Y.		6.75		6.75	Includes materials and labor to install
Install	S.Y.	1.01	5.40	.37	6.78	thin coat veneer plaster.
Demolish and Install	S.Y.	1.01	12.15	.37	13.53	
Clean	S.Y.		3.31		3.31	
Paint	S.Y.	1.91	5.20		7.11	
Minimum Charge	Job		228		228	

Walls / Ceilings

Wall Plaster

Wall Plaster	Unit	Material	Labor	Equip.	Total	Specification
Lath						
Metal						
Demolish	S.Y.		1.58		1.58	Includes material and labor to install
Install	S.Y.	4.94	6.05		10.99	wire lath on steel framed walls.
Demolish and Install	S.Y.	4.94	7.63		12.57	
Clean	S.Y.		3.31		3.31	
Minimum Charge	Job		106		106	
Gypsum						
Demolish	S.Y.		1.58		1.58	Includes material and labor to install
Install	S.Y.	2.97	5.65		8.62	gypsum lath on walls.
Demolish and Install	S.Y.	2.97	7.23		10.20	
Clean	S.Y.		3.31		3.31	
Minimum Charge	Job		228		228	
Repairs						
Fill in Small Cracks						
Install	L.F.	.07	3.65		3.72	Includes labor and material to fill in
Minimum Charge	Job		228		228	hairline cracks up to 1/8" with filler, sand and prep for paint.
Repair up to 1" Stock						
Install	L.F.	.30	4.56		4.86	Includes labor and material to fill in
Minimum Charge	Job		228		228	hairline cracks up to 1" with filler, sand and prep for paint.
Patch Small Hole (2")						
Install	Ea.	.07	4.80		4.87	Includes labor and material to patch
Minimum Charge	Job		228		228	up to a 2" diameter hole in plaster.
Patch Large Hole						
Install	Ea.	8.80	152		160.80	Includes labor and material to patch
Minimum Charge	Job		228		228	up to a 20" diameter hole in plaster.
Patch Plaster Section						
Install	Ea.	2.65	4.56		7.21	Includes labor and material to repair
Minimum Charge	Job		228		228	section of plaster wall, area less than 20 S.F.
Repair Inside Corner						
Install	L.F.	3.52	11.40		14.92	Includes material and labor to patch
Minimum Charge	Job		228		228	plaster corners.

Ceiling Plaster

Ceiling Plaster	Unit	Material	Labor	Equip.	Total	Specification
Interior						
Cement						
Demolish	S.Y.		6.75		6.75	Includes material and labor to install
Install	S.Y.	15.25	29	2.01	46.26	cement plaster on interior wood
Demolish and Install	S.Y.	15.25	35.75	2.01	53.01	framing with metal lath.
Clean	S.Y.		3.46		3.46	
Paint	S.Y.	1.91	5.20		7.11	
Minimum Charge	Job		228		228	
Acoustic						
Demolish	S.Y.		6.75		6.75	Includes material and labor to install
Install	S.Y.	11.65	35	2.01	48.66	acoustical cement plaster on interior
Demolish and Install	S.Y.	11.65	41.75	2.01	55.41	wood framing with metal lath.
Clean	S.Y.		3.46		3.46	
Paint	S.Y.	1.91	5.20		7.11	
Minimum Charge	Job		228		228	

Walls / Ceilings

Ceiling Plaster

	Unit	Material	Labor	Equip.	Total	Specification
Gypsum						
Demolish	S.Y.		6.75		6.75	Includes material and labor to install
Install	S.Y.	10.45	30	1.71	42.16	gypsum wall or ceiling plaster on
Demolish and Install	S.Y.	10.45	36.75	1.71	48.91	interior wood framing with metal lath.
Clean	S.Y.		3.46		3.46	
Paint	S.Y.	1.91	5.20		7.11	
Minimum Charge	Job		228		228	
Stucco						
Demolish	S.Y.		16.15		16.15	Includes material and labor to install
Install	S.Y.	9.30	41.50	2.36	53.16	stucco on wire mesh over wood
Demolish and Install	S.Y.	9.30	57.65	2.36	69.31	framing.
Clean	S.Y.		3.46		3.46	
Paint	S.Y.	1.91	5.20		7.11	
Minimum Charge	Job		228		228	
Thin Coat						
Veneer						
Demolish	S.Y.		6.75		6.75	Includes materials and labor to install
Install	S.Y.	1.01	5.40	.37	6.78	thin coat veneer plaster.
Demolish and Install	S.Y.	1.01	12.15	.37	13.53	
Clean	S.Y.		3.46		3.46	
Paint	S.Y.	1.91	5.20		7.11	
Minimum Charge	Job		228		228	
Lath						
Metal						
Demolish	S.Y.		1.58		1.58	Includes material and labor to install
Install	S.Y.	4.94	7.55		12.49	wire lath on steel framed ceilings.
Demolish and Install	S.Y.	4.94	9.13		14.07	
Minimum Charge	Job		228		228	
Gypsum						
Demolish	S.Y.		1.58		1.58	Includes material and labor to install
Install	S.Y.	2.97	4.71		7.68	gypsum lath on ceilings.
Demolish and Install	S.Y.	2.97	6.29		9.26	
Minimum Charge	Job		228		228	
Repairs						
Fill in Small Cracks						
Install	S.F.	.07	.83		.90	Includes labor and material to fill in
Minimum Charge	Job		455		455	hairline cracks up to 1/8" with filler, sand and prep for paint.

Gypsum Wallboard

	Unit	Material	Labor	Equip.	Total	Specification
Finished						
1/4"						
Demolish	S.F.		.40		.40	Cost includes material and labor to
Install	S.F.	.37	1.45		1.82	install 1/4" drywall, including joint
Demolish and Install	S.F.	.37	1.85		2.22	taping, premixed compound and
Clean	S.F.		.33		.33	screws.
Paint	S.F.	.28	.34		.62	
Minimum Charge	Job		236		236	
3/8"						
Demolish	S.F.		.40		.40	Cost includes material and labor to
Install	S.F.	.34	1.47		1.81	install 3/8" drywall including joint
Demolish and Install	S.F.	.34	1.87		2.21	tape, premixed compound and
Clean	S.F.		.33		.33	screws.
Paint	S.F.	.28	.34		.62	
Minimum Charge	Job		236		236	

For customer support on your Contractor's Pricing Guide: Residential Repair & Remodeling, call 800.448.8182.

Gypsum Wallboard		Unit	Material	Labor	Equip.	Total	Specification
1/2"							
	Demolish	S.F.		.40		.40	Cost includes material and labor to
	Install	S.F.	.32	1.49		1.81	install 1/2" drywall including joint
	Demolish and Install	S.F.	.32	1.89		2.21	tape, premixed compound and
	Clean	S.F.		.33		.33	screws.
	Paint	S.F.	.28	.34		.62	
	Minimum Charge	Job		236		236	
1/2" Water Resistant							
	Demolish	S.F.		.40		.40	Cost includes material and labor to
	Install	S.F.	.36	1.49		1.85	install 1/2" WR drywall including joint
	Demolish and Install	S.F.	.36	1.89		2.25	tape, premixed compound and
	Clean	S.F.		.33		.33	screws.
	Paint	S.F.	.28	.34		.62	
	Minimum Charge	Job		236		236	
1/2" Fire-rated							
	Demolish	S.F.		.40		.40	Cost includes material and labor to
	Install	S.F.	.35	1.49		1.84	install 1/2" FR drywall including
	Demolish and Install	S.F.	.35	1.89		2.24	screws. Taping and finishing not
	Clean	S.F.		.33		.33	included.
	Paint	S.F.	.28	.34		.62	
	Minimum Charge	Job		236		236	
5/8"							
	Demolish	S.F.		.40		.40	Cost includes material and labor to
	Install	S.F.	.37	1.43		1.80	install 5/8" drywall including joint
	Demolish and Install	S.F.	.37	1.83		2.20	tape, premixed compound and
	Clean	S.F.		.33		.33	screws.
	Paint	S.F.	.28	.34		.62	
	Minimum Charge	Job		236		236	
5/8" Fire-rated							
	Demolish	S.F.		.40		.40	Cost includes material and labor to
	Install	S.F.	.37	1.69		2.06	install 5/8" FR drywall, including joint
	Demolish and Install	S.F.	.37	2.09		2.46	taping, premixed compound and
	Clean	S.F.		.33		.33	screws.
	Paint	S.F.	.28	.34		.62	
	Minimum Charge	Job		236		236	
2-hour Fire-rated							
	Demolish	S.F.		1.01		1.01	Cost includes material and labor to
	Install	S.F.	.83	3.35		4.18	install 2 layers of 5/8" type X drywall
	Demolish and Install	S.F.	.83	4.36		5.19	including joint taping, premixed
	Clean	S.F.		.33		.33	compound and screws.
	Paint	S.F.	.28	.34		.62	
	Minimum Charge	Job		236		236	
Unfinished							
1/4"							
	Demolish	S.F.		.40		.40	Cost includes material and labor to
	Install	S.F.	.33	.50		.83	install 1/4" drywall including screws.
	Demolish and Install	S.F.	.33	.90		1.23	Taping and finishing not included.
	Clean	S.F.		.33		.33	
	Minimum Charge	Job		236		236	
3/8"							
	Demolish	S.F.		.40		.40	Cost includes material and labor to
	Install	S.F.	.30	.51		.81	install 3/8" drywall including screws.
	Demolish and Install	S.F.	.30	.91		1.21	Tape, finish or texture is not included.
	Clean	S.F.		.33		.33	
	Minimum Charge	Job		236		236	

Walls / Ceilings

Gypsum Wallboard

Gypsum Wallboard	Unit	Material	Labor	Equip.	Total	Specification
1/2"						
Demolish	S.F.		.40		.40	Cost includes material and labor to
Install	S.F.	.28	.51		.79	install 1/2" drywall including screws.
Demolish and Install	S.F.	.28	.91		1.19	Taping and finishing not included.
Clean	S.F.		.33		.33	
Minimum Charge	Job		236		236	
1/2" Water Resistant						
Demolish	S.F.		.40		.40	Cost includes material and labor to
Install	S.F.	.32	.51		.83	install 1/2" WR drywall including
Demolish and Install	S.F.	.32	.91		1.23	screws. Taping and finishing not
Clean	S.F.		.33		.33	included.
Minimum Charge	Job		236		236	
1/2" Fire-rated						
Demolish	S.F.		.40		.40	Cost includes labor and material to
Install	S.F.	.28	.34		.62	install 1/2" FR drywall including
Demolish and Install	S.F.	.28	.74		1.02	screws. Tape, finish or texture is not
Clean	S.F.		.33		.33	included.
Minimum Charge	Job		236		236	
5/8"						
Demolish	S.F.		.40		.40	Cost includes material and labor to
Install	S.F.	.32	.58		.90	install 5/8" drywall including screws.
Demolish and Install	S.F.	.32	.98		1.30	Taping and finishing not included.
Clean	S.F.		.33		.33	
Minimum Charge	Job		236		236	
5/8" Fire-rated						
Demolish	S.F.		.40		.40	Cost includes material and labor to
Install	S.F.	.33	.58		.91	install 5/8" FR drywall including
Demolish and Install	S.F.	.33	.98		1.31	screws. Taping and finishing not
Clean	S.F.		.33		.33	included.
Minimum Charge	Job		236		236	
2-hour Fire-rated						
Demolish	S.F.		1.01		1.01	Cost includes material and labor to
Install	S.F.	.83	3.35		4.18	install 2 layers of 5/8" type X drywall
Demolish and Install	S.F.	.83	4.36		5.19	including joint taping, premixed
Clean	S.F.		.33		.33	compound and screws.
Minimum Charge	Job		236		236	
Accessories						
Tape & Finish						
Install	S.F.	.06	.47		.53	Includes labor and material for taping,
Minimum Charge	Job		236		236	sanding and finish.
Texture						
Sprayed						
Install	S.F.	.04	.57		.61	Includes labor and material to install
Minimum Charge	Job		236		236	by spray, texture finish.
Trowel						
Install	S.F.	.45	.37		.82	Includes labor and material to apply
Minimum Charge	Job		236		236	hand troweled texture.
Patch Hole in Gypsum Wallboard						
Install	S.F.	.53	64.50		65.03	Includes labor and material to patch a
Minimum Charge	Job		236		236	one square foot area of damaged drywall.
Furring Strips						
Demolish	S.F.		.10		.10	Cost includes material and labor to
Install	S.F.	.22	1.05		1.27	install 1" x 2" furring strips, 16" O.C.
Demolish and Install	S.F.	.22	1.15		1.37	per S.F. of surface area to be covered.
Minimum Charge	Job		118		118	

179

Ceramic Wall Tile

Ceramic Wall Tile	Unit	Material	Labor	Equip.	Total	Specification
Thin Set						
2″ x 2″						
Demolish	S.F.		1.35		1.35	Cost includes material and labor to
Install	S.F.	7.25	7.40		14.65	install 2″ x 2″ ceramic tile in organic
Demolish and Install	S.F.	7.25	8.75		16	adhesive including cutting and grout.
Clean	S.F.		.69		.69	
Minimum Charge	Job		222		222	
4-1/4″ x 4-1/4″						
Demolish	S.F.		1.35		1.35	Cost includes material and labor to
Install	S.F.	2.85	4.14		6.99	install 4-1/4″ x 4-1/4″ ceramic tile in
Demolish and Install	S.F.	2.85	5.49		8.34	organic adhesive including grout.
Clean	S.F.		.69		.69	
Minimum Charge	Job		222		222	
6″ x 6″						
Demolish	S.F.		1.35		1.35	Cost includes material and labor to
Install	S.F.	3.99	4.49		8.48	install 6″ x 6″ ceramic tile in organic
Demolish and Install	S.F.	3.99	5.84		9.83	adhesive including cutting and grout.
Clean	S.F.		.69		.69	
Minimum Charge	Job		222		222	
8″ x 8″						
Demolish	S.F.		1.35		1.35	Cost includes material and labor to
Install	S.F.	5.70	4.63		10.33	install 8″ x 8″ ceramic tile in organic
Demolish and Install	S.F.	5.70	5.98		11.68	adhesive including cutting and grout.
Clean	S.F.		.69		.69	
Minimum Charge	Job		222		222	
Thick Set						
2″ x 2″						
Demolish	S.F.		1.62		1.62	Cost includes material and labor to
Install	S.F.	8.30	8.90		17.20	install 2″ x 2″ ceramic tile in mortar
Demolish and Install	S.F.	8.30	10.52		18.82	bed including cutting and grout.
Clean	S.F.		.69		.69	
Minimum Charge	Job		222		222	
4-1/4″ x 4-1/4″						
Demolish	S.F.		1.62		1.62	Cost includes material and labor to
Install	S.F.	10.50	8.90		19.40	install 4-1/4″ x 4-1/4″ ceramic tile in
Demolish and Install	S.F.	10.50	10.52		21.02	mortar bed including cutting and
Clean	S.F.		.69		.69	grout.
Minimum Charge	Job		222		222	
6″ x 6″						
Demolish	S.F.		1.62		1.62	Cost includes material and labor to
Install	S.F.	11.20	8.10		19.30	install 6″ x 6″ ceramic tile in mortar
Demolish and Install	S.F.	11.20	9.72		20.92	bed including cutting and grout.
Clean	S.F.		.69		.69	
Minimum Charge	Job		222		222	
8″ x 8″						
Demolish	S.F.		1.62		1.62	Cost includes material and labor to
Install	S.F.	10.40	7.40		17.80	install 8″ x 8″ ceramic tile in mortar
Demolish and Install	S.F.	10.40	9.02		19.42	bed including cutting and grout.
Clean	S.F.		.69		.69	
Minimum Charge	Job		222		222	
Re-grout						
Install	S.F.	.17	4.45		4.62	Includes material and labor to regrout
Minimum Charge	Job		222		222	wall tiles.

Walls / Ceilings

Wall Tile

	Unit	Material	Labor	Equip.	Total	Specification
Mirror						
Custom Cut Wall						
Demolish	S.F.		.77		.77	Includes material and labor to install
Install	S.F.	2.95	4.69		7.64	distortion-free float glass mirror, 1/8"
Demolish and Install	S.F.	2.95	5.46		8.41	to 3/16" thick with cut and polished
Clean	S.F.	.01	.21		.22	edges.
Minimum Charge	Job		229		229	
Beveled Wall						
Demolish	S.F.		.77		.77	Includes material and labor to install
Install	S.F.	7.55	8.30		15.85	distortion-free float glass mirror, 1/8"
Demolish and Install	S.F.	7.55	9.07		16.62	to 3/16" thick with beveled edges.
Clean	S.F.	.01	.21		.22	
Minimum Charge	Job		229		229	
Smoked Wall						
Demolish	S.F.		.77		.77	Includes material and labor to install
Install	S.F.	9.35	8.30		17.65	distortion-free smoked float glass
Demolish and Install	S.F.	9.35	9.07		18.42	mirror, 1/8" to 3/16" thick with cut
Clean	S.F.	.01	.21		.22	and polished edges.
Minimum Charge	Job		229		229	
Marble						
Good Grade						
Demolish	S.F.		2.53		2.53	Cost includes material and labor to
Install	S.F.	10	19.35		29.35	install 3/8" x 12" x 12" marble tile in
Demolish and Install	S.F.	10	21.88		31.88	mortar bed with grout.
Clean	S.F.		.69		.69	
Minimum Charge	Job		222		222	
Better Grade						
Demolish	S.F.		2.53		2.53	Cost includes material and labor to
Install	S.F.	11.15	19.35		30.50	install 3/8" x 12" x 12" marble, in
Demolish and Install	S.F.	11.15	21.88		33.03	mortar bed with grout.
Clean	S.F.		.69		.69	
Minimum Charge	Job		222		222	
Premium Grade						
Demolish	S.F.		2.53		2.53	Cost includes material and labor to
Install	S.F.	21.50	19.35		40.85	install 3/8" x 12" x 12" marble, in
Demolish and Install	S.F.	21.50	21.88		43.38	mortar bed with grout.
Clean	S.F.		.69		.69	
Minimum Charge	Job		222		222	
Glass Block						
Demolish	S.F.		2.02		2.02	Cost includes material and labor to
Install	S.F.	26	19.10		45.10	install 3-7/8" thick, 6" x 6" smooth
Demolish and Install	S.F.	26	21.12		47.12	face glass block including mortar and
Clean	S.F.		.69		.69	cleaning.
Minimum Charge	Job		430		430	
Cork						
Demolish	S.F.		.31		.31	Cost includes material and labor to
Install	S.F.	4.68	1.64		6.32	install 3/16" thick cork tile and
Demolish and Install	S.F.	4.68	1.95		6.63	installation.
Clean	S.F.		.69		.69	
Minimum Charge	Job		196		196	
Re-grout						
Install	S.F.	.17	4.45		4.62	Includes material and labor to regrout
Minimum Charge	Job		222		222	wall tiles.

181

Walls / Ceilings

Wall Paneling		Unit	Material	Labor	Equip.	Total	Specification
Plywood							
Good Grade							
	Demolish	S.F.		.40		.40	Cost includes material and labor to
	Install	S.F.	1.64	1.88		3.52	install 4' x 8' x 1/4" paneling,
	Demolish and Install	S.F.	1.64	2.28		3.92	including nails and adhesive.
	Clean	S.F.		.33		.33	
	Paint	S.F.	.34	.75		1.09	
	Minimum Charge	Job		236		236	
Better Grade							
	Demolish	S.F.		.40		.40	Cost includes material and labor to
	Install	S.F.	1.33	2.24		3.57	install 4' x 8' x 1/4" plywood wall
	Demolish and Install	S.F.	1.33	2.64		3.97	panel, including nails and adhesive.
	Clean	S.F.		.33		.33	
	Paint	S.F.	.34	.75		1.09	
	Minimum Charge	Job		236		236	
Premium Grade							
	Demolish	S.F.		.40		.40	Cost includes material and labor to
	Install	S.F.	1.22	2.69		3.91	install 4' x 8' x 1/4" plywood wall
	Demolish and Install	S.F.	1.22	3.09		4.31	panel, including nails and adhesive.
	Clean	S.F.		.33		.33	
	Paint	S.F.	.34	.75		1.09	
	Minimum Charge	Job		236		236	
Hardboard							
Embossed Paneling							
	Demolish	S.F.		.40		.40	Includes material and labor to install
	Install	S.F.	.68	1.88		2.56	hardboard embossed paneling
	Demolish and Install	S.F.	.68	2.28		2.96	including adhesives.
	Clean	S.F.		.33		.33	
	Minimum Charge	Job		236		236	
Plastic Molding							
	Demolish	L.F.		.40		.40	Includes material and labor to install
	Install	L.F.	.87	1.18		2.05	plastic trim for paneling.
	Demolish and Install	L.F.	.87	1.58		2.45	
	Clean	L.F.	.01	.21		.22	
	Minimum Charge	Job		236		236	
Hardboard							
Embossed Brick							
	Demolish	S.F.		.40		.40	Cost includes material and labor to
	Install	S.F.	.89	2.09		2.98	install 4' x 8' non-ceramic mineral
	Demolish and Install	S.F.	.89	2.49		3.38	brick-like veneer wall panels including
	Clean	S.F.		.33		.33	adhesive.
	Paint	S.F.	.21	.59		.80	
	Minimum Charge	Job		236		236	
Embossed Ceramic Tile							
	Demolish	S.F.		.40		.40	Includes material and labor to install
	Install	S.F.	.68	1.88		2.56	hardboard embossed paneling
	Demolish and Install	S.F.	.68	2.28		2.96	including adhesives.
	Clean	S.F.		.33		.33	
	Paint	S.F.	.21	.59		.80	
	Minimum Charge	Job		236		236	
Pegboard							
	Demolish	S.F.		.40		.40	Cost includes material and labor to
	Install	S.F.	.53	1.88		2.41	install 4' x 8' x 1/4" pegboard panel,
	Demolish and Install	S.F.	.53	2.28		2.81	including adhesive.
	Clean	S.F.		.33		.33	
	Paint	S.F.	.21	.59		.80	
	Minimum Charge	Job		236		236	

Walls / Ceilings

Wall Paneling		Unit	Material	Labor	Equip.	Total	Specification
Furring Strips							
	Demolish	S.F.		.10		.10	Cost includes material and labor to
	Install	S.F.	.22	1.05		1.27	install 1" x 2" furring strips, 16" O.C.
	Demolish and Install	S.F.	.22	1.15		1.37	per S.F. of surface area to be covered.
	Minimum Charge	Job		236		236	
Molding							
Unfinished Base							
	Demolish	L.F.		.67		.67	Cost includes material and labor to
	Install	L.F.	.61	1.42		2.03	install 1/2" x 2-1/2" pine or birch
	Demolish and Install	L.F.	.61	2.09		2.70	base molding.
	Reinstall	L.F.		1.14		1.14	
	Clean	L.F.	.01	.21		.22	
	Paint	L.F.	.06	.43		.49	
	Minimum Charge	Job		236		236	
Unfinished Corner							
	Demolish	L.F.		.67		.67	Includes material and labor to install
	Install	L.F.	1.77	1.96		3.73	unfinished pine corner molding.
	Demolish and Install	L.F.	1.77	2.63		4.40	
	Reinstall	L.F.		1.57		1.57	
	Clean	L.F.	.01	.21		.22	
	Paint	L.F.	.17	.63		.80	
	Minimum Charge	Job		236		236	
Unfinished Cove							
	Demolish	L.F.		.67		.67	Cost includes material and labor to
	Install	L.F.	2.09	1.78		3.87	install 11/16" x 2-3/4" pine cove
	Demolish and Install	L.F.	2.09	2.45		4.54	molding.
	Reinstall	L.F.		1.42		1.42	
	Clean	L.F.	.01	.21		.22	
	Paint	L.F.	.17	.63		.80	
	Minimum Charge	Job		236		236	
Unfinished Crown							
	Demolish	L.F.		.81		.81	Includes material and labor to install
	Install	L.F.	2.34	1.88		4.22	unfinished pine crown molding.
	Demolish and Install	L.F.	2.34	2.69		5.03	
	Reinstall	L.F.		1.51		1.51	
	Clean	L.F.	.02	.28		.30	
	Paint	L.F.	.10	.42		.52	
	Minimum Charge	Job		236		236	
Prefinished Base							
	Demolish	L.F.		.67		.67	Cost includes material and labor to
	Install	L.F.	1.37	1.51		2.88	install 9/16" x 3-5/16" prefinished
	Demolish and Install	L.F.	1.37	2.18		3.55	base molding.
	Reinstall	L.F.		1.21		1.21	
	Clean	L.F.	.01	.21		.22	
	Paint	L.F.	.06	.43		.49	
	Minimum Charge	Job		236		236	
Prefinished Corner							
	Demolish	L.F.		.67		.67	Includes material and labor to install
	Install	L.F.	2.30	1.96		4.26	inside or outside prefinished corner
	Demolish and Install	L.F.	2.30	2.63		4.93	molding.
	Reinstall	L.F.		1.57		1.57	
	Clean	L.F.	.01	.21		.22	
	Paint	L.F.	.17	.63		.80	
	Minimum Charge	Job		236		236	

For customer support on your Contractor's Pricing Guide: Residential Repair & Remodeling, call 800.448.8182.

Wall Paneling

	Unit	Material	Labor	Equip.	Total	Specification
Prefinished Crown						
Demolish	L.F.		.81		.81	Cost includes material and labor to
Install	L.F.	1.86	1.88		3.74	install 3/4" x 3-13/16" prefinished
Demolish and Install	L.F.	1.86	2.69		4.55	crown molding.
Reinstall	L.F.		1.51		1.51	
Clean	L.F.	.02	.28		.30	
Paint	L.F.	.10	.42		.52	
Minimum Charge	Job		236		236	

Wallcovering

	Unit	Material	Labor	Equip.	Total	Specification
Surface Prep						
Patch Walls						
Install	S.F.	.13	4.56		4.69	Includes labor and material to repair
Minimum Charge	Job		196		196	section of plaster wall, area less than
						20 S.F.
Sizing						
Install	S.F.	.09	.20		.29	Includes labor and materials to apply
Minimum Charge	Job		196		196	wall sizing.
Good Grade						
Walls						
Demolish	S.F.		.73		.73	Includes material and labor to install
Install	S.F.	1.01	.82		1.83	pre-pasted, vinyl-coated wallcovering.
Demolish and Install	S.F.	1.01	1.55		2.56	
Clean	S.F.		.08		.08	
Minimum Charge	Job		196		196	
Ceiling						
Demolish	S.F.		.73		.73	Includes material and labor to install
Install	S.F.	1.01	.82		1.83	pre-pasted, vinyl-coated wallcovering.
Demolish and Install	S.F.	1.01	1.55		2.56	
Clean	S.F.	.01	.21		.22	
Minimum Charge	Job		196		196	
Better Grade						
Walls						
Demolish	S.F.		.73		.73	Includes material and labor to install
Install	S.F.	1.14	.82		1.96	pre-pasted, vinyl-coated wallcovering.
Demolish and Install	S.F.	1.14	1.55		2.69	
Clean	S.F.		.08		.08	
Minimum Charge	Job		196		196	
Ceiling						
Demolish	S.F.		.73		.73	Includes material and labor to install
Install	S.F.	1.14	.82		1.96	pre-pasted, vinyl-coated wallcovering.
Demolish and Install	S.F.	1.14	1.55		2.69	
Clean	S.F.	.01	.21		.22	
Minimum Charge	Job		196		196	
Premium Grade						
Walls						
Demolish	S.F.		.91		.91	Includes material and labor to install
Install	S.F.	1.22	.82		2.04	pre-pasted, vinyl-coated wallcovering.
Demolish and Install	S.F.	1.22	1.73		2.95	
Clean	S.F.		.08		.08	
Minimum Charge	Job		196		196	
Ceiling						
Demolish	S.F.		.91		.91	Includes material and labor to install
Install	S.F.	1.22	.82		2.04	pre-pasted, vinyl-coated wallcovering.
Demolish and Install	S.F.	1.22	1.73		2.95	
Clean	S.F.	.01	.21		.22	
Minimum Charge	Job		196		196	

Walls / Ceilings

Wallcovering	Unit	Material	Labor	Equip.	Total	Specification
Mural						
Pre-printed						
Demolish	Ea.		126		126	Includes material and labor to install
Install	Ea.	273	295		568	wallcovering with a pre-printed mural.
Demolish and Install	Ea.	273	421		694	
Clean	Ea.		9.10		9.10	
Minimum Charge	Job		196		196	
Hand-painted						
Demolish	Ea.		126		126	Includes material and labor to install
Install	Ea.	545	595		1140	wallcovering with a hand-painted
Demolish and Install	Ea.	545	721		1266	mural.
Clean	Ea.		9.10		9.10	
Minimum Charge	Job		196		196	
Vinyl-coated						
Good Grade						
Demolish	Ea.		22.50		22.50	Includes material and labor to install
Install	Ea.	17.10	27.50		44.60	pre-pasted, vinyl-coated wallcovering.
Demolish and Install	Ea.	17.10	50		67.10	Based upon 30 S.F. rolls.
Clean	Ea.		9.10		9.10	
Minimum Charge	Job		196		196	
Better Grade						
Demolish	Ea.		22.50		22.50	Includes material and labor to install
Install	Ea.	26.50	27.50		54	pre-pasted, vinyl-coated wallcovering.
Demolish and Install	Ea.	26.50	50		76.50	Based upon 30 S.F. rolls.
Clean	Ea.		9.10		9.10	
Minimum Charge	Job		196		196	
Premium Grade						
Demolish	Ea.		22.50		22.50	Includes material and labor to install
Install	Ea.	31.50	27.50		59	pre-pasted, vinyl-coated wallcovering.
Demolish and Install	Ea.	31.50	50		81.50	Based upon 30 S.F. rolls.
Clean	Ea.		9.10		9.10	
Minimum Charge	Job		196		196	
Blank Underliner						
Demolish	Ea.		22.50		22.50	Includes material and labor to install
Install	Ea.	9.70	28		37.70	blank stock liner paper or bridging
Demolish and Install	Ea.	9.70	50.50		60.20	paper. Based upon 30 S.F. rolls.
Clean	Ea.		9.10		9.10	
Minimum Charge	Job		196		196	
Solid Vinyl						
Good Grade						
Demolish	Roll		24.50		24.50	Includes material and labor to install
Install	Ea.	24	27.50		51.50	pre-pasted, vinyl wallcovering. Based
Demolish and Install	Ea.	24	52		76	upon 30 S.F. rolls.
Clean	Ea.		9.10		9.10	
Minimum Charge	Job		196		196	
Better Grade						
Demolish	Roll		24.50		24.50	Includes material and labor to install
Install	Ea.	32	27.50		59.50	pre-pasted, vinyl wallcovering. Based
Demolish and Install	Ea.	32	52		84	upon 30 S.F. rolls.
Clean	Ea.		9.10		9.10	
Minimum Charge	Job		196		196	
Premium Grade						
Demolish	Roll		24.50		24.50	Includes material and labor to install
Install	Ea.	50.50	27.50		78	pre-pasted, vinyl wallcovering. Based
Demolish and Install	Ea.	50.50	52		102.50	upon 30 S.F. rolls.
Clean	Ea.		9.10		9.10	
Minimum Charge	Job		196		196	

Wallcovering		Unit	Material	Labor	Equip.	Total	Specification
Blank Underliner							
	Demolish	Ea.		22.50		22.50	Includes material and labor to install
	Install	Ea.	9.70	28		37.70	blank stock liner paper or bridging
	Demolish and Install	Ea.	9.70	50.50		60.20	paper. Based upon 30 S.F. rolls.
	Clean	Ea.		9.10		9.10	
	Minimum Charge	Job		196		196	
Grasscloth / String							
Good Grade							
	Demolish	Ea.		22.50		22.50	Includes material and labor to install
	Install	Ea.	54.50	35.50		90	grasscloth or string wallcovering.
	Demolish and Install	Ea.	54.50	58		112.50	Based upon 30 S.F. rolls.
	Clean	Ea.		12.10		12.10	
	Minimum Charge	Job		196		196	
Better Grade							
	Demolish	Ea.		22.50		22.50	Includes material and labor to install
	Install	Ea.	61	35.50		96.50	grasscloth or string wallcovering.
	Demolish and Install	Ea.	61	58		119	Based upon 30 S.F. rolls.
	Clean	Ea.		12.10		12.10	
	Minimum Charge	Job		196		196	
Premium Grade							
	Demolish	Ea.		22.50		22.50	Includes material and labor to install
	Install	Ea.	67	35.50		102.50	grasscloth or string wallcovering.
	Demolish and Install	Ea.	67	58		125	Based upon 30 S.F. rolls.
	Clean	Ea.		12.10		12.10	
	Minimum Charge	Job		196		196	
Blank Underliner							
	Demolish	Ea.		22.50		22.50	Includes material and labor to install
	Install	Ea.	9.70	28		37.70	blank stock liner paper or bridging
	Demolish and Install	Ea.	9.70	50.50		60.20	paper. Based upon 30 S.F. rolls.
	Clean	Ea.		9.10		9.10	
	Minimum Charge	Job		196		196	
Designer							
Premium Grade							
	Demolish	S.F.		.91		.91	Includes material and labor to install
	Install	Ea.	127	41.50		168.50	paperback vinyl, untrimmed and
	Demolish and Install	Ea.	127	42.41		169.41	hand-painted. Based upon 30 S.F.
	Clean	Ea.		9.10		9.10	rolls.
	Minimum Charge	Job		196		196	
Blank Underliner							
	Demolish	Ea.		22.50		22.50	Includes material and labor to install
	Install	Ea.	9.70	28		37.70	blank stock liner paper or bridging
	Demolish and Install	Ea.	9.70	50.50		60.20	paper. Based upon 30 S.F. rolls.
	Clean	Ea.		9.10		9.10	
	Minimum Charge	Job		196		196	
Foil							
	Demolish	Ea.		22.50		22.50	Includes material and labor to install
	Install	Ea.	29.50	35.50		65	foil coated wallcovering. Based upon
	Demolish and Install	Ea.	29.50	58		87.50	30 S.F. rolls.
	Clean	Ea.		9.10		9.10	
	Minimum Charge	Job		196		196	
Wallcovering Border							
Good Grade							
	Demolish	L.F.		1.47		1.47	Includes material and labor to install
	Install	L.F.	1.66	1.51		3.17	vinyl-coated border 4" to 6" wide.
	Demolish and Install	L.F.	1.66	2.98		4.64	
	Clean	L.F.		.14		.14	
	Minimum Charge	Job		196		196	

Walls / Ceilings

Wallcovering

	Unit	Material	Labor	Equip.	Total	Specification
Better Grade						
Demolish	L.F.		1.47		1.47	Includes material and labor to install
Install	L.F.	2.98	1.51		4.49	vinyl-coated border 4" to 6" wide.
Demolish and Install	L.F.	2.98	2.98		5.96	
Clean	L.F.		.14		.14	
Minimum Charge	Job		196		196	
Premium Grade						
Demolish	L.F.		1.47		1.47	Includes material and labor to install
Install	L.F.	3.64	1.51		5.15	vinyl-coated border 4" to 6" wide.
Demolish and Install	L.F.	3.64	2.98		6.62	
Clean	L.F.		.14		.14	
Minimum Charge	Job		196		196	

Acoustical Ceiling

	Unit	Material	Labor	Equip.	Total	Specification
Complete System						
2' x 2' Square Edge						
Demolish	S.F.		.67		.67	Includes material and labor to install
Install	S.F.	2.16	1.45		3.61	T-bar grid acoustical ceiling system,
Demolish and Install	S.F.	2.16	2.12		4.28	random pinhole 5/8" thick square
Reinstall	S.F.		1.16		1.16	edge tile.
Clean	S.F.	.02	.28		.30	
Minimum Charge	Job		236		236	
2' x 2' Reveal Edge						
Demolish	S.F.		.67		.67	Includes material and labor to install
Install	S.F.	2.89	1.88		4.77	T-bar grid acoustical ceiling system,
Demolish and Install	S.F.	2.89	2.55		5.44	random pinhole 5/8" thick reveal
Reinstall	S.F.		1.51		1.51	edge tile.
Clean	S.F.	.02	.28		.30	
Minimum Charge	Job		236		236	
2' x 2' Fire-rated						
Demolish	S.F.		.67		.67	Includes material and labor to install
Install	S.F.	4.11	1.45		5.56	T-bar grid FR acoustical ceiling system,
Demolish and Install	S.F.	4.11	2.12		6.23	random pinhole 5/8" thick square
Reinstall	S.F.		1.16		1.16	edge tile.
Clean	S.F.	.02	.28		.30	
Minimum Charge	Job		236		236	
2' x 4' Square Edge						
Demolish	S.F.		.67		.67	Cost includes material and labor to
Install	S.F.	2.51	1.24		3.75	install 2' x 4' T-bar grid acoustical
Demolish and Install	S.F.	2.51	1.91		4.42	ceiling system, random pinhole 5/8"
Reinstall	S.F.		.99		.99	thick square edge tile.
Clean	S.F.	.02	.28		.30	
Minimum Charge	Job		236		236	
2' x 4' Reveal Edge						
Demolish	S.F.		.67		.67	Includes material and labor to install
Install	S.F.	2.59	1.71		4.30	T-bar grid acoustical ceiling system,
Demolish and Install	S.F.	2.59	2.38		4.97	random pinhole 5/8" thick reveal
Reinstall	S.F.		1.37		1.37	edge tile.
Clean	S.F.	.02	.28		.30	
Minimum Charge	Job		236		236	
2' x 4' Fire-rated						
Demolish	S.F.		.67		.67	Includes material and labor to install
Install	S.F.	4.11	1.45		5.56	T-bar grid FR acoustical ceiling system,
Demolish and Install	S.F.	4.11	2.12		6.23	random pinhole 5/8" thick square
Reinstall	S.F.		1.16		1.16	edge tile.
Clean	S.F.	.02	.28		.30	
Minimum Charge	Job		236		236	

For customer support on your Contractor's Pricing Guide: Residential Repair & Remodeling, call 800.448.8182.

Walls / Ceilings

Acoustical Ceiling	Unit	Material	Labor	Equip.	Total	Specification
T-bar Grid						
2′ x 2′						
Demolish	S.F.		.79		.79	Includes material and labor to install
Install	S.F.	1.14	.73		1.87	T-bar grid acoustical ceiling
Demolish and Install	S.F.	1.14	1.52		2.66	suspension system.
Reinstall	S.F.		.58		.58	
Clean	S.F.	.02	.28		.30	
Minimum Charge	Job		228		228	
2′ x 4′						
Demolish	S.F.		.79		.79	Includes material and labor to install
Install	S.F.	.88	.59		1.47	T-bar grid acoustical ceiling
Demolish and Install	S.F.	.88	1.38		2.26	suspension system.
Reinstall	S.F.		.47		.47	
Clean	S.F.	.02	.28		.30	
Minimum Charge	Job		228		228	
Tile Panels						
2′ x 2′ Square Edge						
Demolish	S.F.		.22		.22	Includes material and labor to install
Install	S.F.	.91	.75		1.66	square edge panels 5/8″ thick with
Demolish and Install	S.F.	.91	.97		1.88	pinhole pattern. T-bar grid not
Reinstall	S.F.		.60		.60	included.
Clean	S.F.	.02	.28		.30	
Minimum Charge	Job		236		236	
2′ x 2′ Reveal Edge						
Demolish	S.F.		.22		.22	Includes material and labor to install
Install	S.F.	1.27	1		2.27	reveal edge panels 5/8″ thick with
Demolish and Install	S.F.	1.27	1.22		2.49	pinhole pattern. T-bar grid not
Reinstall	S.F.		.80		.80	included.
Clean	S.F.	.02	.28		.30	
Minimum Charge	Job		236		236	
2′ x 2′ Fire-rated						
Demolish	S.F.		.22		.22	Includes material and labor to install
Install	S.F.	1.44	.70		2.14	FR square edge panels 5/8″ thick with
Demolish and Install	S.F.	1.44	.92		2.36	pinhole pattern. T-bar grid not
Reinstall	S.F.		.56		.56	included.
Clean	S.F.	.02	.28		.30	
Minimum Charge	Job		236		236	
2′ x 4′ Square Edge						
Demolish	S.F.		.22		.22	Includes material and labor to install
Install	S.F.	.91	.75		1.66	square edge panels 5/8″ thick with
Demolish and Install	S.F.	.91	.97		1.88	pinhole pattern. T-bar grid not
Reinstall	S.F.		.60		.60	included.
Clean	S.F.	.02	.28		.30	
Minimum Charge	Job		236		236	
2′ x 4′ Reveal Edge						
Demolish	S.F.		.22		.22	Includes material and labor to install
Install	S.F.	1.27	1		2.27	reveal edge panels 5/8″ thick with
Demolish and Install	S.F.	1.27	1.22		2.49	pinhole pattern. T-bar grid not
Reinstall	S.F.		.80		.80	included.
Clean	S.F.	.02	.28		.30	
Minimum Charge	Job		236		236	
2′ x 4′ Fire-rated						
Demolish	S.F.		.22		.22	Includes material and labor to install
Install	S.F.	1.44	.70		2.14	FR square edge panels 5/8″ thick with
Demolish and Install	S.F.	1.44	.92		2.36	pinhole pattern. T-bar grid not
Reinstall	S.F.		.56		.56	included.
Clean	S.F.	.02	.28		.30	
Minimum Charge	Job		236		236	

Walls / Ceilings

Acoustical Ceiling		Unit	Material	Labor	Equip.	Total	Specification
Bleach Out Stains							
	Install	S.F.	.19	.36		.55	Includes labor and material to apply
	Minimum Charge	Job		236		236	laundry type chlorine bleach to remove light staining of acoustic tile material.
Adhesive Tile							
Good Grade							
	Demolish	S.F.		.90		.90	Cost includes material and labor to
	Install	S.F.	1.25	1.57		2.82	install 12″ x 12″ tile with embossed
	Demolish and Install	S.F.	1.25	2.47		3.72	texture pattern. Furring strips not
	Clean	S.F.	.01	.21		.22	included.
	Minimum Charge	Job		236		236	
Better Grade							
	Demolish	S.F.		.90		.90	Cost includes material and labor to
	Install	S.F.	1.55	1.57		3.12	install 12″ x 12″ tile with embossed
	Demolish and Install	S.F.	1.55	2.47		4.02	texture pattern. Furring strips not
	Clean	S.F.	.01	.21		.22	included.
	Minimum Charge	Job		236		236	
Premium Grade							
	Demolish	S.F.		.90		.90	Cost includes material and labor to
	Install	S.F.	5.65	1.57		7.22	install 12″ x 12″ tile with embossed
	Demolish and Install	S.F.	5.65	2.47		8.12	texture pattern. Furring strips not
	Clean	S.F.	.01	.21		.22	included.
	Minimum Charge	Job		236		236	
Bleach Out Stains							
	Install	S.F.	.19	.36		.55	Includes labor and material to apply
	Minimum Charge	Job		236		236	laundry type chlorine bleach to remove light staining of acoustic tile material.
Stapled Tile							
Good Grade							
	Demolish	S.F.		.54		.54	Cost includes material and labor to
	Install	S.F.	1.25	1.57		2.82	install 12″ x 12″ tile with embossed
	Demolish and Install	S.F.	1.25	2.11		3.36	texture pattern. Furring strips not
	Clean	S.F.	.01	.21		.22	included.
	Minimum Charge	Job		236		236	
Better Grade							
	Demolish	S.F.		.54		.54	Cost includes material and labor to
	Install	S.F.	1.55	1.57		3.12	install 12″ x 12″ tile with embossed
	Demolish and Install	S.F.	1.55	2.11		3.66	texture pattern. Furring strips not
	Clean	S.F.	.01	.21		.22	included.
	Minimum Charge	Job		236		236	
Premium Grade							
	Demolish	S.F.		.54		.54	Cost includes material and labor to
	Install	S.F.	5.65	1.57		7.22	install 12″ x 12″ tile with embossed
	Demolish and Install	S.F.	5.65	2.11		7.76	texture pattern. Furring strips not
	Clean	S.F.	.01	.21		.22	included.
	Minimum Charge	Job		236		236	
Bleach Out Stains							
	Install	S.F.	.19	.36		.55	Includes labor and material to apply
	Minimum Charge	Job		236		236	laundry type chlorine bleach to remove light staining of acoustic tile material.

Acoustical Ceiling

	Unit	Material	Labor	Equip.	Total	Specification
Blown						
Scrape Off Existing						
Install	S.F.		.75		.75	Includes labor and material to remove
Minimum Charge	Job		236		236	existing blown acoustical ceiling material.
Seal Ceiling Gypsum Wallboard						
Install	S.F.	.07	.34		.41	Includes labor and material to seal
Minimum Charge	Job		196		196	existing drywall with shellac-based material.
Reblow Existing						
Install	S.F.	.04	.57		.61	Includes labor and material to install
Minimum Charge	Job		605	69	674	by spray, texture finish.
Bleach Out Stains						
Install	S.F.	.19	.36		.55	Includes labor and material to apply
Minimum Charge	Job		236		236	laundry type chlorine bleach to remove light staining of acoustic tile material.
New Blown						
Install	S.F.	.04	.57		.61	Includes labor and material to install
Minimum Charge	Job		196		196	by spray, texture finish.
Apply Glitter To						
Install	S.F.	.23	.18		.41	Includes labor and material to apply
Minimum Charge	Job		196		196	glitter material to existing or new drywall.
Furring Strips						
Demolish	S.F.		.10		.10	Cost includes material and labor to
Install	S.F.	.22	1.05		1.27	install 1" x 2" furring strips, 16" O.C.
Demolish and Install	S.F.	.22	1.15		1.37	per S.F. of surface area to be covered.
Minimum Charge	Job		118		118	

Finished Stair	Bi-Fold Door	Raised Panel Door

Hollow Core Door	Unit	Material	Labor	Equip.	Total	Specification
Pre-hung Wood						
1' 6" x 6' 8"						
Demolish	Ea.		43.50		43.50	Cost includes material and labor to
Install	Ea.	165	26		191	install 1' 6" x 6' 8" pre-hung 1-3/8"
Demolish and Install	Ea.	165	69.50		234.50	lauan door with split pine jamb
Reinstall	Ea.		26.18		26.18	including casing, stops, and hinges.
Clean	Ea.	1.50	15.15		16.65	
Paint	Ea.	10.85	56		66.85	
Minimum Charge	Job		236		236	
2' x 6' 8"						
Demolish	Ea.		43.50		43.50	Cost includes material and labor to
Install	Ea.	179	27.50		206.50	install 2' x 6' 8" pre-hung 1-3/8"
Demolish and Install	Ea.	179	71		250	lauan door with split pine jamb
Reinstall	Ea.		27.72		27.72	including casing, stops, and hinges.
Clean	Ea.	1.50	15.15		16.65	
Paint	Ea.	10.85	56		66.85	
Minimum Charge	Job		236		236	
2' 4" x 6' 8"						
Demolish	Ea.		43.50		43.50	Cost includes material and labor to
Install	Ea.	184	27.50		211.50	install 2' 4" x 6' 8" pre-hung 1-3/8"
Demolish and Install	Ea.	184	71		255	lauan door with split pine jamb
Reinstall	Ea.		27.72		27.72	including casing, stops, and hinges.
Clean	Ea.	1.50	15.15		16.65	
Paint	Ea.	10.85	56		66.85	
Minimum Charge	Job		236		236	
2' 6" x 6' 8"						
Demolish	Ea.		43.50		43.50	Cost includes material and labor to
Install	Ea.	187	27.50		214.50	install 2' 6" x 6' 8" pre-hung 1-3/8"
Demolish and Install	Ea.	187	71		258	lauan door with split pine jamb
Reinstall	Ea.		27.72		27.72	including casing, stops, and hinges.
Clean	Ea.	1.50	15.15		16.65	
Paint	Ea.	10.85	56		66.85	
Minimum Charge	Job		236		236	
2' 8" x 6' 8"						
Demolish	Ea.		43.50		43.50	Cost includes material and labor to
Install	Ea.	176	27.50		203.50	install 2' 8" x 6' 8" pre-hung 1-3/8"
Demolish and Install	Ea.	176	71		247	lauan door with split pine jamb,
Reinstall	Ea.		27.72		27.72	including casing, stop and hinges.
Clean	Ea.	1.50	15.15		16.65	
Paint	Ea.	10.85	56		66.85	
Minimum Charge	Job		236		236	

Finish Carpentry

Hollow Core Door

	Unit	Material	Labor	Equip.	Total	Specification
3' x 6' 8"						
Demolish	Ea.		43.50		43.50	Cost includes material and labor to
Install	Ea.	184	29.50		213.50	install 3' x 6' 8" pre-hung 1-3/8"
Demolish and Install	Ea.	184	73		257	lauan door with split pine jamb
Reinstall	Ea.		29.45		29.45	including casing, stops, and hinges.
Clean	Ea.	1.50	15.15		16.65	
Paint	Ea.	10.85	56		66.85	
Minimum Charge	Job		236		236	
Pocket Type						
Demolish	Ea.		65.50		65.50	Cost includes material and labor to
Install	Ea.	83	52.50		135.50	install 2' 8" x 6' 8" lauan door.
Demolish and Install	Ea.	83	118		201	
Reinstall	Ea.		52.36		52.36	
Clean	Ea.	1.50	15.15		16.65	
Paint	Ea.	10.85	56		66.85	
Minimum Charge	Job		236		236	
Stain						
Install	Ea.	8.70	32.50		41.20	Includes labor and material to stain
Minimum Charge	Job		196		196	single door and trim on both sides.
Pre-hung Masonite						
1' 6" x 6' 8"						
Demolish	Ea.		43.50		43.50	Cost includes material and labor to
Install	Ea.	119	26		145	install 1' 6" x 6' 8" pre-hung 1-3/8"
Demolish and Install	Ea.	119	69.50		188.50	masonite door with split pine jamb
Reinstall	Ea.		26.18		26.18	including casing, stops, and hinges.
Clean	Ea.	1.50	15.15		16.65	
Paint	Ea.	10.85	56		66.85	
Minimum Charge	Job		236		236	
2' x 6' 8"						
Demolish	Ea.		43.50		43.50	Cost includes material and labor to
Install	Ea.	240	27.50		267.50	install 2' x 6' 8" pre-hung 1-3/8"
Demolish and Install	Ea.	240	71		311	masonite door with split pine jamb
Reinstall	Ea.		27.72		27.72	including casing, stops, and hinges.
Clean	Ea.	1.50	15.15		16.65	
Paint	Ea.	10.85	56		66.85	
Minimum Charge	Job		236		236	
2' 4" x 6' 8"						
Demolish	Ea.		43.50		43.50	Cost includes material and labor to
Install	Ea.	232	27.50		259.50	install 2' 4" x 6' 8" pre-hung 1-3/8"
Demolish and Install	Ea.	232	71		303	masonite door with split pine jamb
Reinstall	Ea.		27.72		27.72	including casing, stops, and hinges.
Clean	Ea.	1.50	15.15		16.65	
Paint	Ea.	10.85	56		66.85	
Minimum Charge	Job		236		236	
2' 8" x 6' 8"						
Demolish	Ea.		43.50		43.50	Cost includes material and labor to
Install	Ea.	234	27.50		261.50	install 2' 8" x 6' 8" pre-hung 1-3/8"
Demolish and Install	Ea.	234	71		305	masonite door with split pine jamb
Reinstall	Ea.		27.72		27.72	including casing, stops, and hinges.
Clean	Ea.	1.50	15.15		16.65	
Paint	Ea.	10.85	56		66.85	
Minimum Charge	Job		236		236	
3' x 6' 8"						
Demolish	Ea.		43.50		43.50	Cost includes material and labor to
Install	Ea.	253	29.50		282.50	install 3' x 6' 8" pre-hung 1-3/8"
Demolish and Install	Ea.	253	73		326	masonite door with split pine jamb
Reinstall	Ea.		29.45		29.45	including casing, stops, and hinges.
Clean	Ea.	1.50	15.15		16.65	
Paint	Ea.	10.85	56		66.85	
Minimum Charge	Job		236		236	

Finish Carpentry

Hollow Core Door

	Unit	Material	Labor	Equip.	Total	Specification
Pocket Type						
Demolish	Ea.		65.50		65.50	Cost includes material and labor to
Install	Ea.	83	52.50		135.50	install 2' 8" x 6' 8" lauan door.
Demolish and Install	Ea.	83	118		201	
Reinstall	Ea.		52.36		52.36	
Clean	Ea.	1.50	15.15		16.65	
Paint	Ea.	10.85	56		66.85	
Minimum Charge	Job		236		236	
Stain						
Install	Ea.	8.70	32.50		41.20	Includes labor and material to stain
Minimum Charge	Job		196		196	single door and trim on both sides.

Hollow Core Door Only
1' 6" x 6' 8"

	Unit	Material	Labor	Equip.	Total	Specification
Demolish	Ea.		9.35		9.35	Cost includes material and labor to
Install	Ea.	77.50	25		102.50	install 1' 6" x 6' 8" lauan hollow core
Demolish and Install	Ea.	77.50	34.35		111.85	door slab.
Reinstall	Ea.		24.80		24.80	
Clean	Ea.	.63	11.35		11.98	
Paint	Ea.	5	49		54	
Minimum Charge	Job		236		236	

2' x 6' 8"

	Unit	Material	Labor	Equip.	Total	Specification
Demolish	Ea.		9.35		9.35	Cost includes material and labor to
Install	Ea.	89	26		115	install 2' x 6' 8" lauan hollow core
Demolish and Install	Ea.	89	35.35		124.35	door slab.
Reinstall	Ea.		26.18		26.18	
Clean	Ea.	.63	11.35		11.98	
Paint	Ea.	5	49		54	
Minimum Charge	Job		236		236	

2' 4" x 6' 8"

	Unit	Material	Labor	Equip.	Total	Specification
Demolish	Ea.		9.35		9.35	Cost includes material and labor to
Install	Ea.	98.50	26		124.50	install 2' 4" x 6' 8" lauan hollow core
Demolish and Install	Ea.	98.50	35.35		133.85	door slab.
Reinstall	Ea.		26.18		26.18	
Clean	Ea.	.63	11.35		11.98	
Paint	Ea.	5	49		54	
Minimum Charge	Job		236		236	

2' 8" x 6' 8"

	Unit	Material	Labor	Equip.	Total	Specification
Demolish	Ea.		9.35		9.35	Cost includes material and labor to
Install	Ea.	101	26		127	install 2' 8" x 6' 8" lauan hollow core
Demolish and Install	Ea.	101	35.35		136.35	door slab.
Reinstall	Ea.		26.18		26.18	
Clean	Ea.	.63	11.35		11.98	
Paint	Ea.	5	49		54	
Minimum Charge	Job		236		236	

3' x 6' 8"

	Unit	Material	Labor	Equip.	Total	Specification
Demolish	Ea.		9.35		9.35	Cost includes material and labor to
Install	Ea.	107	27.50		134.50	install 3' x 6' 8" lauan hollow core
Demolish and Install	Ea.	107	36.85		143.85	door slab.
Reinstall	Ea.		27.72		27.72	
Clean	Ea.	.63	11.35		11.98	
Paint	Ea.	5	49		54	
Minimum Charge	Job		236		236	

Pocket Type

	Unit	Material	Labor	Equip.	Total	Specification
Demolish	Ea.		9.35		9.35	Cost includes material and labor to
Install	Ea.	83	52.50		135.50	install 2' 8" x 6' 8" lauan door.
Demolish and Install	Ea.	83	61.85		144.85	
Reinstall	Ea.		52.36		52.36	
Clean	Ea.	.63	11.35		11.98	
Paint	Ea.	5	49		54	
Minimum Charge	Job		236		236	

193

Hollow Core Door		Unit	Material	Labor	Equip.	Total	Specification
Stain							
	Install	Ea.	4.82	43.50		48.32	Includes labor and materials to stain a
	Minimum Charge	Job		196		196	flush style door on both sides by brush.
Casing Trim							
Single Width							
	Demolish	Ea.		8.30		8.30	Cost includes material and labor to
	Install	Opng.	26.50	39.50		66	install 11/16" x 2-1/2" pine ranch
	Demolish and Install	Opng.	26.50	47.80		74.30	style casing for one side of a standard
	Reinstall	Opng.		31.41		31.41	door opening.
	Clean	Opng.	.21	3.64		3.85	
	Paint	Ea.	5.05	7.80		12.85	
	Minimum Charge	Job		236		236	
Double Width							
	Demolish	Ea.		9.90		9.90	Cost includes material and labor to
	Install	Opng.	31	47		78	install 11/16" x 2-1/2" pine ranch
	Demolish and Install	Opng.	31	56.90		87.90	style door casing for one side of a
	Reinstall	Opng.		37.70		37.70	double door opening.
	Clean	Opng.	.25	7.25		7.50	
	Paint	Ea.	5.05	8.90		13.95	
	Minimum Charge	Job		236		236	
Hardware							
Doorknob w / Lock							
	Demolish	Ea.		18.70		18.70	Includes material and labor to install
	Install	Ea.	42	29.50		71.50	privacy lockset.
	Demolish and Install	Ea.	42	48.20		90.20	
	Reinstall	Ea.		23.56		23.56	
	Clean	Ea.	.15	11.35		11.50	
	Minimum Charge	Job		236		236	
Doorknob							
	Demolish	Ea.		18.70		18.70	Includes material and labor to install
	Install	Ea.	44.50	29.50		74	residential passage lockset, keyless
	Demolish and Install	Ea.	44.50	48.20		92.70	bored type.
	Reinstall	Ea.		23.56		23.56	
	Clean	Ea.	.15	11.35		11.50	
	Minimum Charge	Job		236		236	
Deadbolt							
	Demolish	Ea.		16.35		16.35	Includes material and labor to install a
	Install	Ea.	81.50	33.50		115	deadbolt.
	Demolish and Install	Ea.	81.50	49.85		131.35	
	Reinstall	Ea.		26.93		26.93	
	Clean	Ea.	.15	11.35		11.50	
	Minimum Charge	Job		236		236	
Lever Handle							
	Demolish	Ea.		18.70		18.70	Includes material and labor to install
	Install	Ea.	53.50	47		100.50	residential passage lockset, keyless
	Demolish and Install	Ea.	53.50	65.70		119.20	bored type with lever handle.
	Reinstall	Ea.		37.70		37.70	
	Clean	Ea.	.15	11.35		11.50	
	Minimum Charge	Job		236		236	
Closer							
	Demolish	Ea.		9.80		9.80	Includes material and labor to install
	Install	Ea.	172	72.50		244.50	pneumatic light duty closer for interior
	Demolish and Install	Ea.	172	82.30		254.30	type doors.
	Reinstall	Ea.		57.99		57.99	
	Clean	Ea.	.08	15.15		15.23	
	Minimum Charge	Job		236		236	

Finish Carpentry

Hollow Core Door

		Unit	Material	Labor	Equip.	Total	Specification
Push Plate							
	Demolish	Ea.		14.55		14.55	Includes material and labor to install
	Install	Ea.	29	39.50		68.50	bronze push-pull plate.
	Demolish and Install	Ea.	29	54.05		83.05	
	Reinstall	Ea.		31.41		31.41	
	Clean	Ea.	.15	11.35		11.50	
	Minimum Charge	Job		236		236	
Kickplate							
	Demolish	Ea.		14.55		14.55	Includes material and labor to install
	Install	Ea.	26.50	31.50		58	aluminum kickplate, 10" x 28".
	Demolish and Install	Ea.	26.50	46.05		72.55	
	Reinstall	Ea.		25.13		25.13	
	Clean	Ea.	.15	11.35		11.50	
	Minimum Charge	Job		236		236	
Exit Sign							
	Demolish	Ea.		27.50		27.50	Includes material and labor to install a
	Install	Ea.	113	67.50		180.50	wall mounted interior electric exit sign.
	Demolish and Install	Ea.	113	95		208	
	Reinstall	Ea.		54.04		54.04	
	Clean	Ea.	.01	8.50		8.51	
	Minimum Charge	Job		236		236	
Jamb							
	Demolish	Ea.		32.50		32.50	Includes material and labor to install
	Install	Ea.	116	12.75		128.75	flat pine jamb with square cut heads
	Demolish and Install	Ea.	116	45.25		161.25	and rabbeted sides for 6' 8" high and
	Clean	Ea.	.04	9.10		9.14	3-9/16" door including trim sets for
	Paint	Ea.	7.10	26		33.10	both sides.
	Minimum Charge	Job		236		236	
Shave & Refit							
	Install	Ea.		39.50		39.50	Includes labor to shave and rework
	Minimum Charge	Job		236		236	door to fit opening at the job site.

Solid Core Door

		Unit	Material	Labor	Equip.	Total	Specification
Pre-hung Wood							
2' 4" x 6' 8"							
	Demolish	Ea.		43.50		43.50	Cost includes material and labor to
	Install	Ea.	257	27.50		284.50	install 2' 4" x 6' 8" pre-hung 1-3/4"
	Demolish and Install	Ea.	257	71		328	door, 4-1/2" split jamb, including
	Reinstall	Ea.		27.72		27.72	casing and stop, hinges, aluminum sill,
	Clean	Ea.	1.50	15.15		16.65	weatherstripping.
	Paint	Ea.	11.50	56		67.50	
	Minimum Charge	Job		236		236	
2' 6" x 6' 8"							
	Demolish	Ea.		43.50		43.50	Cost includes material and labor to
	Install	Ea.	250	27.50		277.50	install 2' 6" x 6' 8" pre-hung 1-3/4"
	Demolish and Install	Ea.	250	71		321	door, 4-1/2" split jamb, including
	Reinstall	Ea.		27.72		27.72	casing and stop, hinges, aluminum sill,
	Clean	Ea.	1.50	15.15		16.65	weatherstripping.
	Paint	Ea.	11.50	56		67.50	
	Minimum Charge	Job		236		236	

Solid Core Door

Solid Core Door	Unit	Material	Labor	Equip.	Total	Specification
2' 8" x 6' 8"						
Demolish	Ea.		43.50		43.50	Cost includes material and labor to
Install	Ea.	259	27.50		286.50	install 2' 8" x 6' 8" pre-hung 1-3/4"
Demolish and Install	Ea.	259	71		330	door, 4-1/2" split jamb, including
Reinstall	Ea.		27.72		27.72	casing and stop, hinges, aluminum sill,
Clean	Ea.	1.50	15.15		16.65	weatherstripping.
Paint	Ea.	11.50	56		67.50	
Minimum Charge	Job		236		236	
3' x 6' 8"						
Demolish	Ea.		43.50		43.50	Cost includes material and labor to
Install	Ea.	263	29.50		292.50	install 3' x 6' 8" pre-hung 1-3/4"
Demolish and Install	Ea.	263	73		336	door, including casing and stop,
Reinstall	Ea.		29.45		29.45	hinges, jamb, aluminum sill,
Clean	Ea.	1.50	15.15		16.65	weatherstripped.
Paint	Ea.	11.50	56		67.50	
Minimum Charge	Job		236		236	
3' 6" x 6' 8"						
Demolish	Ea.		43.50		43.50	Cost includes material and labor to
Install	Ea.	415	31.50		446.50	install 3' 6" x 6' 8" pre-hung 1-3/4"
Demolish and Install	Ea.	415	75		490	door, 4-1/2" split jamb, including
Reinstall	Ea.		31.41		31.41	casing and stop, hinges, aluminum sill,
Clean	Ea.	1.50	15.15		16.65	weatherstripping.
Paint	Ea.	11.50	56		67.50	
Minimum Charge	Job		236		236	
Raised Panel						
Demolish	Ea.		43.50		43.50	Cost includes material and labor to
Install	Ea.	505	26		531	install 3' x 6' 8" pre-hung 1-3/4"
Demolish and Install	Ea.	505	69.50		574.50	door, jamb, including casing and
Reinstall	Ea.		26.18		26.18	stop, hinges, aluminum sill,
Clean	Ea.	1.50	15.15		16.65	weatherstripping.
Paint	Ea.	11.50	130		141.50	
Minimum Charge	Job		236		236	
Stain						
Install	Ea.	8.70	32.50		41.20	Includes labor and material to stain
Minimum Charge	Job		196		196	single door and trim on both sides.

Bifold Door

Bifold Door	Unit	Material	Labor	Equip.	Total	Specification
Wood, Single						
3'						
Demolish	Ea.		32.50		32.50	Cost includes material and labor to
Install	Ea.	85.50	72.50		158	install 3' wide solid wood bi-fold door
Demolish and Install	Ea.	85.50	105		190.50	not including jamb, trim, track, and
Reinstall	Ea.		72.49		72.49	hardware.
Clean	Ea.	1.50	15.15		16.65	
Paint	Ea.	10.85	56		66.85	
Minimum Charge	Job		236		236	
5'						
Demolish	Ea.		32.50		32.50	Includes material and labor to install
Install	Ea.	129	85.50		214.50	pair of 2' 6" wide solid wood bi-fold
Demolish and Install	Ea.	129	118		247	doors not including jamb, trim, track,
Reinstall	Ea.		85.67		85.67	and hardware.
Clean	Ea.	.66	26		26.66	
Paint	Ea.	34.50	98		132.50	
Minimum Charge	Job		236		236	

Bifold Door

	Unit	Material	Labor	Equip.	Total	Specification
6'						
Demolish	Ea.		32.50		32.50	Includes material and labor to install
Install	Ea.	146	94		240	pair of 3' wide solid wood bi-fold
Demolish and Install	Ea.	146	126.50		272.50	doors not including jamb, trim, track,
Reinstall	Ea.		94.24		94.24	and hardware.
Clean	Ea.	.75	30.50		31.25	
Paint	Ea.	34.50	98		132.50	
Minimum Charge	Job		236		236	
8'						
Demolish	Ea.		32.50		32.50	Includes material and labor to install
Install	Ea.	232	105		337	pair of 4' wide solid wood bi-fold
Demolish and Install	Ea.	232	137.50		369.50	doors not including jamb, trim, track,
Reinstall	Ea.		104.71		104.71	and hardware.
Clean	Ea.	1.03	36.50		37.53	
Paint	Ea.	18.10	98		116.10	
Minimum Charge	Job		236		236	

Mirrored

	Unit	Material	Labor	Equip.	Total	Specification
6'						
Demolish	Ea.		32.50		32.50	Includes material and labor to install
Install	Ea.	470	52.50		522.50	pair of 3' wide mirrored bi-fold doors
Demolish and Install	Ea.	470	85		555	including jamb, trim, track, and
Reinstall	Ea.		52.36		52.36	hardware.
Clean	Ea.	.75	15.15		15.90	
Minimum Charge	Job		236		236	
8'						
Demolish	Ea.		32.50		32.50	Includes material and labor to install
Install	Ea.	815	78.50		893.50	pair of 4' wide mirrored bi-fold doors
Demolish and Install	Ea.	815	111		926	including jamb, trim, track, and
Reinstall	Ea.		78.53		78.53	hardware.
Clean	Ea.	1.03	18.20		19.23	
Minimum Charge	Job		236		236	

Louvered

	Unit	Material	Labor	Equip.	Total	Specification
3'						
Demolish	Ea.		32.50		32.50	Cost includes material and labor to
Install	Ea.	279	72.50		351.50	install 3' wide louvered bi-fold door
Demolish and Install	Ea.	279	105		384	not including jamb, trim, track, and
Reinstall	Ea.		72.49		72.49	hardware.
Clean	Ea.	1.50	15.15		16.65	
Paint	Ea.	10.85	56		66.85	
Minimum Charge	Job		236		236	
5'						
Demolish	Ea.		32.50		32.50	Includes material and labor to install
Install	Ea.	295	85.50		380.50	pair of 2' 6" wide louvered wood
Demolish and Install	Ea.	295	118		413	bi-fold doors not including jamb, trim,
Reinstall	Ea.		85.67		85.67	track, and hardware.
Clean	Ea.	.66	26		26.66	
Paint	Ea.	34.50	98		132.50	
Minimum Charge	Job		236		236	
6'						
Demolish	Ea.		32.50		32.50	Includes material and labor to install
Install	Ea.	350	94		444	pair of 3' wide louvered wood bi-fold
Demolish and Install	Ea.	350	126.50		476.50	doors not including jamb, trim, track,
Reinstall	Ea.		94.24		94.24	and hardware.
Clean	Ea.	.75	30.50		31.25	
Paint	Ea.	34.50	98		132.50	
Minimum Charge	Job		236		236	

197

Finish Carpentry

Bifold Door	Unit	Material	Labor	Equip.	Total	Specification
8'						
Demolish	Ea.		32.50		32.50	Includes material and labor to install
Install	Ea.	540	94		634	pair of 4' wide louvered wood bi-fold
Demolish and Install	Ea.	540	126.50		666.50	doors not including jamb, trim, track,
Reinstall	Ea.		94.24		94.24	and hardware.
Clean	Ea.	3.55	26		29.55	
Paint	Ea.	18.10	98		116.10	
Minimum Charge	Job		236		236	
Accordion Type						
Demolish	S.F.		1.19		1.19	Includes material and labor to install
Install	Ea.	45.50	94		139.50	vinyl folding accordion closet doors,
Demolish and Install	Ea.	45.50	95.19		140.69	including track and frame.
Reinstall	Ea.		94.24		94.24	
Clean	S.F.		.45		.45	
Minimum Charge	Job		236		236	
Casing Trim						
Single Width						
Demolish	Ea.		8.30		8.30	Cost includes material and labor to
Install	Opng.	26.50	39.50		66	install 11/16" x 2-1/2" pine ranch
Demolish and Install	Opng.	26.50	47.80		74.30	style casing for one side of a standard
Reinstall	Opng.		31.41		31.41	door opening.
Clean	Opng.	.21	3.64		3.85	
Paint	Ea.	5.05	7.80		12.85	
Minimum Charge	Job		236		236	
Double Width						
Demolish	Ea.		9.90		9.90	Cost includes material and labor to
Install	Opng.	31	47		78	install 11/16" x 2-1/2" pine ranch
Demolish and Install	Opng.	31	56.90		87.90	style door casing for one side of a
Reinstall	Opng.		37.70		37.70	double door opening.
Clean	Opng.	.25	7.25		7.50	
Paint	Ea.	5.05	8.90		13.95	
Minimum Charge	Job		236		236	
Jamb						
Demolish	Ea.		32.50		32.50	Includes material and labor to install
Install	Ea.	116	12.75		128.75	flat pine jamb with square cut heads
Demolish and Install	Ea.	116	45.25		161.25	and rabbeted sides for 6' 8" high and
Clean	Ea.	.04	9.10		9.14	3-9/16" door including trim sets for
Paint	Ea.	7.10	26		33.10	both sides.
Minimum Charge	Job		118		118	
Stain						
Install	Ea.	8.70	32.50		41.20	Includes labor and material to stain
Minimum Charge	Job		196		196	single door and trim on both sides.
Shave & Refit						
Install	Ea.		39.50		39.50	Includes labor to shave and rework
Minimum Charge	Job		118		118	door to fit opening at the job site.

Finish Carpentry

Louver Door

	Unit	Material	Labor	Equip.	Total	Specification
Full						
2' x 6' 8"						
Demolish	Ea.		9.35		9.35	Cost includes material and labor to
Install	Ea.	310	47		357	install 2' x 6' 8" pine full louver door,
Demolish and Install	Ea.	310	56.35		366.35	jamb, hinges.
Reinstall	Ea.		47.12		47.12	
Clean	Ea.	2.77	45.50		48.27	
Paint	Ea.	10.85	56		66.85	
Minimum Charge	Job		236		236	
2' 8" x 6' 8"						
Demolish	Ea.		9.35		9.35	Cost includes material and labor to
Install	Ea.	360	47		407	install 2' 8" x 6' 8" pine full louver
Demolish and Install	Ea.	360	56.35		416.35	door, jamb, hinges.
Reinstall	Ea.		47.12		47.12	
Clean	Ea.	2.77	45.50		48.27	
Paint	Ea.	10.85	56		66.85	
Minimum Charge	Job		236		236	
3' x 6' 8"						
Demolish	Ea.		9.35		9.35	Cost includes material and labor to
Install	Ea.	375	49.50		424.50	install 3' x 6' 8" pine full louver door,
Demolish and Install	Ea.	375	58.85		433.85	jamb, hinges.
Reinstall	Ea.		49.60		49.60	
Clean	Ea.	2.77	45.50		48.27	
Paint	Ea.	10.85	56		66.85	
Minimum Charge	Job		236		236	
Half						
2' x 6' 8"						
Demolish	Ea.		9.35		9.35	Cost includes material and labor to
Install	Ea.	400	47		447	install 2' x 6' 8" pine half louver door,
Demolish and Install	Ea.	400	56.35		456.35	jamb, hinges.
Reinstall	Ea.		47.12		47.12	
Clean	Ea.	2.77	45.50		48.27	
Paint	Ea.	10.85	56		66.85	
Minimum Charge	Job		236		236	
2' 8" x 6' 8"						
Demolish	Ea.		9.35		9.35	Cost includes material and labor to
Install	Ea.	420	47		467	install 2' 8" x 6' 8" pine half louver
Demolish and Install	Ea.	420	56.35		476.35	door, jamb, hinges.
Reinstall	Ea.		47.12		47.12	
Clean	Ea.	2.77	45.50		48.27	
Paint	Ea.	10.85	56		66.85	
Minimum Charge	Job		236		236	
3' x 6' 8"						
Demolish	Ea.		9.35		9.35	Cost includes material and labor to
Install	Ea.	440	49.50		489.50	install 3' x 6' 8" pine half louver door,
Demolish and Install	Ea.	440	58.85		498.85	jamb, hinges.
Reinstall	Ea.		49.60		49.60	
Clean	Ea.	2.77	45.50		48.27	
Paint	Ea.	10.85	56		66.85	
Minimum Charge	Job		236		236	
Jamb						
Demolish	Ea.		32.50		32.50	Includes material and labor to install
Install	Ea.	116	12.75		128.75	flat pine jamb with square cut heads
Demolish and Install	Ea.	116	45.25		161.25	and rabbeted sides for 6' 8" high and
Clean	Ea.	.04	9.10		9.14	3-9/16" door including trim sets for
Paint	Ea.	7.10	26		33.10	both sides.
Minimum Charge	Job		118		118	

For customer support on your Contractor's Pricing Guide: Residential Repair & Remodeling, call 800.448.8182.

Finish Carpentry

Louver Door

Louver Door		Unit	Material	Labor	Equip.	Total	Specification
Stain							
	Install	Ea.	8.70	32.50		41.20	Includes labor and material to stain
	Minimum Charge	Job		196		196	single door and trim on both sides.
Shave & Refit							
	Install	Ea.		39.50		39.50	Includes labor to shave and rework
	Minimum Charge	Job		196		196	door to fit opening at the job site.

Bypass Sliding Door

Bypass Sliding Door		Unit	Material	Labor	Equip.	Total	Specification
Wood							
5'							
	Demolish	Ea.		43.50		43.50	Includes material and labor to install
	Install	Opng.	208	85.50		293.50	pair of 2' 6" wide lauan hollow core
	Demolish and Install	Opng.	208	129		337	wood bypass doors with jamb, track
	Reinstall	Opng.		85.67		85.67	hardware.
	Clean	Ea.	.66	26		26.66	
	Paint	Ea.	34.50	98		132.50	
	Minimum Charge	Job		236		236	
6'							
	Demolish	Ea.		43.50		43.50	Includes material and labor to install
	Install	Opng.	194	94		288	pair of 3' wide lauan hollow core
	Demolish and Install	Opng.	194	137.50		331.50	wood bi-pass doors with jamb, track
	Reinstall	Opng.		94.24		94.24	and hardware.
	Clean	Ea.	.75	30.50		31.25	
	Paint	Ea.	34.50	98		132.50	
	Minimum Charge	Job		236		236	
8'							
	Demolish	Ea.		43.50		43.50	Includes material and labor to install
	Install	Opng.	380	118		498	pair of 4' wide lauan hollow core
	Demolish and Install	Opng.	380	161.50		541.50	wood bi-pass doors with jamb, track
	Reinstall	Opng.		117.80		117.80	and hardware.
	Clean	Ea.	1.03	36.50		37.53	
	Paint	Ea.	18.10	98		116.10	
	Minimum Charge	Job		236		236	
Mirrored							
6'							
	Demolish	Ea.		43.50		43.50	Includes material and labor to install
	Install	Opng.	325	94		419	pair of 3' wide mirrored bi-pass doors
	Demolish and Install	Opng.	325	137.50		462.50	with jamb, trim, track hardware.
	Reinstall	Opng.		94.24		94.24	
	Clean	Ea.	.75	15.15		15.90	
	Minimum Charge	Job		236		236	
8'							
	Demolish	Ea.		43.50		43.50	Includes material and labor to install
	Install	Opng.	530	105		635	pair of 4' wide mirrored bi-pass doors
	Demolish and Install	Opng.	530	148.50		678.50	with jamb, trim, track hardware.
	Reinstall	Opng.		104.71		104.71	
	Clean	Ea.	1.03	18.20		19.23	
	Minimum Charge	Job		236		236	

Finish Carpentry

Bypass Sliding Door

	Unit	Material	Labor	Equip.	Total	Specification
Casing Trim						
Single Width						
Demolish	Ea.		8.30		8.30	Cost includes material and labor to
Install	Opng.	26.50	39.50		66	install 11/16" x 2-1/2" pine ranch
Demolish and Install	Opng.	26.50	47.80		74.30	style casing for one side of a standard
Reinstall	Opng.		31.41		31.41	door opening.
Clean	Opng.	.21	3.64		3.85	
Paint	Ea.	5.05	7.80		12.85	
Minimum Charge	Job		236		236	
Double Width						
Demolish	Ea.		9.90		9.90	Cost includes material and labor to
Install	Opng.	31	47		78	install 11/16" x 2-1/2" pine ranch
Demolish and Install	Opng.	31	56.90		87.90	style door casing for one side of a
Reinstall	Opng.		37.70		37.70	double door opening.
Clean	Opng.	.25	7.25		7.50	
Paint	Ea.	5.05	8.90		13.95	
Minimum Charge	Job		236		236	
Jamb						
Demolish	Ea.		32.50		32.50	Includes material and labor to install
Install	Ea.	116	12.75		128.75	flat pine jamb with square cut heads
Demolish and Install	Ea.	116	45.25		161.25	and rabbeted sides for 6' 8" high and
Clean	Ea.	.04	9.10		9.14	3-9/16" door including trim sets for
Paint	Ea.	7.10	26		33.10	both sides.
Minimum Charge	Job		118		118	
Stain						
Install	Ea.	8.70	32.50		41.20	Includes labor and material to stain
Minimum Charge	Job		196		196	single door and trim on both sides.
Shave & Refit						
Install	Ea.		39.50		39.50	Includes labor to shave and rework
Minimum Charge	Job		118		118	door to fit opening at the job site.

Base Molding

	Unit	Material	Labor	Equip.	Total	Specification
One Piece						
1-5/8"						
Demolish	L.F.		.67		.67	Includes material and labor to install
Install	L.F.	.39	1.39		1.78	pine or birch base molding.
Demolish and Install	L.F.	.39	2.06		2.45	
Reinstall	L.F.		1.12		1.12	
Clean	L.F.	.01	.21		.22	
Paint	L.F.	.06	.43		.49	
Minimum Charge	Job		236		236	
2-1/2"						
Demolish	L.F.		.67		.67	Cost includes material and labor to
Install	L.F.	.61	1.42		2.03	install 1/2" x 2-1/2" pine or birch
Demolish and Install	L.F.	.61	2.09		2.70	base molding.
Reinstall	L.F.		1.14		1.14	
Clean	L.F.	.01	.21		.22	
Paint	L.F.	.06	.43		.49	
Minimum Charge	Job		236		236	

Finish Carpentry

Base Molding		Unit	Material	Labor	Equip.	Total	Specification
3-1/2"							
	Demolish	L.F.		.67		.67	Cost includes material and labor to
	Install	L.F.	.93	1.51		2.44	install 9/16" x 3-1/2" pine or birch
	Demolish and Install	L.F.	.93	2.18		3.11	base molding.
	Reinstall	L.F.		1.21		1.21	
	Clean	L.F.	.01	.21		.22	
	Paint	L.F.	.08	.47		.55	
	Minimum Charge	Job		236		236	
6"							
	Demolish	L.F.		.67		.67	Cost includes material and labor to
	Install	L.F.	.56	1.51		2.07	install 1" x 6" pine or birch base
	Demolish and Install	L.F.	.56	2.18		2.74	molding.
	Reinstall	L.F.		1.21		1.21	
	Clean	L.F.	.01	.30		.31	
	Paint	L.F.	.10	.59		.69	
	Minimum Charge	Job		236		236	
8"							
	Demolish	L.F.		.81		.81	Cost includes material and labor to
	Install	L.F.	.72	1.51		2.23	install 1" x 8" pine or birch base
	Demolish and Install	L.F.	.72	2.32		3.04	molding.
	Reinstall	L.F.		1.21		1.21	
	Clean	L.F.	.02	.36		.38	
	Paint	L.F.	.11	.65		.76	
	Minimum Charge	Job		236		236	
Premium / Custom Grade							
	Demolish	L.F.		.81		.81	Includes material and labor to install
	Install	L.F.	4.62	4.53		9.15	custom 2 or 3 piece oak or other
	Demolish and Install	L.F.	4.62	5.34		9.96	hardwood, stain grade trim.
	Reinstall	L.F.		3.62		3.62	
	Clean	L.F.	.02	.36		.38	
	Paint	L.F.	.12	.65		.77	
	Minimum Charge	Job		236		236	
One Piece Oak							
1-5/8"							
	Demolish	L.F.		.67		.67	Cost includes material and labor to
	Install	L.F.	.98	1.39		2.37	install 1/2" x 1-5/8" oak base
	Demolish and Install	L.F.	.98	2.06		3.04	molding.
	Reinstall	L.F.		1.12		1.12	
	Clean	L.F.	.01	.21		.22	
	Paint	L.F.	.14	.58		.72	
	Minimum Charge	Job		236		236	
2-1/2"							
	Demolish	L.F.		.67		.67	Cost includes material and labor to
	Install	L.F.	1.73	1.42		3.15	install 1/2" x 2-1/2" oak base
	Demolish and Install	L.F.	1.73	2.09		3.82	molding.
	Reinstall	L.F.		1.14		1.14	
	Clean	L.F.	.01	.21		.22	
	Paint	L.F.	.14	.59		.73	
	Minimum Charge	Job		236		236	
3-1/2"							
	Demolish	L.F.		.67		.67	Cost includes material and labor to
	Install	L.F.	2.18	1.45		3.63	install 1/2" x 3-1/2" oak base
	Demolish and Install	L.F.	2.18	2.12		4.30	molding.
	Reinstall	L.F.		1.16		1.16	
	Clean	L.F.	.01	.21		.22	
	Paint	L.F.	.14	.60		.74	
	Minimum Charge	Job		236		236	

Finish Carpentry

Base Molding

	Unit	Material	Labor	Equip.	Total	Specification
6″						
Demolish	L.F.		.67		.67	Cost includes material and labor to
Install	L.F.	3.15	1.51		4.66	install 1″ x 6″ oak base molding.
Demolish and Install	L.F.	3.15	2.18		5.33	
Reinstall	L.F.		1.21		1.21	
Clean	L.F.	.02	.36		.38	
Paint	L.F.	.15	.63		.78	
Minimum Charge	Job		236		236	
8″						
Demolish	L.F.		.81		.81	Cost includes material and labor to
Install	L.F.	3.89	1.51		5.40	install 1″ x 8″ oak base molding.
Demolish and Install	L.F.	3.89	2.32		6.21	
Reinstall	L.F.		1.21		1.21	
Clean	L.F.	.02	.36		.38	
Paint	L.F.	.15	.65		.80	
Minimum Charge	Job		236		236	

Vinyl Base Molding

	Unit	Material	Labor	Equip.	Total	Specification
2-1/2″						
Demolish	L.F.		.67		.67	Cost includes material and labor to
Install	L.F.	1.43	1.41		2.84	install 2-1/2″ vinyl or rubber cove
Demolish and Install	L.F.	1.43	2.08		3.51	base including adhesive.
Reinstall	L.F.		1.41		1.41	
Clean	L.F.		.33		.33	
Minimum Charge	Job		226		226	
4″						
Demolish	L.F.		.67		.67	Cost includes material and labor to
Install	Ea.	2.73	1.41		4.14	install 4″ vinyl or rubber cove base
Demolish and Install	Ea.	2.73	2.08		4.81	including adhesive.
Reinstall	Ea.		1.41		1.41	
Clean	L.F.		.33		.33	
Minimum Charge	Job		226		226	
6″						
Demolish	L.F.		.67		.67	Cost includes material and labor to
Install	L.F.	1.30	1.41		2.71	install 6″ vinyl or rubber cove base
Demolish and Install	L.F.	1.30	2.08		3.38	including adhesive.
Reinstall	L.F.		1.41		1.41	
Clean	L.F.		.33		.33	
Minimum Charge	Job		226		226	

Prefinished

	Unit	Material	Labor	Equip.	Total	Specification
Demolish	L.F.		.67		.67	Cost includes material and labor to
Install	L.F.	1.37	1.51		2.88	install 9/16″ x 3-5/16″ prefinished
Demolish and Install	L.F.	1.37	2.18		3.55	base molding.
Reinstall	L.F.		1.21		1.21	
Clean	L.F.	.01	.21		.22	
Paint	L.F.	.08	.47		.55	
Minimum Charge	Job		236		236	

Shoe

	Unit	Material	Labor	Equip.	Total	Specification
Demolish	L.F.		.67		.67	Includes material and labor to install
Install	L.F.	.39	1.39		1.78	pine or birch base molding.
Demolish and Install	L.F.	.39	2.06		2.45	
Reinstall	L.F.		1.12		1.12	
Clean	L.F.	.01	.21		.22	
Paint	L.F.	.17	.63		.80	
Minimum Charge	Job		236		236	

For customer support on your Contractor's Pricing Guide: Residential Repair & Remodeling, call 800.448.8182.

Finish Carpentry

Base Molding

Base Molding		Unit	Material	Labor	Equip.	Total	Specification
Carpeted							
	Demolish	L.F.		.27		.27	Includes material and labor to install
	Install	L.F.	.99	2.70		3.69	carpet baseboard with adhesive.
	Demolish and Install	L.F.	.99	2.97		3.96	
	Reinstall	L.F.		2.16		2.16	
	Clean	L.F.		.14		.14	
	Minimum Charge	Job		222		222	
Ceramic Tile							
	Demolish	L.F.		1.47		1.47	Includes material and labor to install
	Install	L.F.	4.63	6.15		10.78	thin set ceramic tile cove base
	Demolish and Install	L.F.	4.63	7.62		12.25	including grout.
	Clean	L.F.		.69		.69	
	Minimum Charge	Job		222		222	
Quarry Tile							
	Demolish	L.F.		1.47		1.47	Includes material and labor to install
	Install	L.F.	7.05	7.15		14.20	quarry tile base including grout.
	Demolish and Install	L.F.	7.05	8.62		15.67	
	Clean	L.F.		.69		.69	
	Minimum Charge	Job		222		222	
Caulk and Renail							
	Install	L.F.	.01	.74		.75	Includes labor and material to renail
	Minimum Charge	Job		236		236	base and caulk where necessary.

Molding

Molding		Unit	Material	Labor	Equip.	Total	Specification
Oak Base							
	Demolish	L.F.		.67		.67	Cost includes material and labor to
	Install	L.F.	1.73	1.42		3.15	install 1/2" x 2-1/2" oak base
	Demolish and Install	L.F.	1.73	2.09		3.82	molding.
	Reinstall	L.F.		1.14		1.14	
	Clean	L.F.	.01	.21		.22	
	Paint	L.F.	.14	.59		.73	
	Minimum Charge	Job		236		236	
Shoe							
	Demolish	L.F.		.67		.67	Includes material and labor to install
	Install	L.F.	.39	1.39		1.78	pine or birch base molding.
	Demolish and Install	L.F.	.39	2.06		2.45	
	Reinstall	L.F.		1.12		1.12	
	Clean	L.F.	.01	.21		.22	
	Paint	L.F.	.17	.63		.80	
	Minimum Charge	Job		236		236	
Quarter Round							
	Demolish	L.F.		.67		.67	Includes material and labor to install
	Install	L.F.	.58	1.85		2.43	pine base quarter round molding.
	Demolish and Install	L.F.	.58	2.52		3.10	
	Reinstall	L.F.		1.48		1.48	
	Clean	L.F.	.01	.21		.22	
	Paint	L.F.	.12	.98		1.10	
	Minimum Charge	Job		236		236	
Chair Rail							
	Demolish	L.F.		.67		.67	Cost includes material and labor to
	Install	L.F.	1.69	1.75		3.44	install 5/8" x 2-1/2" oak chair rail
	Demolish and Install	L.F.	1.69	2.42		4.11	molding.
	Reinstall	L.F.		1.40		1.40	
	Clean	L.F.	.01	.21		.22	
	Paint	L.F.	.12	.98		1.10	
	Minimum Charge	Job		236		236	

For customer support on your Contractor's Pricing Guide: Residential Repair & Remodeling, call 800.448.8182.

Finish Carpentry

Molding

	Unit	Material	Labor	Equip.	Total	Specification
Crown						
Demolish	L.F.		.67		.67	Includes material and labor to install
Install	L.F.	2.34	1.88		4.22	unfinished pine crown molding.
Demolish and Install	L.F.	2.34	2.55		4.89	
Reinstall	L.F.		1.51		1.51	
Clean	L.F.	.02	.28		.30	
Paint	L.F.	.12	.98		1.10	
Minimum Charge	Job		236		236	
Cove						
Demolish	L.F.		.67		.67	Cost includes material and labor to
Install	L.F.	2.09	1.78		3.87	install 11/16" x 2-3/4" pine cove
Demolish and Install	L.F.	2.09	2.45		4.54	molding.
Reinstall	L.F.		1.42		1.42	
Clean	L.F.	.01	.21		.22	
Paint	L.F.	.12	.98		1.10	
Minimum Charge	Job		236		236	
Corner						
Demolish	L.F.		.67		.67	Includes material and labor to install
Install	L.F.	1.83	.74		2.57	inside or outside pine corner molding.
Demolish and Install	L.F.	1.83	1.41		3.24	
Reinstall	L.F.		.59		.59	
Clean	L.F.	.01	.21		.22	
Paint	L.F.	.12	.98		1.10	
Minimum Charge	Job		236		236	
Picture						
Demolish	L.F.		.67		.67	Cost includes material and labor to
Install	L.F.	1.55	1.96		3.51	install 9/16" x 2-1/2" pine casing.
Demolish and Install	L.F.	1.55	2.63		4.18	
Reinstall	L.F.		1.57		1.57	
Clean	L.F.	.01	.21		.22	
Paint	L.F.	.12	.98		1.10	
Minimum Charge	Job		236		236	
Single Cased Opening						
Demolish	Ea.		32.50		32.50	Includes material and labor to install
Install	Ea.	26	29.50		55.50	flat pine jamb with square cut heads
Demolish and Install	Ea.	26	62		88	and rabbeted sides for 6' 8" high
Clean	Ea.	.10	7.60		7.70	opening including trim sets for both
Paint	Ea.	1.57	9.80		11.37	sides.
Minimum Charge	Job		236		236	
Double Cased Opening						
Demolish	Ea.		32.50		32.50	Includes material and labor to install
Install	Ea.	51	33.50		84.50	flat pine jamb with square cut heads
Demolish and Install	Ea.	51	66		117	and rabbeted sides for 6' 8" high
Clean	Ea.	.12	8.85		8.97	opening including trim sets for both
Paint	Ea.	1.82	11.50		13.32	sides.
Minimum Charge	Job		236		236	

Window Trim Set

	Unit	Material	Labor	Equip.	Total	Specification
Single						
Demolish	Ea.		9		9	Cost includes material and labor to
Install	Opng.	42.50	47		89.50	install 11/16" x 2-1/2" pine ranch
Demolish and Install	Opng.	42.50	56		98.50	style window casing.
Reinstall	Opng.		37.70		37.70	
Clean	Ea.	.15	5.20		5.35	
Paint	Ea.	1.47	9.30		10.77	
Minimum Charge	Job		236		236	

For customer support on your Contractor's Pricing Guide: Residential Repair & Remodeling, call 800.448.8182.

Finish Carpentry

Window Trim Set

	Unit	Material	Labor	Equip.	Total	Specification
Double						
Demolish	Ea.		10.10		10.10	Cost includes material and labor to
Install	Ea.	58	59		117	install 11/16" x 2-1/2" pine ranch
Demolish and Install	Ea.	58	69.10		127.10	style trim for casing one side of a
Reinstall	Ea.		47.12		47.12	double window opening.
Clean	Ea.	.26	6.05		6.31	
Paint	Ea.	2.46	15.65		18.11	
Minimum Charge	Job		236		236	
Triple						
Demolish	Ea.		11.55		11.55	Cost includes material and labor to
Install	Opng.	86	78.50		164.50	install 11/16" x 2-1/2" pine ranch
Demolish and Install	Opng.	86	90.05		176.05	style trim for casing one side of a
Reinstall	Opng.		62.83		62.83	triple window opening.
Clean	Ea.	.37	7.25		7.62	
Paint	Ea.	3.45	21.50		24.95	
Minimum Charge	Job		236		236	
Window Casing Per L.F.						
Demolish	L.F.		.67		.67	Cost includes material and labor to
Install	L.F.	1.56	1.88		3.44	install 11/16" x 2-1/2" pine ranch
Demolish and Install	L.F.	1.56	2.55		4.11	style trim for casing.
Reinstall	L.F.		1.51		1.51	
Clean	L.F.	.01	.21		.22	
Paint	L.F.	.12	.98		1.10	
Minimum Charge	Job		236		236	

Window Sill

	Unit	Material	Labor	Equip.	Total	Specification
Wood						
Demolish	L.F.		.67		.67	Includes material and labor to install
Install	L.F.	2.64	2.36		5	flat, wood window stool.
Demolish and Install	L.F.	2.64	3.03		5.67	
Reinstall	L.F.		1.88		1.88	
Clean	L.F.	.01	.21		.22	
Paint	L.F.	.12	.98		1.10	
Minimum Charge	Job		236		236	
Marble						
Demolish	L.F.		1.01		1.01	Includes material and labor to install
Install	L.F.	13.15	10.15		23.30	marble window sills, 6" x 3/4" thick.
Demolish and Install	L.F.	13.15	11.16		24.31	
Reinstall	L.F.		8.13		8.13	
Clean	L.F.	.01	.21		.22	
Minimum Charge	Job		222		222	

Shelving

	Unit	Material	Labor	Equip.	Total	Specification
Pine						
18"						
Demolish	L.F.		1.15		1.15	Cost includes material and labor to
Install	L.F.	8.90	4.96		13.86	install 1" thick custom shelving board
Demolish and Install	L.F.	8.90	6.11		15.01	18" wide.
Reinstall	L.F.		3.97		3.97	
Clean	L.F.	.02	.61		.63	
Paint	L.F.	1.53	.62		2.15	
Minimum Charge	Job		236		236	

Finish Carpentry

Shelving

Shelving		Unit	Material	Labor	Equip.	Total	Specification
24"							
	Demolish	L.F.		1.15		1.15	Cost includes material and labor to
	Install	L.F.	11.90	5.55		17.45	install 1" thick custom shelving board.
	Demolish and Install	L.F.	11.90	6.70		18.60	
	Reinstall	L.F.		4.43		4.43	
	Clean	L.F.	.03	.91		.94	
	Paint	L.F.	2.09	.82		2.91	
	Minimum Charge	Job		236		236	
Particle Board							
	Demolish	L.F.		1.06		1.06	Cost includes material and labor to
	Install	L.F.	1.43	2.48		3.91	install 3/4" thick particle shelving
	Demolish and Install	L.F.	1.43	3.54		4.97	board 18" wide.
	Reinstall	L.F.		1.98		1.98	
	Clean	L.F.	.02	.61		.63	
	Paint	S.F.	.14	.91		1.05	
	Minimum Charge	Job		236		236	
Plywood							
	Demolish	L.F.		1.06		1.06	Cost includes material and labor to
	Install	L.F.	1.09	7.25		8.34	install 3/4" thick plywood shelving
	Demolish and Install	L.F.	1.09	8.31		9.40	board.
	Reinstall	L.F.		5.80		5.80	
	Clean	L.F.	.02	.61		.63	
	Paint	S.F.	.14	.91		1.05	
	Minimum Charge	Job		236		236	
Hardwood (oak)							
	Demolish	L.F.		1.06		1.06	Includes material and labor to install
	Install	L.F.	6.95	3.25		10.20	oak shelving, 1" x 12", incl. cleats and
	Demolish and Install	L.F.	6.95	4.31		11.26	bracing.
	Reinstall	L.F.		2.60		2.60	
	Clean	L.F.	.02	.61		.63	
	Paint	S.F.	.14	.91		1.05	
	Minimum Charge	Job		236		236	
Custom Bookcase							
	Demolish	L.F.		13.45		13.45	Includes material and labor to install
	Install	L.F.	45.50	47		92.50	custom modular bookcase unit with
	Demolish and Install	L.F.	45.50	60.45		105.95	clear pine faced frames, shelves 12"
	Reinstall	L.F.		37.70		37.70	O.C., 7' high, 8" deep.
	Clean	L.F.	.59	13		13.59	
	Paint	L.F.	1.75	30		31.75	
	Minimum Charge	Job		236		236	
Glass							
	Demolish	L.F.		1.06		1.06	Includes material and labor to install
	Install	L.F.	28	3.25		31.25	glass shelving, 1" x 12", incl. cleats
	Demolish and Install	L.F.	28	4.31		32.31	and bracing.
	Reinstall	L.F.		2.60		2.60	
	Clean	L.F.	.02	.56		.58	
	Minimum Charge	Job		229		229	

Closet Shelving

Closet Shelving		Unit	Material	Labor	Equip.	Total	Specification
Closet Shelf and Rod							
	Demolish	L.F.		1.06		1.06	Cost includes material and labor to
	Install	L.F.	3.52	7.85		11.37	install 1" thick custom shelving board
	Demolish and Install	L.F.	3.52	8.91		12.43	18" wide and 1" diameter clothes pole
	Reinstall	L.F.		6.28		6.28	with brackets 3' O.C.
	Clean	L.F.	.01	.45		.46	
	Paint	L.F.	.24	.98		1.22	
	Minimum Charge	Job		236		236	

207

Finish Carpentry

Closet Shelving

	Unit	Material	Labor	Equip.	Total	Specification
Clothing Rod						Includes material and labor to install
Demolish	L.F.		.70		.70	fir closet pole, 1-5/8" diameter.
Install	L.F.	2.26	2.36		4.62	
Demolish and Install	L.F.	2.26	3.06		5.32	
Reinstall	L.F.		1.88		1.88	
Clean	L.F.	.01	.61		.62	
Paint	L.F.	.23	1.08		1.31	
Minimum Charge	Job		236		236	

Stair Assemblies

	Unit	Material	Labor	Equip.	Total	Specification
Straight Hardwood						Includes material and labor to install
Demolish	Ea.		465		465	factory cut and assembled straight
Install	Flight	1325	315		1640	closed box stairs with oak treads and
Demolish and Install	Flight	1325	780		2105	prefinished stair rail with balusters.
Clean	Flight	1.50	30.50		32	
Paint	Flight	21.50	52		73.50	
Minimum Charge	Job		236		236	
Clear Oak Tread						Includes material and labor to install
Demolish	Ea.		11.90		11.90	1-1/4" thick clear oak treads per riser.
Install	Ea.	107	26		133	
Demolish and Install	Ea.	107	37.90		144.90	
Clean	Ea.	.04	3.03		3.07	
Paint	Ea.	.72	2.90		3.62	
Minimum Charge	Job		236		236	
Landing						Includes material and labor to install
Demolish	S.F.		4.21		4.21	clear oak landing.
Install	S.F.	13.10	26		39.10	
Demolish and Install	S.F.	13.10	30.21		43.31	
Clean	S.F.		.45		.45	
Paint	S.F.	.50	.97		1.47	
Minimum Charge	Job		236		236	
Refinish / Stain						Includes labor and material to refinish
Install	S.F.	.24	2.80		3.04	wood floor.
Minimum Charge	Job		196		196	
Spiral Hardwood						Cost includes material and labor to
Demolish	Ea.		465		465	install 4' - 6' diameter spiral stairs with
Install	Flight	2950	630		3580	oak treads, factory cut and assembled
Demolish and Install	Flight	2950	1095		4045	with double handrails.
Clean	Flight	1.50	30.50		32	
Paint	Flight	21.50	52		73.50	
Minimum Charge	Job		236		236	

Stairs

	Unit	Material	Labor	Equip.	Total	Specification
Disappearing						Includes material and labor to install a
Demolish	Ea.		25.50		25.50	folding staircase.
Install	Ea.	260	135		395	
Demolish and Install	Ea.	260	160.50		420.50	
Reinstall	Ea.		107.70		107.70	
Clean	Ea.	.22	22.50		22.72	
Minimum Charge	Job		236		236	

Stair Components		Unit	Material	Labor	Equip.	Total	Specification
Handrail w / Balusters							
	Demolish	L.F.		3.37		3.37	Includes material and labor to install
	Install	L.F.	53	9.80		62.80	prefinished assembled stair rail with
	Demolish and Install	L.F.	53	13.17		66.17	brackets, turned balusters and newel.
	Clean	L.F.	.30	2.51		2.81	
	Paint	L.F.	2.50	4.60		7.10	
	Minimum Charge	Job		236		236	
Banister							
	Demolish	L.F.		3.37		3.37	Includes material and labor to install
	Install	L.F.	42.50	7.85		50.35	built-up oak railings.
	Demolish and Install	L.F.	42.50	11.22		53.72	
	Clean	L.F.	.30	2.27		2.57	
	Paint	L.F.	2.50	3.56		6.06	
	Minimum Charge	Job		236		236	

Wall and Base Cabinets   **Countertop**

Wall Cabinets

	Unit	Material	Labor	Equip.	Total	Specification
Good Grade Laminated						
12″ Wide, 1 Door						
Demolish	Ea.		10.10		10.10	Includes material and labor to install
Install	Ea.	223	43		266	good modular unit with melamine
Demolish and Install	Ea.	223	53.10		276.10	laminated to particle board, in
Reinstall	Ea.		34.27		34.27	textured colors or wood grain print
Clean	Ea.	.22	7		7.22	finish, hinges and pulls.
Paint	Ea.	.75	13.95		14.70	
Minimum Charge	Job		236		236	
15″ Wide, 1 Door						
Demolish	Ea.		12.65		12.65	Includes material and labor to install
Install	Ea.	234	44		278	good modular unit with melamine
Demolish and Install	Ea.	234	56.65		290.65	laminated to particle board, in
Reinstall	Ea.		35.23		35.23	textured colors or wood grain print
Clean	Ea.	.29	8.65		8.94	finish, hinges and pulls.
Paint	Ea.	.94	17.80		18.74	
Minimum Charge	Job		236		236	
18″ Wide, 1 Door						
Demolish	Ea.		15.25		15.25	Includes material and labor to install
Install	Ea.	257	45		302	good modular unit with melamine
Demolish and Install	Ea.	257	60.25		317.25	laminated to particle board, in
Reinstall	Ea.		36.07		36.07	textured colors or wood grain print
Clean	Ea.	.34	10.40		10.74	finish, hinges and pulls.
Paint	Ea.	1.12	20.50		21.62	
Minimum Charge	Job		236		236	
21″ Wide, 1 Door						
Demolish	Ea.		17.55		17.55	Includes material and labor to install
Install	Ea.	261	46		307	good modular unit with melamine
Demolish and Install	Ea.	261	63.55		324.55	laminated to particle board, in
Reinstall	Ea.		36.60		36.60	textured colors or wood grain print
Clean	Ea.	.40	12.10		12.50	finish, hinges and pulls.
Paint	Ea.	1.31	24.50		25.81	
Minimum Charge	Job		236		236	
24″ w / Blind Corner						
Demolish	Ea.		20		20	Includes material and labor to install
Install	Ea.	300	46.50		346.50	good modular unit with melamine
Demolish and Install	Ea.	300	66.50		366.50	laminated to particle board, in
Reinstall	Ea.		37.14		37.14	textured colors or wood grain print
Clean	Ea.	.45	14		14.45	finish, hinges and pulls.
Paint	Ea.	1.50	28		29.50	
Minimum Charge	Job		236		236	

Cabinets and Countertops

Wall Cabinets	Unit	Material	Labor	Equip.	Total	Specification
27″ Wide, 2 Door						
Demolish	Ea.		22.50		22.50	Includes material and labor to install
Install	Ea.	76	47.50		123.50	good modular unit with melamine
Demolish and Install	Ea.	76	70		146	laminated to particle board, in
Reinstall	Ea.		38.08		38.08	textured colors or wood grain print
Clean	Ea.	.51	15.80		16.31	finish, hinges and pulls.
Paint	Ea.	1.69	32.50		34.19	
Minimum Charge	Job		236		236	
30″ Wide, 2 Door						
Demolish	Ea.		25.50		25.50	Includes material and labor to install
Install	Ea.	350	49		399	good modular unit with melamine
Demolish and Install	Ea.	350	74.50		424.50	laminated to particle board, in
Reinstall	Ea.		39.06		39.06	textured colors or wood grain print
Clean	Ea.	.56	17.30		17.86	finish, hinges and pulls.
Paint	Ea.	1.87	35.50		37.37	
Minimum Charge	Job		236		236	
36″ Wide, 2 Door						
Demolish	Ea.		30		30	Includes material and labor to install
Install	Ea.	395	50		445	good modular unit with melamine
Demolish and Install	Ea.	395	80		475	laminated to particle board, in
Reinstall	Ea.		40.10		40.10	textured colors or wood grain print
Clean	Ea.	.67	21.50		22.17	finish, hinges and pulls.
Paint	Ea.	2.27	43.50		45.77	
Minimum Charge	Job		236		236	
48″ Wide, 2 Door						
Demolish	Ea.		40.50		40.50	Includes material and labor to install
Install	Ea.	490	51		541	good modular unit with melamine
Demolish and Install	Ea.	490	91.50		581.50	laminated to particle board, in
Reinstall	Ea.		40.97		40.97	textured colors or wood grain print
Clean	Ea.	.90	28		28.90	finish, hinges and pulls.
Paint	Ea.	3	56		59	
Minimum Charge	Job		236		236	
Above-Appliance						
Demolish	Ea.		25.50		25.50	Includes material and labor to install
Install	Ea.	234	38		272	good modular unit with melamine
Demolish and Install	Ea.	234	63.50		297.50	laminated to particle board, in
Reinstall	Ea.		30.40		30.40	textured colors or wood grain print
Clean	Ea.	.56	17.30		17.86	finish, hinges and pulls.
Paint	Ea.	1.87	35.50		37.37	
Minimum Charge	Job		236		236	
Better Grade Wood						
12″ Wide, 1 Door						
Demolish	Ea.		10.10		10.10	Includes material and labor to install
Install	Ea.	298	43		341	custom modular unit with solid
Demolish and Install	Ea.	298	53.10		351.10	hardwood faced frames, hardwood
Reinstall	Ea.		34.27		34.27	door frames and drawer fronts,
Clean	Ea.	.22	7		7.22	hardwood veneer on raised door
Paint	Ea.	.75	15.05		15.80	panels, hinges and pulls.
Minimum Charge	Job		236		236	
15″ Wide, 1 Door						
Demolish	Ea.		12.65		12.65	Includes material and labor to install
Install	Ea.	310	44		354	custom modular unit with solid
Demolish and Install	Ea.	310	56.65		366.65	hardwood faced frames, hardwood
Reinstall	Ea.		35.23		35.23	door frames and drawer fronts,
Clean	Ea.	.29	8.65		8.94	hardwood veneer on raised door
Paint	Ea.	.94	18.65		19.59	panels, hinges and pulls.
Minimum Charge	Job		236		236	

Cabinets and Countertops

Wall Cabinets	Unit	Material	Labor	Equip.	Total	Specification
18" Wide, 1 Door						
Demolish	Ea.		15.25		15.25	Includes material and labor to install
Install	Ea.	340	45		385	custom modular unit with solid
Demolish and Install	Ea.	340	60.25		400.25	hardwood faced frames, hardwood
Reinstall	Ea.		36.07		36.07	door frames and drawer fronts,
Clean	Ea.	.34	10.40		10.74	hardwood veneer on raised door
Paint	Ea.	1.12	23		24.12	panels, hinges and pulls.
Minimum Charge	Job		236		236	
21" Wide, 1 Door						
Demolish	Ea.		17.55		17.55	Includes material and labor to install
Install	Ea.	345	46		391	custom modular unit with solid
Demolish and Install	Ea.	345	63.55		408.55	hardwood faced frames, hardwood
Reinstall	Ea.		36.60		36.60	door frames and drawer fronts,
Clean	Ea.	.40	12.10		12.50	hardwood veneer on raised door
Paint	Ea.	1.31	26		27.31	panels, hinges and pulls.
Minimum Charge	Job		236		236	
24" w / Blind Corner						
Demolish	Ea.		20		20	Includes material and labor to install
Install	Ea.	400	46.50		446.50	custom modular unit with solid
Demolish and Install	Ea.	400	66.50		466.50	hardwood faced frames, hardwood
Reinstall	Ea.		37.14		37.14	door frames and drawer fronts,
Clean	Ea.	.45	14		14.45	hardwood veneer on raised door
Paint	Ea.	1.50	30		31.50	panels, hinges and pulls.
Minimum Charge	Job		236		236	
27" Wide, 2 Door						
Demolish	Ea.		22.50		22.50	Includes material and labor to install
Install	Ea.	102	47.50		149.50	custom modular unit with solid
Demolish and Install	Ea.	102	70		172	hardwood faced frames, hardwood
Reinstall	Ea.		38.08		38.08	door frames and drawer fronts,
Clean	Ea.	.51	15.80		16.31	hardwood veneer on raised door
Paint	Ea.	1.69	32.50		34.19	panels, hinges and pulls.
Minimum Charge	Job		236		236	
30" Wide, 2 Door						
Demolish	Ea.		25.50		25.50	Includes material and labor to install
Install	Ea.	470	49		519	custom modular unit with solid
Demolish and Install	Ea.	470	74.50		544.50	hardwood faced frames, hardwood
Reinstall	Ea.		39.06		39.06	door frames and drawer fronts,
Clean	Ea.	.56	17.30		17.86	hardwood veneer on raised door
Paint	Ea.	1.87	39		40.87	panels, hinges and pulls.
Minimum Charge	Job		236		236	
36" Wide, 2 Door						
Demolish	Ea.		30		30	Includes material and labor to install
Install	Ea.	530	50		580	custom modular unit with solid
Demolish and Install	Ea.	530	80		610	hardwood faced frames, hardwood
Reinstall	Ea.		40.10		40.10	door frames and drawer fronts,
Clean	Ea.	.67	21.50		22.17	hardwood veneer on raised door
Paint	Ea.	2.27	43.50		45.77	panels, hinges and pulls.
Minimum Charge	Job		236		236	
48" Wide, 2 Door						
Demolish	Ea.		40.50		40.50	Includes material and labor to install
Install	Ea.	650	51		701	custom modular unit with solid
Demolish and Install	Ea.	650	91.50		741.50	hardwood faced frames, hardwood
Reinstall	Ea.		40.97		40.97	door frames and drawer fronts,
Clean	Ea.	.90	28		28.90	hardwood veneer on raised door
Paint	Ea.	3	56		59	panels, hinges and pulls.
Minimum Charge	Job		236		236	

213

Cabinets and Countertops

Wall Cabinets

	Unit	Material	Labor	Equip.	Total	Specification
Above-appliance						
Demolish	Ea.		25.50		25.50	Includes material and labor to install
Install	Ea.	310	38		348	custom modular unit with solid
Demolish and Install	Ea.	310	63.50		373.50	hardwood faced frames, hardwood
Reinstall	Ea.		30.40		30.40	door frames and drawer fronts,
Clean	Ea.	.56	17.30		17.86	hardwood veneer on raised door
Paint	Ea.	1.87	39		40.87	panels, hinges and pulls.
Minimum Charge	Job		236		236	
Premium Grade Wood						
12″ Wide, 1 Door						
Demolish	Ea.		10.10		10.10	Includes material and labor to install
Install	Ea.	360	43		403	premium modular unit with solid
Demolish and Install	Ea.	360	53.10		413.10	hardwood faced frames, hardwood
Reinstall	Ea.		34.27		34.27	door frames and drawer fronts,
Clean	Ea.	.22	7		7.22	hardwood veneer on raised door
Paint	Ea.	.75	16.30		17.05	panels, hinges and pulls.
Minimum Charge	Job		236		236	
15″ Wide, 1 Door						
Demolish	Ea.		12.65		12.65	Includes material and labor to install
Install	Ea.	420	44		464	premium modular unit with solid
Demolish and Install	Ea.	420	56.65		476.65	hardwood faced frames, hardwood
Reinstall	Ea.		35.23		35.23	door frames and drawer fronts,
Clean	Ea.	.29	8.65		8.94	hardwood veneer on raised door
Paint	Ea.	.94	20.50		21.44	panels, hinges and pulls.
Minimum Charge	Job		236		236	
18″ Wide, 1 Door						
Demolish	Ea.		15.25		15.25	Includes material and labor to install
Install	Ea.	415	45		460	premium modular unit with solid
Demolish and Install	Ea.	415	60.25		475.25	hardwood faced frames, hardwood
Reinstall	Ea.		36.07		36.07	door frames and drawer fronts,
Clean	Ea.	.34	10.40		10.74	hardwood veneer on raised door
Paint	Ea.	1.12	24.50		25.62	panels, hinges and pulls.
Minimum Charge	Job		236		236	
21″ Wide, 1 Door						
Demolish	Ea.		17.55		17.55	Includes material and labor to install
Install	Ea.	420	46		466	premium modular unit with solid
Demolish and Install	Ea.	420	63.55		483.55	hardwood faced frames, hardwood
Reinstall	Ea.		36.60		36.60	door frames and drawer fronts,
Clean	Ea.	.40	12.10		12.50	hardwood veneer on raised door
Paint	Ea.	1.31	28		29.31	panels, hinges and pulls.
Minimum Charge	Job		236		236	
24″ w / Blind Corner						
Demolish	Ea.		20		20	Includes material and labor to install
Install	Ea.	485	46.50		531.50	premium modular unit with solid
Demolish and Install	Ea.	485	66.50		551.50	hardwood faced frames, hardwood
Reinstall	Ea.		37.14		37.14	door frames and drawer fronts,
Clean	Ea.	.45	14		14.45	hardwood veneer on raised door
Paint	Ea.	1.50	32.50		34	panels, hinges and pulls.
Minimum Charge	Job		236		236	
27″ Wide, 2 Door						
Demolish	Ea.		22.50		22.50	Includes material and labor to install
Install	Ea.	445	47.50		492.50	premium modular unit with solid
Demolish and Install	Ea.	445	70		515	hardwood faced frames, hardwood
Reinstall	Ea.		38.08		38.08	door frames and drawer fronts,
Clean	Ea.	.51	15.80		16.31	hardwood veneer on raised door
Paint	Ea.	1.69	35.50		37.19	panels, hinges and pulls.
Minimum Charge	Job		236		236	

Cabinets and Countertops

Wall Cabinets

	Unit	Material	Labor	Equip.	Total	Specification
30" Wide, 2 Door						
Demolish	Ea.		25.50		25.50	Includes material and labor to install
Install	Ea.	560	49		609	premium modular unit with solid
Demolish and Install	Ea.	560	74.50		634.50	hardwood faced frames, hardwood
Reinstall	Ea.		39.06		39.06	door frames and drawer fronts,
Clean	Ea.	.56	17.30		17.86	hardwood veneer on raised door
Paint	Ea.	1.87	39		40.87	panels, hinges and pulls.
Minimum Charge	Job		236		236	
36" Wide, 2 Door						
Demolish	Ea.		30		30	Includes material and labor to install
Install	Ea.	635	50		685	premium modular unit with solid
Demolish and Install	Ea.	635	80		715	hardwood faced frames, hardwood
Reinstall	Ea.		40.10		40.10	door frames and drawer fronts,
Clean	Ea.	.67	21.50		22.17	hardwood veneer on raised door
Paint	Ea.	2.27	49		51.27	panels, hinges and pulls.
Minimum Charge	Job		236		236	
48" Wide, 2 Door						
Demolish	Ea.		40.50		40.50	Includes material and labor to install
Install	Ea.	780	51		831	premium modular unit with solid
Demolish and Install	Ea.	780	91.50		871.50	hardwood faced frames, hardwood
Reinstall	Ea.		40.97		40.97	door frames and drawer fronts,
Clean	Ea.	.90	28		28.90	hardwood veneer on raised door
Paint	Ea.	3	65		68	panels, hinges and pulls.
Minimum Charge	Job		236		236	
Above-appliance						
Demolish	Ea.		25.50		25.50	Includes material and labor to install
Install	Ea.	375	38		413	premium modular unit with solid
Demolish and Install	Ea.	375	63.50		438.50	hardwood faced frames, hardwood
Reinstall	Ea.		30.40		30.40	door frames and drawer fronts,
Clean	Ea.	.56	17.30		17.86	hardwood veneer on raised door
Paint	Ea.	1.87	39		40.87	panels, hinges, pulls.
Minimum Charge	Job		236		236	

Premium Hardwood

	Unit	Material	Labor	Equip.	Total	Specification
12" Wide, 1 Door						
Demolish	Ea.		10.10		10.10	Includes material and labor to install
Install	Ea.	390	43		433	premium hardwood modular unit with
Demolish and Install	Ea.	390	53.10		443.10	solid hardwood faced frames, drawer
Reinstall	Ea.		34.27		34.27	fronts, and door panels, steel drawer
Clean	Ea.	.22	7		7.22	guides, hinges, pulls, laminated
Paint	Ea.	.75	17.80		18.55	interior.
Minimum Charge	Job		236		236	
15" Wide, 1 Door						
Demolish	Ea.		12.65		12.65	Includes material and labor to install
Install	Ea.	545	44		589	premium hardwood modular unit with
Demolish and Install	Ea.	545	56.65		601.65	solid hardwood faced frames, drawer
Reinstall	Ea.		35.23		35.23	fronts, and door panels, steel drawer
Clean	Ea.	.29	8.65		8.94	guides, hinges, pulls, laminated
Paint	Ea.	.94	21.50		22.44	interior.
Minimum Charge	Job		236		236	
18" Wide, 1 Door						
Demolish	Ea.		15.25		15.25	Includes material and labor to install
Install	Ea.	450	45		495	premium hardwood modular unit with
Demolish and Install	Ea.	450	60.25		510.25	solid hardwood faced frames, drawer
Reinstall	Ea.		36.07		36.07	fronts, and door panels, steel drawer
Clean	Ea.	.34	10.40		10.74	guides, hinges, pulls, laminated
Paint	Ea.	1.12	26		27.12	interior.
Minimum Charge	Job		236		236	

215

Cabinets and Countertops

Wall Cabinets	Unit	Material	Labor	Equip.	Total	Specification
21" Wide, 1 Door						
Demolish	Ea.		17.55		17.55	Includes material and labor to install
Install	Ea.	455	46		501	premium hardwood modular unit with
Demolish and Install	Ea.	455	63.55		518.55	solid hardwood faced frames, drawer
Reinstall	Ea.		36.60		36.60	fronts, and door panels, steel drawer
Clean	Ea.	.40	12.10		12.50	guides, hinges, pulls, laminated
Paint	Ea.	1.31	30		31.31	interior.
Minimum Charge	Job		236		236	
24" w / Blind Corner						
Demolish	Ea.		20		20	Includes material and labor to install
Install	Ea.	530	46.50		576.50	premium hardwood modular unit with
Demolish and Install	Ea.	530	66.50		596.50	solid hardwood faced frames, drawer
Reinstall	Ea.		37.14		37.14	fronts, and door panels, steel drawer
Clean	Ea.	.45	14		14.45	guides, hinges, pulls, laminated
Paint	Ea.	1.50	35.50		37	interior.
Minimum Charge	Job		236		236	
27" Wide, 2 Door						
Demolish	Ea.		22.50		22.50	Includes material and labor to install
Install	Ea.	133	47.50		180.50	premium hardwood modular unit with
Demolish and Install	Ea.	133	70		203	solid hardwood faced frames, drawer
Reinstall	Ea.		38.08		38.08	fronts, and door panels, steel drawer
Clean	Ea.	.51	15.80		16.31	guides, hinges, pulls, laminated
Paint	Ea.	1.69	39		40.69	interior.
Minimum Charge	Job		236		236	
30" Wide, 2 Door						
Demolish	Ea.		25.50		25.50	Includes material and labor to install
Install	Ea.	610	49		659	premium hardwood modular unit with
Demolish and Install	Ea.	610	74.50		684.50	solid hardwood faced frames, drawer
Reinstall	Ea.		39.06		39.06	fronts, and door panels, steel drawer
Clean	Ea.	.56	17.30		17.86	guides, hinges, pulls, laminated
Paint	Ea.	1.87	43.50		45.37	interior.
Minimum Charge	Job		236		236	
36" Wide, 2 Door						
Demolish	Ea.		30		30	Includes material and labor to install
Install	Ea.	695	50		745	premium hardwood modular unit with
Demolish and Install	Ea.	695	80		775	solid hardwood faced frames, drawer
Reinstall	Ea.		40.10		40.10	fronts, and door panels, steel drawer
Clean	Ea.	.67	21.50		22.17	guides, hinges, pulls, laminated
Paint	Ea.	2.27	56		58.27	interior.
Minimum Charge	Job		236		236	
48" Wide, 2 Door						
Demolish	Ea.		40.50		40.50	Includes material and labor to install
Install	Ea.	855	51		906	premium hardwood modular unit with
Demolish and Install	Ea.	855	91.50		946.50	solid hardwood faced frames, drawer
Reinstall	Ea.		40.97		40.97	fronts, and door panels, steel drawer
Clean	Ea.	.90	28		28.90	guides, hinges, pulls, laminated
Paint	Ea.	3	65		68	interior.
Minimum Charge	Job		236		236	
Above-appliance						
Demolish	Ea.		25.50		25.50	Includes material and labor to install
Install	Ea.	405	38		443	premium hardwood modular unit with
Demolish and Install	Ea.	405	63.50		468.50	solid hardwood faced frames, drawer
Reinstall	Ea.		30.40		30.40	fronts, and door panels, steel drawer
Clean	Ea.	.56	17.30		17.86	guides, hinges, pulls, laminated
Paint	Ea.	1.87	43.50		45.37	interior.
Minimum Charge	Job		236		236	

Cabinets and Countertops

Wall Cabinets	Unit	Material	Labor	Equip.	Total	Specification
Medicine Cabinet						
Good Grade						
Demolish	Ea.		16.85		16.85	Cost includes material and labor to
Install	Ea.	161	33.50		194.50	install 14″ x 18″ recessed medicine
Demolish and Install	Ea.	161	50.35		211.35	cabinet with stainless steel frame,
Reinstall	Ea.		26.93		26.93	mirror. Light not included.
Clean	Ea.	.01	17.05		17.06	
Paint	Ea.	2.09	24.50		26.59	
Minimum Charge	Job		236		236	
Good Grade Lighted						
Demolish	Ea.		16.85		16.85	Cost includes material and labor to
Install	Ea.	425	39.50		464.50	install 14″ x 18″ recessed medicine
Demolish and Install	Ea.	425	56.35		481.35	cabinet with stainless steel frame,
Reinstall	Ea.		31.41		31.41	mirror and light.
Clean	Ea.	.01	17.05		17.06	
Paint	Ea.	2.09	24.50		26.59	
Minimum Charge	Job		236		236	
Better Grade						
Demolish	Ea.		16.85		16.85	Includes material and labor to install
Install	Ea.	283	33.50		316.50	recessed medicine cabinet with
Demolish and Install	Ea.	283	50.35		333.35	polished chrome frame and mirror.
Reinstall	Ea.		26.93		26.93	Light not included.
Clean	Ea.	.01	17.05		17.06	
Paint	Ea.	2.09	24.50		26.59	
Minimum Charge	Job		236		236	
Better Grade Lighted						
Demolish	Ea.		16.85		16.85	Includes material and labor to install
Install	Ea.	545	39.50		584.50	recessed medicine cabinet with
Demolish and Install	Ea.	545	56.35		601.35	polished chrome frame, mirror and
Reinstall	Ea.		31.41		31.41	light.
Clean	Ea.	.01	17.05		17.06	
Paint	Ea.	2.09	24.50		26.59	
Minimum Charge	Job		236		236	
Premium Grade						
Demolish	Ea.		16.85		16.85	Cost includes material and labor to
Install	Ea.	340	33.50		373.50	install 14″ x 24″ recessed medicine
Demolish and Install	Ea.	340	50.35		390.35	cabinet with beveled mirror, frameless
Reinstall	Ea.		26.93		26.93	swing door. Light not included.
Clean	Ea.	.01	17.05		17.06	
Paint	Ea.	2.09	24.50		26.59	
Minimum Charge	Job		236		236	
Premium Grade Lighted						
Demolish	Ea.		16.85		16.85	Cost includes material and labor to
Install	Ea.	605	39.50		644.50	install 36″ x 30″ surface mounted
Demolish and Install	Ea.	605	56.35		661.35	medicine cabinet with beveled tri-view
Reinstall	Ea.		31.41		31.41	mirror, frameless swing door and light.
Clean	Ea.	.01	17.05		17.06	
Paint	Ea.	2.09	24.50		26.59	
Minimum Charge	Job		236		236	
Bath Mirror						
Demolish	Ea.		12.65		12.65	Cost includes material and labor to
Install	Ea.	51.50	23.50		75	install 18″ x 24″ surface mount mirror
Demolish and Install	Ea.	51.50	36.15		87.65	with stainless steel frame.
Reinstall	Ea.		18.85		18.85	
Clean	Ea.	.01	1.13		1.14	
Minimum Charge	Job		236		236	

Cabinets and Countertops

Wall Cabinets		Unit	Material	Labor	Equip.	Total	Specification
Garage							
	Demolish	L.F.		10.10		10.10	Includes material and labor to install
	Install	L.F.	114	23.50		137.50	good modular unit with melamine
	Demolish and Install	L.F.	114	33.60		147.60	laminated to particle board, in
	Reinstall	L.F.		18.85		18.85	textured colors or wood grain print
	Clean	L.F.	.22	7		7.22	finish, hinges and pulls.
	Paint	L.F.	.75	13.95		14.70	
	Minimum Charge	Job		236		236	
Strip and Refinish							
	Install	L.F.	5.50	49		54.50	Includes labor and material to strip,
	Minimum Charge	Job		196		196	prep and refinish exterior of cabinets.

Tall Cabinets		Unit	Material	Labor	Equip.	Total	Specification
Good Grade Laminated							
	Demolish	L.F.		17.95		17.95	Includes material and labor to install
	Install	L.F.	282	52.50		334.50	custom modular unit with solid
	Demolish and Install	L.F.	282	70.45		352.45	hardwood faced frames, hardwood
	Reinstall	L.F.		41.88		41.88	door frames and drawer fronts,
	Clean	L.F.	.59	13		13.59	hardwood veneer on raised door
	Paint	L.F.	1.75	28		29.75	panels, hinges and pulls.
	Minimum Charge	Job		236		236	
Better Grade Wood							
	Demolish	L.F.		17.95		17.95	Includes material and labor to install
	Install	L.F.	400	52.50		452.50	custom modular unit with solid
	Demolish and Install	L.F.	400	70.45		470.45	hardwood faced frames, hardwood
	Reinstall	L.F.		41.88		41.88	door frames and drawer fronts,
	Clean	L.F.	.59	13		13.59	hardwood veneer on raised door
	Paint	L.F.	1.75	30		31.75	panels, hinges and pulls.
	Minimum Charge	Job		236		236	
Premium Grade Wood							
	Demolish	L.F.		17.95		17.95	Includes material and labor to install
	Install	L.F.	580	52.50		632.50	premium modular unit with solid
	Demolish and Install	L.F.	580	70.45		650.45	hardwood faced frames, hardwood
	Reinstall	L.F.		41.88		41.88	door frames and drawer fronts,
	Clean	L.F.	.59	13		13.59	hardwood veneer on raised door
	Paint	L.F.	1.75	32.50		34.25	panels, hinges and pulls.
	Minimum Charge	Job		236		236	
Premium Hardwood							
	Demolish	L.F.		17.95		17.95	Includes material and labor to install
	Install	L.F.	810	52.50		862.50	premium hardwood modular unit with
	Demolish and Install	L.F.	810	70.45		880.45	solid hardwood faced frames, drawer
	Reinstall	L.F.		41.88		41.88	fronts, and door panels, steel drawer
	Clean	L.F.	.59	13		13.59	guides, hinges and pulls, laminated
	Paint	L.F.	1.75	35.50		37.25	interior.
	Minimum Charge	Job		236		236	

Cabinets and Countertops

Tall Cabinets	Unit	Material	Labor	Equip.	Total	Specification
Single Oven, 27" Wide						
Good Grade Laminated						
Demolish	Ea.		40.50		40.50	Includes material and labor to install
Install	Ea.	975	118		1093	good modular unit with melamine
Demolish and Install	Ea.	975	158.50		1133.50	laminated to particle board, in
Reinstall	Ea.		94.24		94.24	textured colors or wood grain print
Clean	Ea.	1.87	22.50		24.37	finish, hinges and pulls.
Paint	Ea.	2.93	62.50		65.43	
Minimum Charge	Job		236		236	
Better Grade Veneer						
Demolish	Ea.		40.50		40.50	Includes material and labor to install
Install	Ea.	1725	118		1843	custom modular unit with solid
Demolish and Install	Ea.	1725	158.50		1883.50	hardwood faced frames, hardwood
Reinstall	Ea.		94.24		94.24	door frames and drawer fronts,
Clean	Ea.	1.87	22.50		24.37	hardwood veneer on raised door
Paint	Ea.	2.93	67.50		70.43	panels, hinges and pulls.
Minimum Charge	Job		236		236	
Premium Grade Veneer						
Demolish	Ea.		40.50		40.50	Includes material and labor to install
Install	Ea.	1575	118		1693	premium modular unit with solid
Demolish and Install	Ea.	1575	158.50		1733.50	hardwood faced frames, hardwood
Reinstall	Ea.		94.24		94.24	door frames and drawer fronts,
Clean	Ea.	1.87	22.50		24.37	hardwood veneer on raised door
Paint	Ea.	2.93	74		76.93	panels, hinges and pulls.
Minimum Charge	Job		236		236	
Premium Hardwood						
Demolish	Ea.		40.50		40.50	Includes material and labor to install
Install	Ea.	1700	118		1818	premium hardwood modular unit with
Demolish and Install	Ea.	1700	158.50		1858.50	solid hardwood faced frames, drawer
Reinstall	Ea.		94.24		94.24	fronts, and door panels, steel drawer
Clean	Ea.	1.87	22.50		24.37	guides, hinges and pulls, laminated
Paint	Ea.	2.93	80		82.93	interior.
Minimum Charge	Job		236		236	
Single Oven, 30" Wide						
Good Grade Laminated						
Demolish	Ea.		40.50		40.50	Includes material and labor to install
Install	Ea.	1125	118		1243	good modular unit with melamine
Demolish and Install	Ea.	1125	158.50		1283.50	laminated to particle board, in
Reinstall	Ea.		94.24		94.24	textured colors or wood grain print
Clean	Ea.	1.87	22.50		24.37	finish, hinges and pulls.
Paint	Ea.	2.93	62.50		65.43	
Minimum Charge	Job		236		236	
Better Grade Veneer						
Demolish	Ea.		40.50		40.50	Includes material and labor to install
Install	Ea.	1325	118		1443	custom modular unit with solid
Demolish and Install	Ea.	1325	158.50		1483.50	hardwood faced frames, hardwood
Reinstall	Ea.		94.24		94.24	door frames and drawer fronts,
Clean	Ea.	1.87	22.50		24.37	hardwood veneer on raised door
Paint	Ea.	2.93	62.50		65.43	panels, hinges and pulls.
Minimum Charge	Job		236		236	

Tall Cabinets	Unit	Material	Labor	Equip.	Total	Specification
Premium Grade Veneer						
Demolish	Ea.		40.50		40.50	Includes material and labor to install
Install	Ea.	1475	118		1593	premium modular unit with solid
Demolish and Install	Ea.	1475	158.50		1633.50	hardwood faced frames, hardwood
Reinstall	Ea.		94.24		94.24	door frames and drawer fronts,
Clean	Ea.	1.87	22.50		24.37	hardwood veneer on raised door
Paint	Ea.	2.93	62.50		65.43	panels, hinges and pulls.
Minimum Charge	Job		236		236	
Premium Hardwood						
Demolish	Ea.		40.50		40.50	Includes material and labor to install
Install	Ea.	1575	118		1693	premium hardwood modular unit with
Demolish and Install	Ea.	1575	158.50		1733.50	solid hardwood faced frames, drawer
Reinstall	Ea.		94.24		94.24	fronts, and door panels, steel drawer
Clean	Ea.	1.87	22.50		24.37	guides, hinges and pulls, laminated
Paint	Ea.	2.93	62.50		65.43	interior.
Minimum Charge	Job		236		236	
Utility 18″ W x 12″ D						
Good Grade Laminated						
Demolish	Ea.		40.50		40.50	Includes material and labor to install
Install	Ea.	670	94		764	good modular unit with melamine
Demolish and Install	Ea.	670	134.50		804.50	laminated to particle board, in
Reinstall	Ea.		75.39		75.39	textured colors or wood grain print
Clean	Ea.	.44	11		11.44	finish, hinges and pulls.
Paint	Ea.	1.12	24.50		25.62	
Minimum Charge	Job		236		236	
Better Grade Veneer						
Demolish	Ea.		40.50		40.50	Includes material and labor to install
Install	Ea.	895	94		989	custom modular unit with solid
Demolish and Install	Ea.	895	134.50		1029.50	hardwood faced frames, hardwood
Reinstall	Ea.		75.39		75.39	door frames and drawer fronts,
Clean	Ea.	.44	11		11.44	hardwood veneer on raised door
Paint	Ea.	1.12	26		27.12	panels, hinges and pulls.
Minimum Charge	Job		236		236	
Premium Grade Veneer						
Demolish	Ea.		40.50		40.50	Includes material and labor to install
Install	Ea.	1075	94		1169	premium modular unit with solid
Demolish and Install	Ea.	1075	134.50		1209.50	hardwood faced frames, hardwood
Reinstall	Ea.		75.39		75.39	door frames and drawer fronts,
Clean	Ea.	.44	11		11.44	hardwood veneer on raised door
Paint	Ea.	1.12	30		31.12	panels, hinges and pulls.
Minimum Charge	Job		236		236	
Premium Hardwood						
Demolish	Ea.		40.50		40.50	Includes material and labor to install
Install	Ea.	1175	94		1269	premium hardwood modular unit with
Demolish and Install	Ea.	1175	134.50		1309.50	solid hardwood faced frames, drawer
Reinstall	Ea.		75.39		75.39	fronts, and door panels, steel drawer
Clean	Ea.	.44	11		11.44	guides, hinges and pulls, laminated
Paint	Ea.	1.12	32.50		33.62	interior.
Minimum Charge	Job		236		236	

Tall Cabinets	Unit	Material	Labor	Equip.	Total	Specification
Utility 24″ W x 24″ D						
Good Grade Laminated						
Demolish	Ea.		40.50		40.50	Includes material and labor to install
Install	Ea.	780	118		898	good modular unit with melamine
Demolish and Install	Ea.	780	158.50		938.50	laminated to particle board, in
Reinstall	Ea.		94.24		94.24	textured colors or wood grain print
Clean	Ea.	.57	14.55		15.12	finish, hinges and pulls.
Paint	Ea.	1.50	32.50		34	
Minimum Charge	Job		236		236	
Better Grade Veneer						
Demolish	Ea.		40.50		40.50	Includes material and labor to install
Install	Ea.	1050	118		1168	custom modular unit with solid
Demolish and Install	Ea.	1050	158.50		1208.50	hardwood faced frames, hardwood
Reinstall	Ea.		94.24		94.24	door frames and drawer fronts,
Clean	Ea.	.57	14.55		15.12	hardwood veneer on raised door
Paint	Ea.	1.50	35.50		37	panels, hinges and pulls.
Minimum Charge	Job		236		236	
Premium Grade Veneer						
Demolish	Ea.		40.50		40.50	Includes material and labor to install
Install	Ea.	1250	118		1368	premium modular unit with solid
Demolish and Install	Ea.	1250	158.50		1408.50	hardwood faced frames, hardwood
Reinstall	Ea.		94.24		94.24	door frames and drawer fronts,
Clean	Ea.	.57	14.55		15.12	hardwood veneer on raised door
Paint	Ea.	1.50	39		40.50	panels, hinges and pulls.
Minimum Charge	Job		236		236	
Premium Hardwood						
Demolish	Ea.		40.50		40.50	Includes material and labor to install
Install	Ea.	1375	118		1493	premium hardwood modular unit with
Demolish and Install	Ea.	1375	158.50		1533.50	solid hardwood faced frames, drawer
Reinstall	Ea.		94.24		94.24	fronts, and door panels, steel drawer
Clean	Ea.	.57	14.55		15.12	guides, hinges and pulls, laminated
Paint	Ea.	1.50	43.50		45	interior.
Minimum Charge	Job		236		236	
Plain Shelves						
Install	Ea.	33.50	26		59.50	Includes labor and material to install
Rotating Shelves						tall cabinet shelving, per set.
Install	Ea.	113	23.50		136.50	Includes labor and material to install
Strip and Refinish						lazy susan shelving for wall cabinet.
Install	L.F.	10.95	78		88.95	Includes labor and material to strip,
Minimum Charge	Job		196		196	prep and refinish exterior of cabinets.
Stain						
Install	L.F.	.34	3.56		3.90	Includes labor and material to replace
Minimum Charge	Job		196		196	normal prep and stain on doors and
						exterior cabinets.

Base Cabinets	Unit	Material	Labor	Equip.	Total	Specification
Good Grade Laminated						
12″ w, 1 Door, 1 Drawer						
Demolish	Ea.		10.10		10.10	Includes material and labor to install
Install	Ea.	262	38		300	good modular unit with melamine
Demolish and Install	Ea.	262	48.10		310.10	laminate, textured colors or wood
Reinstall	Ea.		30.40		30.40	grain print finish including hinges and
Clean	Ea.	.30	7.25		7.55	pulls.
Paint	Ea.	.75	16.30		17.05	
Minimum Charge	Job		236		236	
15″ w, 1 Door, 1 Drawer						
Demolish	Ea.		12.65		12.65	Includes material and labor to install
Install	Ea.	271	39.50		310.50	good modular unit with melamine
Demolish and Install	Ea.	271	52.15		323.15	laminate, textured colors or wood
Reinstall	Ea.		31.41		31.41	grain print finish including hinges and
Clean	Ea.	.37	9.10		9.47	pulls.
Paint	Ea.	.94	20.50		21.44	
Minimum Charge	Job		236		236	
18″ w, 1 Door, 1 Drawer						
Demolish	Ea.		14.95		14.95	Includes material and labor to install
Install	Ea.	288	40.50		328.50	good modular unit with melamine
Demolish and Install	Ea.	288	55.45		343.45	laminate, textured colors or wood
Reinstall	Ea.		32.36		32.36	grain print finish including hinges and
Clean	Ea.	.44	11		11.44	pulls.
Paint	Ea.	1.12	24.50		25.62	
Minimum Charge	Job		236		236	
21″ w, 1 Door, 1 Drawer						
Demolish	Ea.		17.55		17.55	Includes material and labor to install
Install	Ea.	295	41.50		336.50	good modular unit with melamine
Demolish and Install	Ea.	295	59.05		354.05	laminate, textured colors or wood
Reinstall	Ea.		33.21		33.21	grain print finish including hinges and
Clean	Ea.	.54	13		13.54	pulls.
Paint	Ea.	1.31	28		29.31	
Minimum Charge	Job		236		236	
24″ w, 1 Door, 1 Drawer						
Demolish	Ea.		20		20	Includes material and labor to install
Install	Ea.	360	42.50		402.50	good modular unit with melamine
Demolish and Install	Ea.	360	62.50		422.50	laminate, textured colors or wood
Reinstall	Ea.		33.81		33.81	grain print finish including hinges and
Clean	Ea.	.57	14.55		15.12	pulls.
Paint	Ea.	1.50	32.50		34	
Minimum Charge	Job		236		236	
30″ Wide Sink Base						
Demolish	Ea.		25.50		25.50	Includes material and labor to install
Install	Ea.	345	44		389	good modular unit with melamine
Demolish and Install	Ea.	345	69.50		414.50	laminate, textured colors or wood
Reinstall	Ea.		35.23		35.23	grain print finish including hinges and
Clean	Ea.	.75	18.20		18.95	pulls.
Paint	Ea.	1.87	39		40.87	
Minimum Charge	Job		236		236	
36″ Wide Sink Base						
Demolish	Ea.		31		31	Includes material and labor to install
Install	Ea.	390	46.50		436.50	good modular unit with melamine
Demolish and Install	Ea.	390	77.50		467.50	laminate, textured colors or wood
Reinstall	Ea.		37.14		37.14	grain print finish including hinges and
Clean	Ea.	.84	21.50		22.34	pulls.
Paint	Ea.	2.27	49		51.27	
Minimum Charge	Job		236		236	

Cabinets and Countertops

Base Cabinets	Unit	Material	Labor	Equip.	Total	Specification
36" Blind Corner						
Demolish	Ea.		31		31	Includes material and labor to install
Install	Ea.	615	52.50		667.50	good modular unit with melamine
Demolish and Install	Ea.	615	83.50		698.50	laminate, textured colors or wood
Reinstall	Ea.		41.88		41.88	grain print finish including hinges and
Clean	Ea.	.84	21.50		22.34	pulls.
Paint	Ea.	2.27	49		51.27	
Minimum Charge	Job		236		236	
42" w, 2 Door, 2 Drawer						
Demolish	Ea.		36.50		36.50	Includes material and labor to install
Install	Ea.	480	47.50		527.50	good modular unit with melamine
Demolish and Install	Ea.	480	84		564	laminate, textured colors or wood
Reinstall	Ea.		38.08		38.08	grain print finish including hinges and
Clean	Ea.	1.07	26		27.07	pulls.
Paint	Ea.	2.60	56		58.60	
Minimum Charge	Job		236		236	
Clean Interior						
Clean	L.F.	.30	5.50		5.80	Includes labor and materials to clean
Minimum Charge	Job		182		182	the interior of a 27" cabinet.
Better Grade Wood						
12" w, 1 Door, 1 Drawer						
Demolish	Ea.		10.10		10.10	Includes material and labor to install
Install	Ea.	345	38		383	custom modular unit including solid
Demolish and Install	Ea.	345	48.10		393.10	hardwood faced frames, hardwood
Reinstall	Ea.		30.40		30.40	door frames and drawer fronts,
Clean	Ea.	.30	7.25		7.55	hardwood veneer on raised door
Paint	Ea.	.75	17.80		18.55	panels, hinges and pulls.
Minimum Charge	Job		236		236	
15" w, 1 Door, 1 Drawer						
Demolish	Ea.		12.65		12.65	Includes material and labor to install
Install	Ea.	365	39.50		404.50	custom modular unit including solid
Demolish and Install	Ea.	365	52.15		417.15	hardwood faced frames, hardwood
Reinstall	Ea.		31.41		31.41	door frames and drawer fronts,
Clean	Ea.	.37	9.10		9.47	hardwood veneer on raised door
Paint	Ea.	.94	23		23.94	panels, hinges and pulls.
Minimum Charge	Job		236		236	
18" w, 1 Door, 1 Drawer						
Demolish	Ea.		14.95		14.95	Includes material and labor to install
Install	Ea.	385	40.50		425.50	custom modular unit including solid
Demolish and Install	Ea.	385	55.45		440.45	hardwood faced frames, hardwood
Reinstall	Ea.		32.36		32.36	door frames and drawer fronts,
Clean	Ea.	.44	11		11.44	hardwood veneer on raised door
Paint	Ea.	1.12	26		27.12	panels, hinges and pulls.
Minimum Charge	Job		236		236	
21" w, 1 Door, 1 Drawer						
Demolish	Ea.		17.55		17.55	Includes material and labor to install
Install	Ea.	395	41.50		436.50	custom modular unit including solid
Demolish and Install	Ea.	395	59.05		454.05	hardwood faced frames, hardwood
Reinstall	Ea.		33.21		33.21	door frames and drawer fronts,
Clean	Ea.	.54	13		13.54	hardwood veneer on raised door
Paint	Ea.	1.31	30		31.31	panels, hinges and pulls.
Minimum Charge	Job		236		236	

For customer support on your Contractor's Pricing Guide: Residential Repair & Remodeling, call 800.448.8182.

Cabinets and Countertops

Base Cabinets	Unit	Material	Labor	Equip.	Total	Specification
24" w, 1 Door, 1 Drawer						
Demolish	Ea.		20		20	Includes material and labor to install
Install	Ea.	480	42.50		522.50	custom modular unit including solid
Demolish and Install	Ea.	480	62.50		542.50	hardwood faced frames, hardwood
Reinstall	Ea.		33.81		33.81	door frames and drawer fronts,
Clean	Ea.	.57	14.55		15.12	hardwood veneer on raised door
Paint	Ea.	1.50	35.50		37	panels, hinges and pulls.
Minimum Charge	Job		236		236	
30" Wide Sink Base						
Demolish	Ea.		25.50		25.50	Includes material and labor to install
Install	Ea.	460	44		504	custom modular unit including solid
Demolish and Install	Ea.	460	69.50		529.50	hardwood faced frames, hardwood
Reinstall	Ea.		35.23		35.23	door frames and drawer fronts,
Clean	Ea.	.75	18.20		18.95	hardwood veneer on raised door
Paint	Ea.	1.87	43.50		45.37	panels, hinges and pulls.
Minimum Charge	Job		236		236	
36" Wide Sink Base						
Demolish	Ea.		31		31	Includes material and labor to install
Install	Ea.	525	46.50		571.50	custom modular unit including solid
Demolish and Install	Ea.	525	77.50		602.50	hardwood faced frames, hardwood
Reinstall	Ea.		37.14		37.14	door frames and drawer fronts,
Clean	Ea.	.84	21.50		22.34	hardwood veneer on raised door
Paint	Ea.	2.27	56		58.27	panels, hinges and pulls.
Minimum Charge	Job		236		236	
36" Blind Corner						
Demolish	Ea.		31		31	Includes material and labor to install
Install	Ea.	820	52.50		872.50	custom modular unit including solid
Demolish and Install	Ea.	820	83.50		903.50	hardwood faced frames, hardwood
Reinstall	Ea.		41.88		41.88	door frames and drawer fronts,
Clean	Ea.	.84	21.50		22.34	hardwood veneer on raised door
Paint	Ea.	2.27	56		58.27	panels, hinges and pulls.
Minimum Charge	Job		236		236	
42" w, 2 Door, 2 Drawer						
Demolish	Ea.		36.50		36.50	Includes material and labor to install
Install	Ea.	640	47.50		687.50	modular unit including solid hardwood
Demolish and Install	Ea.	640	84		724	faced frames, hardwood door frames
Reinstall	Ea.		38.08		38.08	and drawer fronts, hardwood veneer
Clean	Ea.	1.07	26		27.07	on raised door panels, hinges and
Paint	Ea.	2.60	65		67.60	pulls.
Minimum Charge	Job		236		236	
Clean Interior						
Clean	L.F.	.30	5.50		5.80	Includes labor and materials to clean
Minimum Charge	Job		182		182	the interior of a 27" cabinet.
Premium Grade Wood						
12" w, 1 Door, 1 Drawer						
Demolish	Ea.		10.10		10.10	Includes material and labor to install
Install	Ea.	420	19		439	premium hardwood modular unit with
Demolish and Install	Ea.	420	29.10		449.10	solid hardwood faced frames, drawer
Reinstall	Ea.		15.20		15.20	fronts, and door panels, steel drawer
Clean	Ea.	.30	7.25		7.55	guides, hinges, pulls and laminated
Paint	Ea.	.75	19.55		20.30	interior.
Minimum Charge	Job		236		236	

For customer support on your Contractor's Pricing Guide: Residential Repair & Remodeling, call 800.448.8182.

Cabinets and Countertops

Base Cabinets	Unit	Material	Labor	Equip.	Total	Specification
15" w, 1 Door, 1 Drawer						
Demolish	Ea.		12.65		12.65	Includes material and labor to install
Install	Ea.	435	39.50		474.50	premium modular unit including solid
Demolish and Install	Ea.	435	52.15		487.15	hardwood faced frames, hardwood
Reinstall	Ea.		31.41		31.41	door frames and drawer fronts,
Clean	Ea.	.37	9.10		9.47	hardwood veneer on raised door
Paint	Ea.	.94	24.50		25.44	panels, hinges and pulls.
Minimum Charge	Job		236		236	
18" w, 1 Door, 1 Drawer						
Demolish	Ea.		14.95		14.95	Includes material and labor to install
Install	Ea.	460	40.50		500.50	premium modular unit including solid
Demolish and Install	Ea.	460	55.45		515.45	hardwood faced frames, hardwood
Reinstall	Ea.		32.36		32.36	door frames and drawer fronts,
Clean	Ea.	.44	11		11.44	hardwood veneer on raised door
Paint	Ea.	1.12	30		31.12	panels, hinges and pulls.
Minimum Charge	Job		236		236	
21" w, 1 Door, 1 Drawer						
Demolish	Ea.		17.55		17.55	Includes material and labor to install
Install	Ea.	480	41.50		521.50	premium modular unit including solid
Demolish and Install	Ea.	480	59.05		539.05	hardwood faced frames, hardwood
Reinstall	Ea.		33.21		33.21	door frames and drawer fronts,
Clean	Ea.	.54	13		13.54	hardwood veneer on raised door
Paint	Ea.	1.31	32.50		33.81	panels, hinges and pulls.
Minimum Charge	Job		236		236	
24" w, 1 Door, 1 Drawer						
Demolish	Ea.		20		20	Includes material and labor to install
Install	Ea.	570	42.50		612.50	premium modular unit including solid
Demolish and Install	Ea.	570	62.50		632.50	hardwood faced frames, hardwood
Reinstall	Ea.		33.81		33.81	door frames and drawer fronts,
Clean	Ea.	.57	14.55		15.12	hardwood veneer on raised door
Paint	Ea.	1.50	39		40.50	panels, hinges and pulls.
Minimum Charge	Job		236		236	
30" Wide Sink Base						
Demolish	Ea.		25.50		25.50	Includes material and labor to install
Install	Ea.	555	44		599	premium modular unit including solid
Demolish and Install	Ea.	555	69.50		624.50	hardwood faced frames, hardwood
Reinstall	Ea.		35.23		35.23	door frames and drawer fronts,
Clean	Ea.	.75	18.20		18.95	hardwood veneer on raised door
Paint	Ea.	1.87	49		50.87	panels, hinges and pulls.
Minimum Charge	Job		236		236	
36" Wide Sink Base						
Demolish	Ea.		31		31	Includes material and labor to install
Install	Ea.	625	46.50		671.50	premium modular unit including solid
Demolish and Install	Ea.	625	77.50		702.50	hardwood faced frames, hardwood
Reinstall	Ea.		37.14		37.14	door frames and drawer fronts,
Clean	Ea.	.84	21.50		22.34	hardwood veneer on raised door
Paint	Ea.	2.27	56		58.27	panels, hinges and pulls.
Minimum Charge	Job		236		236	
36" Blind Corner						
Demolish	Ea.		31		31	Includes material and labor to install
Install	Ea.	985	52.50		1037.50	premium modular unit including solid
Demolish and Install	Ea.	985	83.50		1068.50	hardwood faced frames, hardwood
Reinstall	Ea.		41.88		41.88	door frames and drawer fronts,
Clean	Ea.	.84	21.50		22.34	hardwood veneer on raised door
Paint	Ea.	2.27	56		58.27	panels, hinges and pulls.
Minimum Charge	Job		236		236	

Cabinets and Countertops

Base Cabinets		Unit	Material	Labor	Equip.	Total	Specification
42" w, 2 Door, 2 Drawer							Includes material and labor to install
	Demolish	Ea.		36.50		36.50	premium modular unit including solid
	Install	Ea.	845	47.50		892.50	hardwood faced frames, hardwood
	Demolish and Install	Ea.	845	84		929	door frames and drawer fronts,
	Reinstall	Ea.		38.08		38.08	hardwood veneer on raised door
	Clean	Ea.	1.07	26		27.07	panels, hinges and pulls.
	Paint	Ea.	2.60	65		67.60	
	Minimum Charge	Job		236		236	
Clean Interior							
	Clean	L.F.	.30	5.50		5.80	Includes labor and materials to clean
	Minimum Charge	Job		182		182	the interior of a 27" cabinet.
Premium Hardwood							
12" w, 1 Door, 1 Drawer							Includes material and labor to install
	Demolish	Ea.		10.10		10.10	premium hardwood modular unit with
	Install	Ea.	455	38		493	solid hardwood faced frames, drawer
	Demolish and Install	Ea.	455	48.10		503.10	fronts, and door panels, steel drawer
	Reinstall	Ea.		30.40		30.40	guides, hinges, pulls and laminated
	Clean	Ea.	.30	7.25		7.55	interior.
	Paint	Ea.	.75	21.50		22.25	
	Minimum Charge	Job		236		236	
15" w, 1 Door, 1 Drawer							Includes material and labor to install
	Demolish	Ea.		12.65		12.65	premium hardwood modular unit with
	Install	Ea.	475	39.50		514.50	solid hardwood faced frames, drawer
	Demolish and Install	Ea.	475	52.15		527.15	fronts, and door panels, steel drawer
	Reinstall	Ea.		31.41		31.41	guides, hinges, pulls and laminated
	Clean	Ea.	.37	9.10		9.47	interior.
	Paint	Ea.	.94	28		28.94	
	Minimum Charge	Job		236		236	
18" w, 1 Door, 1 Drawer							Includes material and labor to install
	Demolish	Ea.		14.95		14.95	premium hardwood modular unit with
	Install	Ea.	505	40.50		545.50	solid hardwood faced frames, drawer
	Demolish and Install	Ea.	505	55.45		560.45	fronts, and door panels, steel drawer
	Reinstall	Ea.		32.36		32.36	guides, hinges, pulls and laminated
	Clean	Ea.	.44	11		11.44	interior.
	Paint	Ea.	1.12	32.50		33.62	
	Minimum Charge	Job		236		236	
21" w, 1 Door, 1 Drawer							Includes material and labor to install
	Demolish	Ea.		17.55		17.55	premium hardwood modular unit with
	Install	Ea.	515	41.50		556.50	solid hardwood faced frames, drawer
	Demolish and Install	Ea.	515	59.05		574.05	fronts, and door panels, steel drawer
	Reinstall	Ea.		33.21		33.21	guides, hinges, pulls and laminated
	Clean	Ea.	.54	13		13.54	interior.
	Paint	Ea.	1.31	35.50		36.81	
	Minimum Charge	Job		236		236	
24" w, 1 Door, 1 Drawer							Includes material and labor to install
	Demolish	Ea.		20		20	premium hardwood modular unit with
	Install	Ea.	625	42.50		667.50	solid hardwood faced frames, drawer
	Demolish and Install	Ea.	625	62.50		687.50	fronts, and door panels, steel drawer
	Reinstall	Ea.		33.81		33.81	guides, hinges and pulls, laminated
	Clean	Ea.	.57	14.55		15.12	interior.
	Paint	Ea.	1.50	43.50		45	
	Minimum Charge	Job		236		236	

Cabinets and Countertops

Base Cabinets	Unit	Material	Labor	Equip.	Total	Specification
30″ Wide Sink Base						
Demolish	Ea.		25.50		25.50	Includes material and labor to install
Install	Ea.	605	44		649	premium hardwood modular unit with
Demolish and Install	Ea.	605	69.50		674.50	solid hardwood faced frames, drawer
Reinstall	Ea.		35.23		35.23	fronts, and door panels, steel drawer
Clean	Ea.	.75	18.20		18.95	guides, hinges, pulls and laminated
Paint	Ea.	1.87	49		50.87	interior.
Minimum Charge	Job		236		236	
36″ Wide Sink Base						
Demolish	Ea.		31		31	Includes material and labor to install
Install	Ea.	680	46.50		726.50	premium hardwood modular unit with
Demolish and Install	Ea.	680	77.50		757.50	solid hardwood faced frames, drawer
Reinstall	Ea.		37.14		37.14	fronts, and door panels, steel drawer
Clean	Ea.	.84	21.50		22.34	guides, hinges, pulls and laminated
Paint	Ea.	2.27	65		67.27	interior.
Minimum Charge	Job		236		236	
36″ Blind Corner						
Demolish	Ea.		31		31	Includes material and labor to install
Install	Ea.	1075	52.50		1127.50	premium hardwood modular unit with
Demolish and Install	Ea.	1075	83.50		1158.50	solid hardwood faced frames, drawer
Reinstall	Ea.		41.88		41.88	fronts, and door panels, steel drawer
Clean	Ea.	.84	21.50		22.34	guides, hinges, pulls and laminated
Paint	Ea.	2.27	65		67.27	interior.
Minimum Charge	Job		236		236	
42″ w, 2 Door, 2 Drawer						
Demolish	Ea.		36.50		36.50	Includes material and labor to install
Install	Ea.	1100	47.50		1147.50	premium hardwood modular unit with
Demolish and Install	Ea.	1100	84		1184	solid hardwood faced frames, drawer
Reinstall	Ea.		38.08		38.08	fronts, and door panels, steel drawer
Clean	Ea.	1.07	26		27.07	guides, hinges, pulls and laminated
Paint	Ea.	2.60	78		80.60	interior.
Minimum Charge	Job		236		236	
Clean Interior						
Clean	L.F.	.30	5.50		5.80	Includes labor and materials to clean
Minimum Charge	Job		182		182	the interior of a 27″ cabinet.
Island						
24″ Wide						
Demolish	Ea.		20		20	Includes material and labor to install
Install	Ea.	705	39.50		744.50	custom modular unit with solid
Demolish and Install	Ea.	705	59.50		764.50	hardwood faced frames, hardwood
Reinstall	Ea.		31.41		31.41	door frames and drawer fronts,
Clean	Ea.	.84	21.50		22.34	hardwood veneer on raised door
Paint	Ea.	2.23	56		58.23	panels, hinges and pulls.
Minimum Charge	Job		236		236	
30″ Wide						
Demolish	Ea.		25.50		25.50	Includes material and labor to install
Install	Ea.	810	49		859	custom modular unit with solid
Demolish and Install	Ea.	810	74.50		884.50	hardwood faced frames, hardwood
Reinstall	Ea.		39.27		39.27	door frames and drawer fronts,
Clean	Ea.	1.07	26		27.07	hardwood veneer on raised door
Paint	Ea.	2.82	65		67.82	panels, hinges and pulls.
Minimum Charge	Job		236		236	

Cabinets and Countertops

Base Cabinets		Unit	Material	Labor	Equip.	Total	Specification
36" Wide							
	Demolish	Ea.		31		31	Includes material and labor to install
	Install	Ea.	935	59		994	custom modular unit with solid
	Demolish and Install	Ea.	935	90		1025	hardwood faced frames, hardwood
	Reinstall	Ea.		47.12		47.12	door frames and drawer fronts,
	Clean	Ea.	1.25	33		34.25	hardwood veneer on raised door
	Paint	Ea.	3.40	78		81.40	panels, hinges and pulls.
	Minimum Charge	Job		236		236	
48" Wide							
	Demolish	Ea.		40.50		40.50	Includes material and labor to install
	Install	Ea.	1400	78.50		1478.50	custom modular unit with solid
	Demolish and Install	Ea.	1400	119		1519	hardwood faced frames, hardwood
	Reinstall	Ea.		62.83		62.83	door frames and drawer fronts,
	Clean	Ea.	1.50	40.50		42	hardwood veneer on raised door
	Paint	Ea.	3.89	98		101.89	panels, hinges and pulls.
	Minimum Charge	Job		236		236	
60" Wide							
	Demolish	Ea.		45		45	Includes material and labor to install
	Install	Ea.	1625	98		1723	custom modular unit with solid
	Demolish and Install	Ea.	1625	143		1768	hardwood faced frames, hardwood
	Reinstall	Ea.		78.53		78.53	door frames and drawer fronts,
	Clean	Ea.	1.87	52		53.87	hardwood veneer on raised door
	Paint	Ea.	3.89	130		133.89	panels, hinges and pulls.
	Minimum Charge	Job		236		236	
Clean Interior							
	Clean	L.F.	.30	5.50		5.80	Includes labor and materials to clean
	Minimum Charge	Job		182		182	the interior of a 27" cabinet.
Liquor Bar							
	Demolish	L.F.		10.10		10.10	Includes material and labor to install
	Install	L.F.	222	47		269	custom modular unit with solid
	Demolish and Install	L.F.	222	57.10		279.10	hardwood faced frames, hardwood
	Reinstall	L.F.		37.70		37.70	door frames and drawer fronts,
	Clean	L.F.	.30	7.60		7.90	hardwood veneer on raised door
	Paint	L.F.	6	17.80		23.80	panels, hinges and pulls.
	Minimum Charge	Job		236		236	
Clean Interior							
	Clean	L.F.	.30	5.50		5.80	Includes labor and materials to clean
	Minimum Charge	Job		182		182	the interior of a 27" cabinet.
Seal Interior							
	Install	L.F.	.57	6.40		6.97	Includes labor and materials to seal a
	Minimum Charge	Job		196		196	cabinet interior.
Back Bar							
	Demolish	L.F.		10.10		10.10	Includes material and labor to install
	Install	L.F.	222	47		269	custom modular unit with solid
	Demolish and Install	L.F.	222	57.10		279.10	hardwood faced frames, hardwood
	Reinstall	L.F.		37.70		37.70	door frames and drawer fronts,
	Clean	L.F.	.30	7.60		7.90	hardwood veneer on raised door
	Paint	L.F.	6	17.80		23.80	panels, hinges and pulls.
	Minimum Charge	Job		236		236	
Clean Interior							
	Clean	L.F.	.30	5.50		5.80	Includes labor and materials to clean
	Minimum Charge	Job		182		182	the interior of a 27" cabinet.

Cabinets and Countertops

Base Cabinets

Base Cabinets		Unit	Material	Labor	Equip.	Total	Specification
Seal Interior							
	Install	L.F.	.57	6.40		6.97	Includes labor and materials to seal a
	Minimum Charge	Job		196		196	cabinet interior.
Storage (garage)							
	Demolish	L.F.		10.10		10.10	Includes material and labor to install
	Install	L.F.	166	47		213	good modular unit with melamine
	Demolish and Install	L.F.	166	57.10		223.10	laminate, textured colors or wood
	Reinstall	L.F.		37.70		37.70	grain print finish including hinges and
	Clean	L.F.	.30	7.60		7.90	pulls.
	Paint	L.F.	6	17.80		23.80	
	Minimum Charge	Job		236		236	
Clean Interior							
	Clean	L.F.	.30	5.50		5.80	Includes labor and materials to clean
	Minimum Charge	Job		182		182	the interior of a 27" cabinet.
Seal Interior							
	Install	L.F.	.57	6.40		6.97	Includes labor and materials to seal a
	Minimum Charge	Job		196		196	cabinet interior.
Built-in Desk							
	Demolish	L.F.		10.10		10.10	Includes material and labor to install
	Install	L.F.	222	47		269	custom modular unit with solid
	Demolish and Install	L.F.	222	57.10		279.10	hardwood faced frames, hardwood
	Reinstall	L.F.		37.70		37.70	door frames and drawer fronts,
	Clean	L.F.	.30	7.60		7.90	hardwood veneer on raised door
	Paint	L.F.	6	17.80		23.80	panels, hinges and pulls.
	Minimum Charge	Job		236		236	
Built-in Bookcase							
	Demolish	L.F.		13.45		13.45	Includes material and labor to install
	Install	L.F.	45.50	47		92.50	custom modular bookcase unit with
	Demolish and Install	L.F.	45.50	60.45		105.95	clear pine faced frames, shelves 12"
	Reinstall	L.F.		37.70		37.70	O.C., 7' high, 8" deep.
	Clean	L.F.	.59	13		13.59	
	Paint	L.F.	1.75	30		31.75	
	Minimum Charge	Job		236		236	
Strip and Refinish							
	Install	L.F.	5.50	49		54.50	Includes labor and material to strip,
	Minimum Charge	Job		196		196	prep and refinish exterior of cabinets.
Stain							
	Install	L.F.	.76	1.82		2.58	Includes labor and material to stain
	Minimum Charge	Job		236		236	door and exterior of cabinets.

Countertop

Countertop		Unit	Material	Labor	Equip.	Total	Specification
Laminated							
With Splash							
	Demolish	L.F.		6.75		6.75	Includes material and labor to install
	Install	L.F.	48.50	16.85		65.35	one piece laminated top with 4"
	Demolish and Install	L.F.	48.50	23.60		72.10	backsplash.
	Reinstall	L.F.		13.46		13.46	
	Clean	L.F.	.03	.91		.94	
	Minimum Charge	Job		236		236	

229

Cabinets and Countertops

Countertop		Unit	Material	Labor	Equip.	Total	Specification
Roll Top							
	Demolish	L.F.		6.75		6.75	Includes material and labor to install
	Install	L.F.	13.85	15.70		29.55	one piece laminated top with rolled
	Demolish and Install	L.F.	13.85	22.45		36.30	drip edge (post formed) and
	Reinstall	L.F.		12.57		12.57	backsplash.
	Clean	L.F.	.06	1.38		1.44	
	Minimum Charge	Job		236		236	
Ceramic Tile							
	Demolish	S.F.		3.37		3.37	Cost includes material and labor to
	Install	S.F.	23	9.40		32.40	install 4-1/4" x 4-1/4" to 6" x 6"
	Demolish and Install	S.F.	23	12.77		35.77	glazed tile set in mortar bed and grout
	Clean	S.F.	.02	.69		.71	on particle board substrate.
	Minimum Charge	Job		222		222	
Cultured Marble							
	Demolish	L.F.		6.75		6.75	Includes material and labor to install
	Install	L.F.	60.50	47		107.50	marble countertop, 24" wide, no
	Demolish and Install	L.F.	60.50	53.75		114.25	backsplash.
	Reinstall	L.F.		37.60		37.60	
	Clean	L.F.	.06	1.38		1.44	
	Minimum Charge	Job		222		222	
With Splash							
	Demolish	L.F.		6.75		6.75	Includes material and labor to install
	Install	L.F.	70	47		117	marble countertop, 24" wide, with
	Demolish and Install	L.F.	70	53.75		123.75	backsplash.
	Reinstall	L.F.		37.60		37.60	
	Clean	L.F.	.06	1.38		1.44	
	Minimum Charge	Job		222		222	
Quarry Tile							
	Demolish	S.F.		3.37		3.37	Includes material and labor to install
	Install	S.F.	8.60	9.40		18	quarry tile countertop with no
	Demolish and Install	S.F.	8.60	12.77		21.37	backsplash.
	Clean	S.F.	.02	.69		.71	
	Minimum Charge	Job		222		222	
With Splash							
	Demolish	S.F.		3.37		3.37	Includes material and labor to install
	Install	S.F.	11.65	10.45		22.10	quarry tile countertop with backsplash.
	Demolish and Install	S.F.	11.65	13.82		25.47	
	Clean	S.F.	.02	.69		.71	
	Minimum Charge	Job		222		222	
Butcher Block							
	Demolish	S.F.		3.37		3.37	Includes material and labor to install
	Install	S.F.	50.50	8.40		58.90	solid laminated maple countertop with
	Demolish and Install	S.F.	50.50	11.77		62.27	no backsplash.
	Reinstall	S.F.		6.73		6.73	
	Clean	S.F.	.04	.28		.32	
	Minimum Charge	Job		236		236	
Solid Surface							
	Demolish	L.F.		6.75		6.75	Includes material and labor to install
	Install	L.F.	59	33.50		92.50	solid surface countertop 22" deep.
	Demolish and Install	L.F.	59	40.25		99.25	
	Reinstall	L.F.		26.93		26.93	
	Clean	L.F.	.06	1.38		1.44	
	Minimum Charge	Job		236		236	

Vanity Cabinets	Unit	Material	Labor	Equip.	Total	Specification
Good Grade Laminated						
24" Wide, 2 Door						
Demolish	Ea.		20		20	Includes material and labor to install
Install	Ea.	305	47		352	modular unit with melamine laminated
Demolish and Install	Ea.	305	67		372	to particle board, textured colors or
Reinstall	Ea.		37.70		37.70	wood grain print finish, hinges and
Clean	Ea.	.57	14.55		15.12	pulls.
Paint	Ea.	1.50	32.50		34	
Minimum Charge	Job		236		236	
30" Wide, 2 Door						
Demolish	Ea.		25.50		25.50	Includes material and labor to install
Install	Ea.	365	59		424	modular unit with melamine laminated
Demolish and Install	Ea.	365	84.50		449.50	to particle board, textured colors or
Reinstall	Ea.		47.12		47.12	wood grain print finish, hinges and
Clean	Ea.	.75	18.20		18.95	pulls.
Paint	Ea.	1.87	39		40.87	
Minimum Charge	Job		236		236	
36" Wide, 2 Door						
Demolish	Ea.		31		31	Includes material and labor to install
Install	Ea.	330	70.50		400.50	modular unit with melamine laminated
Demolish and Install	Ea.	330	101.50		431.50	to particle board, textured colors or
Reinstall	Ea.		56.56		56.56	wood grain print finish, hinges and
Clean	Ea.	.84	21.50		22.34	pulls.
Paint	Ea.	2.27	49		51.27	
Minimum Charge	Job		236		236	
42" w, 3 Doors / Drawer						
Demolish	Ea.		36.50		36.50	Includes material and labor to install
Install	Ea.	600	76		676	modular unit with melamine laminated
Demolish and Install	Ea.	600	112.50		712.50	to particle board, textured colors or
Reinstall	Ea.		60.90		60.90	wood grain print finish, hinges and
Clean	Ea.	1.07	26		27.07	pulls.
Paint	Ea.	2.60	56		58.60	
Minimum Charge	Job		236		236	
48" w, 3 Doors / Drawer						
Demolish	Ea.		40.50		40.50	Includes material and labor to install
Install	Ea.	490	82.50		572.50	modular unit with melamine laminated
Demolish and Install	Ea.	490	123		613	to particle board, textured colors or
Reinstall	Ea.		65.96		65.96	wood grain print finish, hinges and
Clean	Ea.	1.50	40.50		42	pulls.
Paint	Ea.	3.89	98		101.89	
Minimum Charge	Job		236		236	
Clean Interior						
Install	L.F.	.30	7.60		7.90	Includes labor and materials to clean
Minimum Charge	Job		182		182	base cabinetry.
Better Grade Wood						
24" Wide, 2 Door						
Demolish	Ea.		20		20	Includes material and labor to install
Install	Ea.	405	47		452	custom modular unit with solid
Demolish and Install	Ea.	405	67		472	hardwood faced frames, hardwood
Reinstall	Ea.		37.70		37.70	door frames and drawer fronts,
Clean	Ea.	.57	14.55		15.12	hardwood veneer on raised door
Paint	Ea.	1.50	35.50		37	panels, hinges and pulls.
Minimum Charge	Job		236		236	

Vanity Cabinets	Unit	Material	Labor	Equip.	Total	Specification
30" Wide, 2 Door						
Demolish	Ea.		25.50		25.50	Includes material and labor to install
Install	Ea.	490	59		549	custom modular unit with solid
Demolish and Install	Ea.	490	84.50		574.50	hardwood faced frames, hardwood
Reinstall	Ea.		47.12		47.12	door frames and drawer fronts,
Clean	Ea.	.75	18.20		18.95	hardwood veneer on raised door
Paint	Ea.	1.87	43.50		45.37	panels, hinges and pulls.
Minimum Charge	Job		236		236	
36" Wide, 2 Door						
Demolish	Ea.		31		31	Includes material and labor to install
Install	Ea.	440	70.50		510.50	custom modular unit with solid
Demolish and Install	Ea.	440	101.50		541.50	hardwood faced frames, hardwood
Reinstall	Ea.		56.56		56.56	door frames and drawer fronts,
Clean	Ea.	.84	21.50		22.34	hardwood veneer on raised door
Paint	Ea.	2.27	56		58.27	panels, hinges and pulls.
Minimum Charge	Job		236		236	
42" w, 3 Doors / Drawer						
Demolish	Ea.		36.50		36.50	Includes material and labor to install
Install	Ea.	600	76		676	custom modular unit with solid
Demolish and Install	Ea.	600	112.50		712.50	hardwood faced frames, hardwood
Reinstall	Ea.		60.90		60.90	door frames and drawer fronts,
Clean	Ea.	1.07	26		27.07	hardwood veneer on raised door
Paint	Ea.	2.60	65		67.60	panels, hinges and pulls.
Minimum Charge	Job		236		236	
48" w, 3 Doors / Drawer						
Demolish	Ea.		40.50		40.50	Includes material and labor to install
Install	Ea.	655	82.50		737.50	custom modular unit with solid
Demolish and Install	Ea.	655	123		778	hardwood faced frames, hardwood
Reinstall	Ea.		65.96		65.96	door frames and drawer fronts,
Clean	Ea.	1.50	40.50		42	hardwood veneer on raised door
Paint	Ea.	3.89	98		101.89	panels, hinges and pulls.
Minimum Charge	Job		236		236	
Clean Interior						
Install	L.F.	.30	7.60		7.90	Includes labor and materials to clean
Minimum Charge	Job		182		182	base cabinetry.
Premium Grade Wood						
24" Wide, 2 Door						
Demolish	Ea.		20		20	Includes material and labor to install
Install	Ea.	440	47		487	premium modular unit with solid
Demolish and Install	Ea.	440	67		507	hardwood faced frames, hardwood
Reinstall	Ea.		37.70		37.70	door frames and drawer fronts,
Clean	Ea.	.57	14.55		15.12	hardwood veneer on raised door
Paint	Ea.	1.50	39		40.50	panels, hinges and pulls.
Minimum Charge	Job		236		236	
30" Wide, 2 Door						
Demolish	Ea.		25.50		25.50	Includes material and labor to install
Install	Ea.	600	59		659	premium modular unit with solid
Demolish and Install	Ea.	600	84.50		684.50	hardwood faced frames, hardwood
Reinstall	Ea.		47.12		47.12	door frames and drawer fronts,
Clean	Ea.	.75	18.20		18.95	hardwood veneer on raised door
Paint	Ea.	1.87	49		50.87	panels, hinges and pulls.
Minimum Charge	Job		236		236	

Cabinets and Countertops

Vanity Cabinets	Unit	Material	Labor	Equip.	Total	Specification
36″ Wide, 2 Door						
Demolish	Ea.		31		31	Includes material and labor to install
Install	Ea.	610	70.50		680.50	premium modular unit with solid
Demolish and Install	Ea.	610	101.50		711.50	hardwood faced frames, hardwood
Reinstall	Ea.		56.56		56.56	door frames and drawer fronts,
Clean	Ea.	.84	21.50		22.34	hardwood veneer on raised door
Paint	Ea.	2.27	56		58.27	panels, hinges and pulls.
Minimum Charge	Job		236		236	
42″ w, 3 Doors / Drawer						
Demolish	Ea.		36.50		36.50	Includes material and labor to install
Install	Ea.	640	76		716	premium modular unit with solid
Demolish and Install	Ea.	640	112.50		752.50	hardwood faced frames, hardwood
Reinstall	Ea.		60.90		60.90	door frames and drawer fronts,
Clean	Ea.	1.07	26		27.07	hardwood veneer on raised door
Paint	Ea.	2.60	65		67.60	panels, hinges and pulls.
Minimum Charge	Job		236		236	
48″ w, 3 Doors / Drawer						
Demolish	Ea.		40.50		40.50	Includes material and labor to install
Install	Ea.	700	82.50		782.50	premium modular unit with solid
Demolish and Install	Ea.	700	123		823	hardwood faced frames, hardwood
Reinstall	Ea.		65.96		65.96	door frames and drawer fronts,
Clean	Ea.	1.50	40.50		42	hardwood veneer on raised door
Paint	Ea.	3.89	98		101.89	panels, hinges and pulls.
Minimum Charge	Job		236		236	
Clean Interior						
Install	L.F.	.30	7.60		7.90	Includes labor and materials to clean
Minimum Charge	Job		182		182	base cabinetry.
Premium Hardwood						
24″ Wide, 2 Door						
Demolish	Ea.		20		20	Includes material and labor to install
Install	Ea.	475	47		522	premium modular unit with solid
Demolish and Install	Ea.	475	67		542	hardwood faced frames, hardwood
Reinstall	Ea.		37.70		37.70	door frames and drawer fronts,
Clean	Ea.	.57	14.55		15.12	hardwood veneer on raised door
Paint	Ea.	1.50	43.50		45	panels, hinges and pulls.
Minimum Charge	Job		236		236	
30″ Wide, 2 Door						
Demolish	Ea.		25.50		25.50	Includes material and labor to install
Install	Ea.	750	59		809	premium modular unit with solid
Demolish and Install	Ea.	750	84.50		834.50	hardwood faced frames, hardwood
Reinstall	Ea.		47.12		47.12	door frames and drawer fronts,
Clean	Ea.	.75	18.20		18.95	hardwood veneer on raised door
Paint	Ea.	1.87	49		50.87	panels, hinges and pulls.
Minimum Charge	Job		236		236	
36″ Wide, 2 Door						
Demolish	Ea.		31		31	Includes material and labor to install
Install	Ea.	1050	70.50		1120.50	premium modular unit with solid
Demolish and Install	Ea.	1050	101.50		1151.50	hardwood faced frames, hardwood
Reinstall	Ea.		56.56		56.56	door frames and drawer fronts,
Clean	Ea.	.84	21.50		22.34	hardwood veneer on raised door
Paint	Ea.	2.27	65		67.27	panels, hinges and pulls.
Minimum Charge	Job		236		236	

Cabinets and Countertops

Vanity Cabinets

	Unit	Material	Labor	Equip.	Total	Specification
42″ w, 3 Doors / Drawer						Includes material and labor to install
Demolish	Ea.		36.50		36.50	premium modular unit with solid
Install	Ea.	670	76		746	hardwood faced frames, hardwood
Demolish and Install	Ea.	670	112.50		782.50	door frames and drawer fronts,
Reinstall	Ea.		60.90		60.90	hardwood veneer on raised door
Clean	Ea.	1.07	26		27.07	panels, hinges and pulls.
Paint	Ea.	2.60	78		80.60	
Minimum Charge	Job		236		236	
48″ w, 3 Doors / Drawer						Includes material and labor to install
Demolish	Ea.		40.50		40.50	premium modular unit with solid
Install	Ea.	750	82.50		832.50	hardwood faced frames, hardwood
Demolish and Install	Ea.	750	123		873	door frames and drawer fronts,
Reinstall	Ea.		65.96		65.96	hardwood veneer on raised door
Clean	Ea.	1.50	40.50		42	panels, hinges and pulls.
Paint	Ea.	3.89	98		101.89	
Minimum Charge	Job		236		236	
Clean Interior						Includes labor and materials to clean
Install	L.F.	.30	7.60		7.90	base cabinetry.
Minimum Charge	Job		182		182	
Strip and Refinish						Includes labor and material to strip,
Install	L.F.	5.50	49		54.50	prep and refinish exterior of cabinets.
Minimum Charge	Job		196		196	

Vanity Cabinet Tops

	Unit	Material	Labor	Equip.	Total	Specification
Cultured Marble Top						Includes material and labor to install
Demolish	L.F.		6.75		6.75	cultured marble vanity top with
Install	L.F.	56.50	27.50		84	integral sink, 22″ deep.
Demolish and Install	L.F.	56.50	34.25		90.75	
Reinstall	L.F.		22.12		22.12	
Clean	L.F.	.06	1.38		1.44	
Minimum Charge	Job		222		222	
Laminated Top						Includes material and labor to install
Demolish	L.F.		4.04		4.04	one piece laminated top with 4″
Install	L.F.	48.50	16.85		65.35	backsplash.
Demolish and Install	L.F.	48.50	20.89		69.39	
Reinstall	L.F.		13.46		13.46	
Clean	L.F.	.03	.91		.94	
Minimum Charge	Job		236		236	
Solid Surface						Includes material and labor to install
Demolish	L.F.		6.75		6.75	solid surface vanity top with integral
Install	L.F.	87.50	35		122.50	sink installed, 22″ deep.
Demolish and Install	L.F.	87.50	41.75		129.25	
Reinstall	L.F.		27.92		27.92	
Clean	L.F.	.06	1.38		1.44	
Minimum Charge	Job		236		236	

Roller Handle

Paint Brush

Paint Preparation, Walls		Unit	Material	Labor	Equip.	Total	Specification
Cover / Protect Floors							
	Install	S.F.	.03	.15		.18	Includes material and labor to install
	Minimum Charge	Job		196		196	plastic masking sheet on large
							uninterrupted areas.
Cover / Protect Walls							
	Install	S.F.	.03	.15		.18	Includes material and labor to install
	Minimum Charge	Job		196		196	plastic masking sheet on walls.
Clean Walls							
Light							
	Clean	S.F.		.12		.12	Includes labor to wash gypsum
	Minimum Charge	Job		196		196	drywall or plaster wall surfaces.
Heavy							
	Clean	S.F.		.16		.16	Includes labor and material for heavy
	Minimum Charge	Job		196		196	cleaning (multiple applications) with
							detergent and solvent.
Prep Walls							
	Paint	S.F.	.04	.20		.24	Includes labor and material to prepare
	Minimum Charge	Job		196		196	for painting including scraping,
							patching and puttying.
Sand Walls							
	Paint	S.F.		.13		.13	Includes labor to sand gypsum drywall
	Minimum Charge	Job		196		196	wall surfaces.
Seal Walls							
Large Area							
	Paint	S.F.	.07	.29		.36	Includes labor and material to paint
	Minimum Charge	Job		196		196	with primer / sealer by roller.
Spot							
	Paint	S.F.	.07	.34		.41	Includes labor and material to seal
	Minimum Charge	Job		196		196	existing drywall with shellac-based
							material.

For customer support on your Contractor's Pricing Guide: Residential Repair & Remodeling, call 800.448.8182.

Painting

Paint Preparation, Ceilings		Unit	Material	Labor	Equip.	Total	Specification
Cover / Protect Floors							
	Install	S.F.	.03	.15		.18	Includes material and labor to install
	Minimum Charge	Job		196		196	plastic masking sheet on large uninterrupted areas.
Cover / Protect Walls							
	Install	S.F.	.03	.15		.18	Includes material and labor to install
	Minimum Charge	Job		196		196	plastic masking sheet on walls.
Prep Ceiling							
	Paint	S.F.	.07	.83		.90	Includes labor and material to fill in
	Minimum Charge	Job		196		196	hairline cracks up to 1/8" with filler, sand and prep for paint.
Sand Ceiling							
	Paint	S.F.		.18		.18	Includes labor to sand gypsum drywall
	Minimum Charge	Job		196		196	ceiling surfaces.
Seal Ceiling							
Large Area							
	Paint	S.F.	.08	.30		.38	Includes labor and material to paint
	Minimum Charge	Job		196		196	ceiling, one coat flat latex.
Spot							
	Paint	S.F.	.07	.34		.41	Includes labor and material to seal
	Minimum Charge	Job		196		196	existing drywall with shellac-based material.

Paint / Texture, Walls		Unit	Material	Labor	Equip.	Total	Specification
Texture Walls							
Spray							
	Install	S.F.	.04	.57		.61	Includes labor and material to install
	Minimum Charge	Job		196		196	by spray, texture finish.
Trowel							
	Install	S.F.	.45	.37		.82	Includes labor and material to apply
	Minimum Charge	Job		196		196	hand troweled texture.
Paint Walls 1 Coat							
	Paint	S.F.	.08	.30		.38	Includes labor and material to paint,
	Minimum Charge	Job		196		196	one coat flat latex.
Paint Walls 2 Coats							
	Paint	S.F.	.17	.49		.66	Includes labor and material to paint,
	Minimum Charge	Job		196		196	two coats flat latex.
Paint Walls 3 Coats							
	Paint	S.F.	.24	.60		.84	Includes labor and material to paint,
	Minimum Charge	Job		196		196	three coats flat latex.
Clean and Seal							
	Paint	S.F.	.42	1.04		1.46	Includes labor and material to do light
	Minimum Charge	Job		196		196	cleaning, sealing with 2 coats of latex paint.

For customer support on your Contractor's Pricing Guide: Residential Repair & Remodeling, call 800.448.8182.

Paint / Texture, Ceilings		Unit	Material	Labor	Equip.	Total	Specification
Texture Ceiling							
Spray							
	Install	S.F.	.04	.57		.61	Includes labor and material to install
	Minimum Charge	Job		196		196	by spray, texture finish.
Trowel							
	Install	S.F.	.45	.37		.82	Includes labor and material to apply
	Minimum Charge	Job		196		196	hand troweled texture.
Paint Ceiling, 1 Coat							
	Paint	S.F.	.08	.30		.38	Includes labor and material to paint
	Minimum Charge	Job		196		196	ceiling, one coat flat latex.
Paint Ceiling, 2 Coats							
	Paint	S.F.	.25	.49		.74	Includes labor and material to paint
	Minimum Charge	Job		196		196	two coats flat latex paint.
Paint Ceiling, 3 Coats							
	Paint	S.F.	.39	.58		.97	Includes labor and material to paint
	Minimum Charge	Job		196		196	ceiling, three coats flat latex.
Clean and Seal							
	Paint	S.F.	.39	.82		1.21	Includes labor and material for light
	Minimum Charge	Job		196		196	cleaning, sealing and painting 3 coats
							of flat latex paint.

237

For customer support on your Contractor's Pricing Guide: Residential Repair & Remodeling, call 800.448.8182.

Plank Flooring	**Wood Strip Flooring**	**Tile Flooring**

Wood Plank Flooring	Unit	Material	Labor	Equip.	Total	Specification
Maple						
2-1/4"						
Demolish	S.F.		1.62		1.62	Includes material and labor to install
Install	S.F.	4.56	2.35		6.91	unfinished T&G maple flooring,
Demolish and Install	S.F.	4.56	3.97		8.53	25/32" thick, 2-1/4" wide, random
Reinstall	S.F.		1.88		1.88	3' to 16' lengths including felt
Clean	S.F.		.45		.45	underlayment, nailed in place over
Paint	S.F.	.24	2.80		3.04	prepared subfloor.
Minimum Charge	Job		236		236	
3-1/4"						
Demolish	S.F.		1.62		1.62	Includes material and labor to install
Install	S.F.	5.25	2.35		7.60	unfinished T&G maple flooring,
Demolish and Install	S.F.	5.25	3.97		9.22	33/32" thick, 3-1/4" wide, random
Reinstall	S.F.		1.88		1.88	3' to 16' lengths including felt
Clean	S.F.		.45		.45	underlayment, nailed in place over
Paint	S.F.	.24	2.80		3.04	prepared subfloor.
Minimum Charge	Job		236		236	
Sand and Finish						
Install	S.F.	.24	.91		1.15	Includes labor and material to refinish
Minimum Charge	Job		236		236	wood floor.
Oak						
2-1/4"						
Demolish	S.F.		1.62		1.62	Includes material and labor to install
Install	S.F.	5.10	2.35		7.45	unfinished T&G oak flooring, 25/32"
Demolish and Install	S.F.	5.10	3.97		9.07	thick, 2-1/4" wide, random 3' to 16'
Reinstall	S.F.		1.88		1.88	lengths including felt underlayment,
Clean	S.F.		.45		.45	nailed in place over prepared
Paint	S.F.	.24	2.80		3.04	subfloor.
Minimum Charge	Job		236		236	
3-1/4"						
Demolish	S.F.		1.62		1.62	Includes material and labor to install
Install	S.F.	4.74	1.66		6.40	unfinished T&G oak flooring, 25/32"
Demolish and Install	S.F.	4.74	3.28		8.02	thick, 3-1/4" wide, random 3' to 16'
Reinstall	S.F.		1.33		1.33	lengths including felt underlayment,
Clean	S.F.		.45		.45	nailed in place over prepared
Paint	S.F.	.24	2.80		3.04	subfloor.
Minimum Charge	Job		236		236	
Sand and Finish						
Install	S.F.	.24	2.80		3.04	Includes labor and material to refinish
Minimum Charge	Job		236		236	wood floor.

239

Flooring

Wood Plank Flooring

Wood Plank Flooring	Unit	Material	Labor	Equip.	Total	Specification
Walnut						
3" to 7"						
Demolish	S.F.		1.62		1.62	Includes material and labor to install
Install	S.F.	14.30	1.66		15.96	unfinished T&G walnut flooring,
Demolish and Install	S.F.	14.30	3.28		17.58	25/32" thick, random 3' to 16'
Reinstall	S.F.		1.33		1.33	lengths including felt underlayment,
Clean	S.F.		.45		.45	nailed in place over prepared
Paint	S.F.	.24	2.80		3.04	subfloor.
Minimum Charge	Job		236		236	
Sand and Finish						
Install	S.F.	.24	2.80		3.04	Includes labor and material to refinish
Minimum Charge	Job		236		236	wood floor.
Pine						
2-1/4"						
Demolish	S.F.		1.62		1.62	Includes material and labor to install
Install	S.F.	8.35	1.77		10.12	unfinished T&G pine flooring, 25/32"
Demolish and Install	S.F.	8.35	3.39		11.74	thick, 2-1/4" wide, random 3' to 16'
Reinstall	S.F.		1.42		1.42	lengths including felt underlayment,
Clean	S.F.		.45		.45	nailed in place over prepared
Paint	S.F.	.24	.91		1.15	subfloor.
Minimum Charge	Job		236		236	
Sand and Finish						
Install	S.F.	.24	2.80		3.04	Includes labor and material to refinish
Minimum Charge	Job		236		236	wood floor.
Prefinished Oak						
2-1/4"						
Demolish	S.F.		1.62		1.62	Includes material and labor to install
Install	S.F.	5.30	2.35		7.65	prefinished T&G oak flooring, 2-1/4"
Demolish and Install	S.F.	5.30	3.97		9.27	wide, random 3' to 16' lengths
Reinstall	S.F.		1.88		1.88	including felt underlayment, nailed in
Clean	S.F.		.45		.45	place over prepared subfloor.
Minimum Charge	Job		236		236	
3-1/4"						
Demolish	S.F.		1.62		1.62	Includes material and labor to install
Install	S.F.	5.15	2.16		7.31	prefinished T&G oak flooring, 25/32"
Demolish and Install	S.F.	5.15	3.78		8.93	thick, 3" to 7" wide, random 3' to 16'
Reinstall	S.F.		1.73		1.73	lengths including felt underlayment,
Clean	S.F.		.45		.45	nailed in place over prepared
Minimum Charge	Job		236		236	subfloor.
Sand and Finish New						
Install	S.F.	.24	1.23		1.47	Includes labor and material to refinish
Minimum Charge	Job		236		236	wood floor.
Refinish Existing						
Install	S.F.	.24	.91		1.15	Includes labor and material to refinish
Minimum Charge	Job		236		236	wood floor.

Flooring

Wood Plank Flooring

Wood Plank Flooring		Unit	Material	Labor	Equip.	Total	Specification
Urethane Coat							
	Install	S.F.	.17	.21		.38	Includes labor and material to install
	Minimum Charge	Job		236		236	two coats of urethane, on existing floor.
Clean and Wax							
	Clean	S.F.	.04	.28		.32	Includes labor and material to wash
	Minimum Charge	Job		236		236	and wax hardwood.
Underlayment Per S.F.							
	Demolish	S.F.		.69		.69	Cost includes material and labor to
	Install	S.F.	1.05	.63		1.68	install 1/4" lauan subfloor, standard
	Demolish and Install	S.F.	1.05	1.32		2.37	interior grade, nailed every 6".
	Clean	S.F.		.35		.35	
	Minimum Charge	Job		236		236	

Wood Parquet Tile Floor

Wood Parquet Tile Floor		Unit	Material	Labor	Equip.	Total	Specification
Oak							
9" x 9"							
	Demolish	S.F.		1.16		1.16	Includes material and labor to install
	Install	S.F.	7.25	3.77		11.02	prefinished oak 9" x 9" parquet block
	Demolish and Install	S.F.	7.25	4.93		12.18	flooring, 5/16" thick, installed in
	Reinstall	S.F.		3.02		3.02	mastic.
	Clean	S.F.		.45		.45	
	Paint	S.F.	.24	2.80		3.04	
	Minimum Charge	Job		236		236	
13" x 13"							
	Demolish	S.F.		1.16		1.16	Includes material and labor to install
	Install	S.F.	10.60	2.95		13.55	prefinished oak 13" x 13" parquet
	Demolish and Install	S.F.	10.60	4.11		14.71	block flooring, 5/16" thick, installed in
	Reinstall	S.F.		2.36		2.36	mastic.
	Clean	S.F.		.45		.45	
	Paint	S.F.	.24	2.80		3.04	
	Minimum Charge	Job		236		236	
Cherry							
9" x 9"							
	Demolish	S.F.		1.16		1.16	Includes material and labor to install
	Install	S.F.	19.80	3.77		23.57	prefinished cherry 9" x 9" parquet
	Demolish and Install	S.F.	19.80	4.93		24.73	block flooring, 5/16" thick, installed in
	Reinstall	S.F.		3.77		3.77	mastic.
	Clean	S.F.		.45		.45	
	Paint	S.F.	.24	2.80		3.04	
	Minimum Charge	Job		236		236	
13" x 13"							
	Demolish	S.F.		1.16		1.16	Includes material and labor to install
	Install	S.F.	19.80	2.95		22.75	prefinished cherry 13" x 13" parquet
	Demolish and Install	S.F.	19.80	4.11		23.91	block flooring, 5/16" thick, installed in
	Reinstall	S.F.		2.95		2.95	mastic.
	Clean	S.F.		.45		.45	
	Paint	S.F.	.24	2.80		3.04	
	Minimum Charge	Job		236		236	

241

Wood Parquet Tile Floor		Unit	Material	Labor	Equip.	Total	Specification
Walnut							
9" x 9"							
	Demolish	S.F.		1.16		1.16	Includes material and labor to install
	Install	S.F.	7.15	3.77		10.92	prefinished walnut 9" x 9" parquet
	Demolish and Install	S.F.	7.15	4.93		12.08	flooring in mastic.
	Reinstall	S.F.		3.02		3.02	
	Clean	S.F.		.45		.45	
	Paint	S.F.	.24	2.80		3.04	
	Minimum Charge	Job		236		236	
13" x 13"							
	Demolish	S.F.		1.16		1.16	Includes material and labor to install
	Install	S.F.	12.70	2.95		15.65	prefinished walnut 13" x 13" parquet
	Demolish and Install	S.F.	12.70	4.11		16.81	block flooring, 5/16" thick, installed in
	Reinstall	S.F.		2.95		2.95	mastic.
	Clean	S.F.		.45		.45	
	Paint	S.F.	.24	2.80		3.04	
	Minimum Charge	Job		236		236	
Acrylic Impregnated							
Oak 12" x 12"							
	Demolish	S.F.		1.16		1.16	Includes material and labor to install
	Install	S.F.	11.75	3.37		15.12	red oak 12" x 12" acrylic
	Demolish and Install	S.F.	11.75	4.53		16.28	impregnated parquet block flooring,
	Reinstall	S.F.		3.37		3.37	5/16" thick, installed in mastic.
	Clean	S.F.		.45		.45	
	Minimum Charge	Job		236		236	
Cherry 12" x 12"							
	Demolish	S.F.		1.16		1.16	Includes material and labor to install
	Install	S.F.	23	3.37		26.37	prefinished cherry 12" x 12" acrylic
	Demolish and Install	S.F.	23	4.53		27.53	impregnated parquet block flooring,
	Reinstall	S.F.		3.37		3.37	5/16" thick, installed in mastic.
	Clean	S.F.		.45		.45	
	Minimum Charge	Job		236		236	
Ash 12" x 12"							
	Demolish	S.F.		1.16		1.16	Includes material and labor to install
	Install	S.F.	21	3.37		24.37	ash 12" x 12" acrylic impregnated
	Demolish and Install	S.F.	21	4.53		25.53	parquet block flooring, 5/16" thick in
	Reinstall	S.F.		3.37		3.37	mastic.
	Clean	S.F.		.45		.45	
	Minimum Charge	Job		236		236	
Teak							
	Demolish	S.F.		1.16		1.16	Includes material and labor to install
	Install	S.F.	9.75	2.95		12.70	prefinished teak parquet block
	Demolish and Install	S.F.	9.75	4.11		13.86	flooring, 5/16" thick, and installed in
	Reinstall	S.F.		2.95		2.95	mastic.
	Clean	S.F.		.45		.45	
	Paint	S.F.	.24	2.80		3.04	
	Minimum Charge	Job		236		236	
Clean and Wax							
	Clean	S.F.	.04	.28		.32	Includes labor and material to wash
	Minimum Charge	Job		236		236	and wax hardwood.

Flooring

Ceramic Tile Flooring	Unit	Material	Labor	Equip.	Total	Specification
Economy Grade						
Demolish	S.F.		1.20		1.20	Cost includes material and labor to
Install	S.F.	9	8.70		17.70	install 4-1/4″ x 4-1/4″ ceramic tile in
Demolish and Install	S.F.	9	9.90		18.90	mortar bed including grout.
Clean	S.F.		.69		.69	
Minimum Charge	Job		222		222	
Average Grade						
Demolish	S.F.		1.20		1.20	Cost includes material and labor to
Install	S.F.	11.40	8.90		20.30	install 4-1/4″ x 4-1/4″ ceramic tile in
Demolish and Install	S.F.	11.40	10.10		21.50	mortar bed including grout.
Clean	S.F.		.69		.69	
Minimum Charge	Job		222		222	
Premium Grade						
Demolish	S.F.		1.20		1.20	Cost includes material and labor to
Install	S.F.	14.85	8.90		23.75	install 4-1/4″ x 4-1/4″ ceramic tile in
Demolish and Install	S.F.	14.85	10.10		24.95	mortar bed including grout.
Clean	S.F.		.69		.69	
Minimum Charge	Job		222		222	
Re-grout						
Install	S.F.	.07	3.56		3.63	Includes material and labor to regrout
Minimum Charge	Job		222		222	tile floors.

Hard Tile Flooring	Unit	Material	Labor	Equip.	Total	Specification
Thick Set Paver						
Brick						
Demolish	S.F.		2.53		2.53	Cost includes material and labor to
Install	S.F.	1.61	20.50		22.11	install 6″ x 12″ adobe brick paver with
Demolish and Install	S.F.	1.61	23.03		24.64	1/2″ mortar joints.
Clean	S.F.		.69		.69	
Minimum Charge	Job		222		222	
Mexican Red						
Demolish	S.F.		2.53		2.53	Cost includes material and labor to
Install	S.F.	2.05	9.25		11.30	install 12″ x 12″ red Mexican paver
Demolish and Install	S.F.	2.05	11.78		13.83	tile in mortar bed with grout.
Clean	S.F.		.69		.69	
Minimum Charge	Job		222		222	
Saltillo						
Demolish	S.F.		1.20		1.20	Cost includes material and labor to
Install	S.F.	1.65	9.25		10.90	install 12″ x 12″ saltillo tile in mortar
Demolish and Install	S.F.	1.65	10.45		12.10	bed with grout.
Clean	S.F.		.69		.69	
Minimum Charge	Job		222		222	
Marble						
Premium Grade						
Demolish	S.F.		1.20		1.20	Cost includes material and labor to
Install	S.F.	21.50	19.35		40.85	install 3/8″ x 12″ x 12″ marble, in
Demolish and Install	S.F.	21.50	20.55		42.05	mortar bed with grout.
Clean	S.F.		.69		.69	
Minimum Charge	Job		222		222	

Hard Tile Flooring	Unit	Material	Labor	Equip.	Total	Specification
Terrazzo						
Gray Cement						
Demolish	S.F.		1.60		1.60	Cost includes material and labor to
Install	S.F.	4.08	10.90	4.65	19.63	install 1-3/4" terrazzo , #1 and #2
Demolish and Install	S.F.	4.08	12.50	4.65	21.23	chips in gray Portland cement.
Clean	S.F.		.69		.69	
Minimum Charge	Job		410	174	584	
White Cement						
Demolish	S.F.		1.60		1.60	Cost includes material and labor to
Install	S.F.	4.84	10.90	4.65	20.39	install 1-3/4" terrazzo, #1 and #2
Demolish and Install	S.F.	4.84	12.50	4.65	21.99	chips in white Portland cement.
Clean	S.F.		.69		.69	
Minimum Charge	Job		410	174	584	
Non-skid Gray						
Demolish	S.F.		1.60		1.60	Cost includes material and labor to
Install	S.F.	7.55	13.65	5.80	27	install 1-3/4" terrazzo, #1 and #2
Demolish and Install	S.F.	7.55	15.25	5.80	28.60	chips in gray Portland cement with
Clean	S.F.		.69		.69	light non-skid abrasive.
Minimum Charge	Job		410	174	584	
Non-skid White						
Demolish	S.F.		1.60		1.60	Cost includes material and labor to
Install	S.F.	8.40	13.65	5.80	27.85	install 1-3/4" terrazzo, #1 and #2
Demolish and Install	S.F.	8.40	15.25	5.80	29.45	chips in white Portland cement with
Clean	S.F.		.69		.69	light non-skid abrasive.
Minimum Charge	Job		410	174	584	
Gray w / Brass Divider						
Demolish	S.F.		1.62		1.62	Cost includes material and labor to
Install	S.F.	11.50	14.90	6.35	32.75	install 1-3/4" terrazzo, #1 and #2
Demolish and Install	S.F.	11.50	16.52	6.35	34.37	chips in gray Portland cement with
Clean	S.F.		.69		.69	brass strips 2' O.C. each way.
Minimum Charge	Job		410	174	584	
Clean						
Clean	S.F.	.07	.46		.53	Includes labor and material to clean
Minimum Charge	Job		410	174	584	terrazzo flooring.
Slate						
Demolish	S.F.		1.44		1.44	Includes material and labor to install
Install	S.F.	8.60	4.37		12.97	slate tile flooring in thin set with grout.
Demolish and Install	S.F.	8.60	5.81		14.41	
Clean	S.F.		.69		.69	
Minimum Charge	Job		222		222	
Granite						
Demolish	S.F.		1.19		1.19	Includes material and labor to install
Install	S.F.	29	4.45		33.45	granite tile flooring in thin set with
Demolish and Install	S.F.	29	5.64		34.64	grout.
Reinstall	S.F.		4.45		4.45	
Clean	S.F.		.69		.69	
Minimum Charge	Job		222		222	
Re-grout						
Install	S.F.	.07	3.56		3.63	Includes material and labor to regrout
Minimum Charge	Job		222		222	tile floors.

Flooring

Sheet Vinyl Per S.F.

Sheet Vinyl Per S.F.	Unit	Material	Labor	Equip.	Total	Specification
Economy Grade						
Demolish	S.F.		.58		.58	Includes material and labor to install
Install	S.F.	2.50	1.39		3.89	resilient sheet vinyl flooring.
Demolish and Install	S.F.	2.50	1.97		4.47	
Clean	S.F.		.45		.45	
Minimum Charge	Job		222		222	
Average Grade						
Demolish	S.F.		.58		.58	Includes material and labor to install
Install	S.F.	3.76	1.39		5.15	resilient sheet vinyl flooring.
Demolish and Install	S.F.	3.76	1.97		5.73	
Clean	S.F.		.45		.45	
Minimum Charge	Job		222		222	
Premium Grade						
Demolish	S.F.		.58		.58	Includes material and labor to install
Install	S.F.	3.40	1.39		4.79	resilient sheet vinyl flooring.
Demolish and Install	S.F.	3.40	1.97		5.37	
Clean	S.F.		.45		.45	
Minimum Charge	Job		222		222	
Clean and Wax						
Clean	S.F.	.04	.50		.54	Includes labor and material to wash
Minimum Charge	Job		222		222	and wax a vinyl tile floor.

Sheet Vinyl

Sheet Vinyl	Unit	Material	Labor	Equip.	Total	Specification
Economy Grade						
Demolish	S.Y.		5.20		5.20	Includes material and labor to install
Install	S.Y.	22.50	12.50		35	no wax sheet vinyl flooring.
Demolish and Install	S.Y.	22.50	17.70		40.20	
Clean	S.Y.		4.09		4.09	
Minimum Charge	Job		222		222	
Average Grade						
Demolish	S.Y.		5.20		5.20	Includes material and labor to install
Install	S.Y.	34	12.50		46.50	no wax sheet vinyl flooring.
Demolish and Install	S.Y.	34	17.70		51.70	
Clean	S.Y.		4.09		4.09	
Minimum Charge	Job		222		222	
Premium Grade						
Demolish	S.Y.		5.20		5.20	Includes material and labor to install
Install	S.Y.	30.50	12.50		43	no wax sheet vinyl flooring.
Demolish and Install	S.Y.	30.50	17.70		48.20	
Clean	S.Y.		4.09		4.09	
Minimum Charge	Job		222		222	
Clean and Wax						
Clean	S.F.	.04	.50		.54	Includes labor and material to wash
Minimum Charge	Job		222		222	and wax a vinyl tile floor.

Vinyl Tile

Vinyl Tile	Unit	Material	Labor	Equip.	Total	Specification
Economy Grade						
Demolish	S.F.		.81		.81	Includes material and labor to install
Install	S.F.	4.09	.89		4.98	no wax 12" x 12" vinyl tile.
Demolish and Install	S.F.	4.09	1.70		5.79	
Clean	S.F.		.45		.45	
Minimum Charge	Job		222		222	

For customer support on your Contractor's Pricing Guide: Residential Repair & Remodeling, call 800.448.8182.

Flooring

Vinyl Tile

Vinyl Tile		Unit	Material	Labor	Equip.	Total	Specification
Average Grade							Includes material and labor to install
	Demolish	S.F.		.81		.81	no wax 12″ x 12″ vinyl tile.
	Install	S.F.	3.54	.89		4.43	
	Demolish and Install	S.F.	3.54	1.70		5.24	
	Clean	S.F.		.45		.45	
	Minimum Charge	Job		222		222	
Premium Grade							Includes material and labor to install
	Demolish	S.F.		.81		.81	no wax 12″ x 12″ vinyl tile.
	Install	S.F.	7.20	.89		8.09	
	Demolish and Install	S.F.	7.20	1.70		8.90	
	Clean	S.F.		.45		.45	
	Minimum Charge	Job		222		222	
Clean and Wax							Includes labor and material to wash
	Clean	S.F.	.04	.50		.54	and wax a vinyl tile floor.
	Minimum Charge	Job		222		222	

Carpeting Per S.F.

Carpeting Per S.F.		Unit	Material	Labor	Equip.	Total	Specification
Economy Grade							Includes material and labor to install
	Demolish	S.F.		.09		.09	carpet including tack strips and hot
	Install	S.F.	1.33	.66		1.99	melt tape on seams.
	Demolish and Install	S.F.	1.33	.75		2.08	
	Clean	S.F.		.63		.63	
	Minimum Charge	Job		222		222	
Average Grade							Includes material and labor to install
	Demolish	S.F.		.09		.09	carpet including tack strips and hot
	Install	S.F.	1.67	.66		2.33	melt tape on seams.
	Demolish and Install	S.F.	1.67	.75		2.42	
	Clean	S.F.		.63		.63	
	Minimum Charge	Job		222		222	
Premium Grade							Includes material and labor to install
	Demolish	S.F.		.09		.09	carpet including tack strips and hot
	Install	S.F.	2.37	.66		3.03	melt tape on seams.
	Demolish and Install	S.F.	2.37	.75		3.12	
	Clean	S.F.		.63		.63	
	Minimum Charge	Job		222		222	
Indoor/Outdoor							Includes material and labor to install
	Demolish	S.F.		.09		.09	indoor-outdoor carpet.
	Install	S.F.	1.18	.66		1.84	
	Demolish and Install	S.F.	1.18	.75		1.93	
	Clean	S.F.		.63		.63	
	Minimum Charge	Job		222		222	
100% Wool							
Average Grade							Includes material and labor to install
	Demolish	S.F.		.09		.09	carpet including tack strips and hot
	Install	S.F.	8.40	.88		9.28	melt tape on seams.
	Demolish and Install	S.F.	8.40	.97		9.37	
	Clean	S.F.		.63		.63	
	Minimum Charge	Job		222		222	
Premium Grade							Includes material and labor to install
	Demolish	S.F.		.09		.09	carpet including tack strips and hot
	Install	S.F.	11.70	.88		12.58	melt tape on seams.
	Demolish and Install	S.F.	11.70	.97		12.67	
	Clean	S.F.		.63		.63	
	Minimum Charge	Job		222		222	

Flooring

Carpeting Per S.F.

Carpeting Per S.F.		Unit	Material	Labor	Equip.	Total	Specification
Luxury Grade							
	Demolish	S.F.		.09		.09	Includes material and labor to install
	Install	S.F.	12.55	.88		13.43	carpet including tack strips and hot
	Demolish and Install	S.F.	12.55	.97		13.52	melt tape on seams.
	Clean	S.F.		.63		.63	
	Minimum Charge	Job		222		222	
Berber							
Average Grade							
	Demolish	S.F.		.09		.09	Includes material and labor to install
	Install	S.F.	6.15	1.05		7.20	carpet including tack strips and hot
	Demolish and Install	S.F.	6.15	1.14		7.29	melt tape on seams.
	Clean	S.F.		.63		.63	
	Minimum Charge	Job		222		222	
Premium Grade							
	Demolish	S.F.		.09		.09	Includes material and labor to install
	Install	S.F.	8.40	1.05		9.45	carpet including tack strips and hot
	Demolish and Install	S.F.	8.40	1.14		9.54	melt tape on seams.
	Clean	S.F.		.63		.63	
	Minimum Charge	Job		222		222	

Carpeting

Carpeting		Unit	Material	Labor	Equip.	Total	Specification
Indoor/Outdoor							
	Demolish	S.Y.		1.35		1.35	Includes material and labor to install
	Install	S.Y.	10.55	5.95		16.50	indoor-outdoor carpet.
	Demolish and Install	S.Y.	10.55	7.30		17.85	
	Reinstall	S.Y.		4.74		4.74	
	Clean	S.Y.		5.70		5.70	
	Minimum Charge	Job		222		222	
Economy Grade							
	Demolish	S.Y.		1.35		1.35	Includes material and labor to install
	Install	S.Y.	11.95	5.95		17.90	carpet including tack strips and hot
	Demolish and Install	S.Y.	11.95	7.30		19.25	melt tape on seams.
	Reinstall	S.Y.		4.74		4.74	
	Clean	S.Y.		5.70		5.70	
	Minimum Charge	Job		222		222	
Average Grade							
	Demolish	S.Y.		1.35		1.35	Includes material and labor to install
	Install	S.Y.	15	5.95		20.95	carpet including tack strips and hot
	Demolish and Install	S.Y.	15	7.30		22.30	melt tape on seams.
	Reinstall	S.Y.		4.74		4.74	
	Clean	S.Y.		5.70		5.70	
	Minimum Charge	Job		222		222	
Premium Grade							
	Demolish	S.Y.		1.35		1.35	Includes material and labor to install
	Install	S.Y.	21.50	5.95		27.45	carpet including tack strips and hot
	Demolish and Install	S.Y.	21.50	7.30		28.80	melt tape on seams.
	Reinstall	S.Y.		4.74		4.74	
	Clean	S.Y.		5.70		5.70	
	Minimum Charge	Job		222		222	

For customer support on your Contractor's Pricing Guide: Residential Repair & Remodeling, call 800.448.8182.

Flooring

Carpeting	Unit	Material	Labor	Equip.	Total	Specification
100% Wool						
Average Grade						
Demolish	S.Y.		1.35		1.35	Includes material and labor to install
Install	S.Y.	75.50	7.95		83.45	wool carpet including tack strips and
Demolish and Install	S.Y.	75.50	9.30		84.80	hot melt tape on seams.
Reinstall	S.Y.		6.35		6.35	
Clean	S.Y.		5.70		5.70	
Minimum Charge	Job		222		222	
Premium Grade						
Demolish	S.Y.		1.35		1.35	Includes material and labor to install
Install	S.Y.	106	7.95		113.95	carpet including tack strips and hot
Demolish and Install	S.Y.	106	9.30		115.30	melt tape on seams.
Reinstall	S.Y.		6.35		6.35	
Clean	S.Y.		5.70		5.70	
Minimum Charge	Job		222		222	
Luxury Grade						
Demolish	S.Y.		1.35		1.35	Includes material and labor to install
Install	S.Y.	113	7.95		120.95	carpet including tack strips and hot
Demolish and Install	S.Y.	113	9.30		122.30	melt tape on seams.
Reinstall	S.Y.		6.35		6.35	
Clean	S.Y.		5.70		5.70	
Minimum Charge	Job		222		222	
Berber						
Average Grade						
Demolish	S.Y.		1.35		1.35	Includes material and labor to install
Install	S.Y.	55.50	9.45		64.95	berber carpet including tack strips and
Demolish and Install	S.Y.	55.50	10.80		66.30	hot melt tape on seams.
Reinstall	S.Y.		7.57		7.57	
Clean	S.Y.		5.70		5.70	
Minimum Charge	Job		222		222	
Premium Grade						
Demolish	S.Y.		1.35		1.35	Includes material and labor to install
Install	S.Y.	75.50	9.45		84.95	carpet including tack strips and hot
Demolish and Install	S.Y.	75.50	10.80		86.30	melt tape on seams.
Reinstall	S.Y.		7.57		7.57	
Clean	S.Y.		5.70		5.70	
Minimum Charge	Job		222		222	

Carpet Pad Per S.F.	Unit	Material	Labor	Equip.	Total	Specification
Urethane						
Demolish	S.F.		.03		.03	Includes material and labor to install
Install	S.F.	.80	.33		1.13	urethane carpet pad.
Demolish and Install	S.F.	.80	.36		1.16	
Minimum Charge	Job		222		222	
Foam Rubber Slab						
Demolish	S.F.		.03		.03	Includes material and labor to install
Install	S.F.	.65	.33		.98	carpet pad, 3/8" thick foam rubber.
Demolish and Install	S.F.	.65	.36		1.01	
Minimum Charge	Job		222		222	
Waffle						
Demolish	S.F.		.03		.03	Includes material and labor to install
Install	S.F.	.87	.04		.91	rubber waffle carpet pad.
Demolish and Install	S.F.	.87	.07		.94	
Minimum Charge	Job		222		222	

For customer support on your Contractor's Pricing Guide: Residential Repair & Remodeling, call 800.448.8182.

Flooring

Carpet Pad Per S.F.

	Unit	Material	Labor	Equip.	Total	Specification
Jute						
Demolish	S.F.		.03		.03	Includes material and labor to install
Install	S.F.	.85	.29		1.14	jute hair carpet pad.
Demolish and Install	S.F.	.85	.32		1.17	
Minimum Charge	Job		222		222	
Rebound						
Demolish	S.F.		.03		.03	Includes material and labor to install
Install	S.F.	.51	.25		.76	rebound carpet pad.
Demolish and Install	S.F.	.51	.28		.79	
Minimum Charge	Job		222		222	

Carpet Pad

	Unit	Material	Labor	Equip.	Total	Specification
Urethane						
Demolish	S.Y.		.23		.23	Includes material and labor to install
Install	S.Y.	7.25	2.97		10.22	urethane carpet pad.
Demolish and Install	S.Y.	7.25	3.20		10.45	
Reinstall	S.Y.		2.37		2.37	
Minimum Charge	Job		222		222	
Foam Rubber Slab						
Demolish	S.Y.		.23		.23	Includes material and labor to install
Install	S.Y.	5.85	2.97		8.82	rubber slab carpet pad.
Demolish and Install	S.Y.	5.85	3.20		9.05	
Reinstall	S.Y.		2.37		2.37	
Minimum Charge	Job		222		222	
Waffle						
Demolish	S.Y.		.23		.23	Includes material and labor to install
Install	S.Y.	7.85	.33		8.18	rubber waffle carpet pad.
Demolish and Install	S.Y.	7.85	.56		8.41	
Reinstall	S.Y.		.26		.26	
Minimum Charge	Job		222		222	
Jute						
Demolish	S.Y.		.23		.23	Includes material and labor to install
Install	S.Y.	7.60	2.62		10.22	jute hair carpet pad.
Demolish and Install	S.Y.	7.60	2.85		10.45	
Reinstall	S.Y.		2.09		2.09	
Minimum Charge	Job		222		222	
Rebound						
Demolish	S.Y.		.23		.23	Includes material and labor to install
Install	S.Y.	4.59	2.22		6.81	rebound carpet pad.
Demolish and Install	S.Y.	4.59	2.45		7.04	
Reinstall	S.Y.		1.77		1.77	
Minimum Charge	Job		222		222	

Stair Components

	Unit	Material	Labor	Equip.	Total	Specification
Carpeting						
Demolish	Ea.		2.45		2.45	Includes material and labor to install
Install	Riser	59	11.70		70.70	carpet and pad on stairs.
Demolish and Install	Riser	59	14.15		73.15	
Reinstall	Riser		9.36		9.36	
Clean	Ea.		5.70		5.70	
Minimum Charge	Job		222		222	

Tub/Shower

Stainless Steel Double Sink

Water Closet

Commode	Unit	Material	Labor	Equip.	Total	Specification
Floor Mounted w / Tank						
Demolish	Ea.		53		53	Includes material and labor to install a
Install	Ea.	223	180		403	toilet with valve, seat and cover.
Demolish and Install	Ea.	223	233		456	
Reinstall	Ea.		144.06		144.06	
Clean	Ea.	.15	18.20		18.35	
Minimum Charge	Job		265		265	
Wall w / Flush Valve						
Demolish	Ea.		53		53	Includes material and labor to install a
Install	Ea.	1125	165		1290	wall hung toilet with flush valve, seat
Demolish and Install	Ea.	1125	218		1343	and cover.
Reinstall	Ea.		131.64		131.64	
Clean	Ea.	.15	18.20		18.35	
Minimum Charge	Job		265		265	
Designer Floor Mount						
Demolish	Ea.		53		53	Includes material and labor to install a
Install	Ea.	925	180		1105	toilet with valve, seat and cover.
Demolish and Install	Ea.	925	233		1158	
Reinstall	Ea.		144.06		144.06	
Clean	Ea.	.15	18.20		18.35	
Minimum Charge	Job		265		265	
Urinal						
Demolish	Ea.		53		53	Includes material and labor to install a
Install	Ea.	320	320		640	wall hung urinal with flush valve.
Demolish and Install	Ea.	320	373		693	
Reinstall	Ea.		254.51		254.51	
Clean	Ea.	.15	18.20		18.35	
Minimum Charge	Job		265		265	
Bidet						
Vitreous China						
Demolish	Ea.		53		53	Includes material and labor to install a
Install	Ea.	710	191		901	vitreous china bidet complete with
Demolish and Install	Ea.	710	244		954	trim.
Reinstall	Ea.		152.70		152.70	
Clean	Ea.	.15	18.20		18.35	
Minimum Charge	Job		265		265	

Finish Mechanical

Commode		Unit	Material	Labor	Equip.	Total	Specification
Chrome Fittings							
	Install	Ea.	288	66.50		354.50	Includes labor and materials to install chrome fittings for a bidet.
Brass Fittings							
	Install	Ea.	315	66.50		381.50	Includes labor and materials to install brass fittings for a bidet.
Seat							
	Demolish	Ea.		22		22	Includes material and labor to install a toilet seat and cover.
	Install	Ea.	32.50	22		54.50	
	Demolish and Install	Ea.	32.50	44		76.50	
	Reinstall	Ea.		17.68		17.68	
	Clean	Ea.	.08	7.60		7.68	
	Minimum Charge	Job		265		265	
Rough-in							
	Install	Ea.	445	335		780	Includes labor and materials to install water supply pipe with valves and drain, waste and vent pipe with all couplings, hangers and fasteners necessary.

Sink (assembly)		Unit	Material	Labor	Equip.	Total	Specification
Single							
Porcelain							
	Demolish	Ea.		33		33	Includes material and labor to install a single bowl enamel finished cast iron kitchen sink with faucet and drain.
	Install	Ea.	875	170		1045	
	Demolish and Install	Ea.	875	203		1078	
	Reinstall	Ea.		136.34		136.34	
	Clean	Ea.	.01	1.32		1.33	
	Minimum Charge	Job		265		265	
Stainless Steel							
	Demolish	Ea.		33		33	Includes material and labor to install a single bowl stainless steel kitchen sink with faucet and drain.
	Install	Ea.	750	170		920	
	Demolish and Install	Ea.	750	203		953	
	Reinstall	Ea.		136.34		136.34	
	Clean	Ea.	.01	1.32		1.33	
	Minimum Charge	Job		265		265	
Double							
Porcelain							
	Demolish	Ea.		38		38	Includes material and labor to install a double bowl enamel finished cast iron kitchen sink with faucet and drain.
	Install	Ea.	420	199		619	
	Demolish and Install	Ea.	420	237		657	
	Reinstall	Ea.		159.07		159.07	
	Clean	Ea.	.15	18.20		18.35	
	Minimum Charge	Job		265		265	
Stainless Steel							
	Demolish	Ea.		38		38	Includes material and labor to install a double bowl stainless steel kitchen sink with faucet and drain.
	Install	Ea.	475	199		674	
	Demolish and Install	Ea.	475	237		712	
	Reinstall	Ea.		159.07		159.07	
	Clean	Ea.	.15	18.20		18.35	
	Minimum Charge	Job		265		265	

Finish Mechanical

Sink (assembly)

	Unit	Material	Labor	Equip.	Total	Specification
Bar						
Demolish	Ea.		33		33	Includes material and labor to install a
Install	Ea.	196	133		329	small single bowl stainless steel bar /
Demolish and Install	Ea.	196	166		362	vegetable / kitchen sink with faucet
Reinstall	Ea.		106.08		106.08	and drain.
Clean	Ea.	.01	1.32		1.33	
Minimum Charge	Job		265		265	
Floor						
Demolish	Ea.		106		106	Includes material and labor to install
Install	Ea.	1150	217		1367	an enameled cast iron floor mounted
Demolish and Install	Ea.	1150	323		1473	corner service sink with faucet and
Reinstall	Ea.		173.53		173.53	drain.
Clean	Ea.	.15	9.10		9.25	
Minimum Charge	Job		265		265	
Pedestal						
Demolish	Ea.		33		33	Includes material and labor to install a
Install	Ea.	350	149		499	vitreous china pedestal lavatory with
Demolish and Install	Ea.	350	182		532	faucet set and pop-up drain.
Reinstall	Ea.		119.30		119.30	
Clean	Ea.	.01	1.32		1.33	
Minimum Charge	Job		265		265	
Vanity Lavatory						
Demolish	Ea.		33		33	Includes material and labor to install a
Install	Ea.	232	177		409	vitreous china lavatory with faucet set
Demolish and Install	Ea.	232	210		442	and pop-up drain.
Reinstall	Ea.		141.39		141.39	
Clean	Ea.	.01	1.32		1.33	
Minimum Charge	Job		265		265	
Rough-in						
Install	Ea.	315	445		760	Includes labor and materials to install copper supply pipe with valves and drain, waste and vent pipe with all couplings, hangers and fasteners necessary.
Laundry						
Demolish	Ea.		36.50		36.50	Includes material and labor to install a
Install	Ea.	161	147		308	laundry sink (wall mounted or with
Demolish and Install	Ea.	161	183.50		344.50	legs) including faucet and pop up
Clean	Ea.	.01	1.32		1.33	drain.
Minimum Charge	Job		265		265	

Sink Only

	Unit	Material	Labor	Equip.	Total	Specification
Single						
Porcelain						
Demolish	Ea.		33		33	Includes material and labor to install a
Install	Ea.	765	189		954	single bowl enamel finished cast iron
Demolish and Install	Ea.	765	222		987	kitchen sink.
Reinstall	Ea.		151.54		151.54	
Clean	Ea.	.01	1.32		1.33	
Minimum Charge	Job		265		265	

For customer support on your Contractor's Pricing Guide: Residential Repair & Remodeling, call 800.448.8182.

Sink Only		Unit	Material	Labor	Equip.	Total	Specification
Stainless Steel							
	Demolish	Ea.		33		33	Includes material and labor to install a
	Install	Ea.	645	133		778	single bowl stainless steel kitchen sink.
	Demolish and Install	Ea.	645	166		811	
	Reinstall	Ea.		106.08		106.08	
	Clean	Ea.	.01	1.32		1.33	
	Minimum Charge	Job		265		265	
Double							
Porcelain							
	Demolish	Ea.		38		38	Includes material and labor to install a
	Install	Ea.	296	221		517	double bowl enamel finished cast iron
	Demolish and Install	Ea.	296	259		555	kitchen sink.
	Reinstall	Ea.		176.80		176.80	
	Clean	Ea.	.15	18.20		18.35	
	Minimum Charge	Job		265		265	
Stainless Steel							
	Demolish	Ea.		38		38	Includes material and labor to install a
	Install	Ea.	955	171		1126	double bowl stainless steel kitchen
	Demolish and Install	Ea.	955	209		1164	sink.
	Reinstall	Ea.		136.88		136.88	
	Clean	Ea.	.15	18.20		18.35	
	Minimum Charge	Job		265		265	
Bar							
	Demolish	Ea.		33		33	Includes material and labor to install a
	Install	Ea.	86.50	133		219.50	small single bowl stainless steel bar /
	Demolish and Install	Ea.	86.50	166		252.50	vegetable / kitchen sink.
	Reinstall	Ea.		106.08		106.08	
	Clean	Ea.	.01	1.32		1.33	
	Minimum Charge	Job		265		265	
Floor							
	Demolish	Ea.		106		106	Includes material and labor to install
	Install	Ea.	1025	241		1266	an enameled cast iron floor mounted
	Demolish and Install	Ea.	1025	347		1372	corner service sink.
	Reinstall	Ea.		192.87		192.87	
	Clean	Ea.	.15	9.10		9.25	
	Minimum Charge	Job		265		265	
Pedestal							
	Demolish	Ea.		33		33	Includes material and labor to install a
	Install	Ea.	296	166		462	vitreous china pedestal lavatory.
	Demolish and Install	Ea.	296	199		495	
	Reinstall	Ea.		132.60		132.60	
	Clean	Ea.	.01	1.32		1.33	
	Minimum Charge	Job		265		265	
Vanity Lavatory							
	Demolish	Ea.		33		33	Includes material and labor to install a
	Install	Ea.	171	166		337	vitreous china lavatory.
	Demolish and Install	Ea.	171	199		370	
	Reinstall	Ea.		132.60		132.60	
	Clean	Ea.	.01	1.32		1.33	
	Minimum Charge	Job		265		265	
Rough-in							
	Install	Ea.	274	415		689	Includes labor and materials to install copper supply pipe with valves and drain, waste and vent pipe with all couplings, hangers and fasteners necessary.

Finish Mechanical

Sink Only

	Unit	Material	Labor	Equip.	Total	Specification
Faucet Sink						
Good Grade						
Demolish	Ea.		26.50		26.50	Includes material and labor to install a
Install	Ea.	92.50	36.50		129	sink faucet and fittings.
Demolish and Install	Ea.	92.50	63		155.50	
Reinstall	Ea.		29.28		29.28	
Clean	Ea.	.06	11.35		11.41	
Minimum Charge	Job		265		265	
Better Grade						
Demolish	Ea.		26.50		26.50	Includes material and labor to install a
Install	Ea.	155	36.50		191.50	sink faucet and fittings.
Demolish and Install	Ea.	155	63		218	
Reinstall	Ea.		29.28		29.28	
Clean	Ea.	.06	11.35		11.41	
Minimum Charge	Job		265		265	
Premium Grade						
Demolish	Ea.		26.50		26.50	Includes material and labor to install a
Install	Ea.	300	45.50		345.50	sink faucet and fittings.
Demolish and Install	Ea.	300	72		372	
Reinstall	Ea.		36.52		36.52	
Clean	Ea.	.06	11.35		11.41	
Minimum Charge	Job		265		265	
Drain & Basket						
Demolish	Ea.		16.60		16.60	Includes material and labor to install a
Install	Ea.	16.55	33		49.55	sink drain and basket.
Demolish and Install	Ea.	16.55	49.60		66.15	
Reinstall	Ea.		26.52		26.52	
Clean	Ea.	.06	11.35		11.41	
Minimum Charge	Job		265		265	

Bathtub

	Unit	Material	Labor	Equip.	Total	Specification
Enameled						
Steel						
Demolish	Ea.		95.50		95.50	Includes labor and material to install a
Install	Ea.	555	174		729	formed steel bathtub with spout,
Demolish and Install	Ea.	555	269.50		824.50	mixing valve, shower head and
Reinstall	Ea.		138.82		138.82	pop-up drain.
Clean	Ea.	.30	30.50		30.80	
Minimum Charge	Job		265		265	
Cast Iron						
Demolish	Ea.		95.50		95.50	Includes material and labor to install a
Install	Ea.	1325	217		1542	cast iron bathtub with spout, mixing
Demolish and Install	Ea.	1325	312.50		1637.50	valve, shower head and pop-up drain.
Reinstall	Ea.		173.53		173.53	
Clean	Ea.	.30	30.50		30.80	
Minimum Charge	Job		265		265	
Fiberglass						
Demolish	Ea.		95.50		95.50	Includes material and labor to install
Install	Ea.	1850	174		2024	an acrylic soaking tub with spout,
Demolish and Install	Ea.	1850	269.50		2119.50	mixing valve, shower head and
Reinstall	Ea.		138.82		138.82	pop-up drain.
Clean	Ea.	.30	30.50		30.80	
Minimum Charge	Job		265		265	

Bathtub	Unit	Material	Labor	Equip.	Total	Specification
Institutional						
Demolish	Ea.		95.50		95.50	Includes material and labor to install a
Install	Ea.	1400	320		1720	hospital / institutional type bathtub
Demolish and Install	Ea.	1400	415.50		1815.50	with spout, mixing valve, shower head
Reinstall	Ea.		254.51		254.51	and pop-up drain.
Clean	Ea.	.30	30.50		30.80	
Minimum Charge	Job		265		265	
Whirlpool (Acrylic)						
Demolish	Ea.		295		295	Includes material and labor to install a
Install	Ea.	1375	955		2330	molded fiberglass whirlpool tub.
Demolish and Install	Ea.	1375	1250		2625	
Reinstall	Ea.		763.52		763.52	
Clean	Ea.	.30	30.50		30.80	
Minimum Charge	Job		265		265	
Pump / Motor						
Demolish	Ea.		73.50		73.50	Includes material and labor to install a
Install	Ea.	340	159		499	whirlpool tub pump / motor.
Demolish and Install	Ea.	340	232.50		572.50	
Reinstall	Ea.		127.42		127.42	
Clean	Ea.	.08	22.50		22.58	
Minimum Charge	Job		265		265	
Heater / Motor						
Demolish	Ea.		73.50		73.50	Includes material and labor to install a
Install	Ea.	204	236		440	whirlpool tub heater.
Demolish and Install	Ea.	204	309.50		513.50	
Reinstall	Ea.		188.59		188.59	
Clean	Ea.	.08	22.50		22.58	
Minimum Charge	Job		265		265	
Thermal Cover						
Demolish	Ea.		11.80		11.80	Includes material and labor to install a
Install	Ea.	415	33		448	whirlpool tub thermal cover.
Demolish and Install	Ea.	415	44.80		459.80	
Reinstall	Ea.		26.52		26.52	
Clean	Ea.	.75	22.50		23.25	
Minimum Charge	Job		265		265	
Tub / Shower Combination						
Demolish	Ea.		84		84	Includes material and labor to install a
Install	Ea.	845	239		1084	fiberglass module tub with shower
Demolish and Install	Ea.	845	323		1168	surround with spout, mixing valve,
Reinstall	Ea.		190.88		190.88	shower head and pop-up drain.
Clean	Ea.	.37	45.50		45.87	
Minimum Charge	Job		265		265	
Accessories						
Sliding Door						
Demolish	Ea.		25.50		25.50	Includes material and labor to install a
Install	Ea.	665	109		774	48″ aluminum framed tempered glass
Demolish and Install	Ea.	665	134.50		799.50	shower door.
Reinstall	Ea.		87.17		87.17	
Clean	Ea.	.63	15.15		15.78	
Minimum Charge	Job		265		265	
Shower Head						
Demolish	Ea.		6.65		6.65	Includes material and labor to install a
Install	Ea.	54.50	22		76.50	water saving shower head.
Demolish and Install	Ea.	54.50	28.65		83.15	
Reinstall	Ea.		17.68		17.68	
Clean	Ea.	.01	4.55		4.56	
Minimum Charge	Job		265		265	

Finish Mechanical

Bathtub

Bathtub		Unit	Material	Labor	Equip.	Total	Specification
Faucet Set							
	Demolish	Ea.		33		33	Includes material and labor to install a
	Install	Ea.	95	66.50		161.50	combination spout / diverter for a
	Demolish and Install	Ea.	95	99.50		194.50	bathtub.
	Reinstall	Ea.		53.04		53.04	
	Clean	Ea.	.06	11.35		11.41	
	Minimum Charge	Job		265		265	
Shower Rod							
	Demolish	Ea.		4.21		4.21	Includes material and labor to install a
	Install	Ea.	30.50	36.50		67	chrome plated shower rod.
	Demolish and Install	Ea.	30.50	40.71		71.21	
	Reinstall	Ea.		29		29	
	Clean	Ea.		4.55		4.55	
	Minimum Charge	Job		265		265	
Rough-in							
	Install	Ea.	480	460		940	Includes labor and materials to install copper supply pipe with valves and drain, waste and vent pipe with all couplings, hangers and fasteners necessary.

Shower

Shower		Unit	Material	Labor	Equip.	Total	Specification
Fiberglass							
32" x 32"							
	Demolish	Ea.		199		199	Includes material and labor to install a
	Install	Ea.	380	174		554	fiberglass shower stall with door,
	Demolish and Install	Ea.	380	373		753	mixing valve and shower head and
	Reinstall	Ea.		138.82		138.82	drain fitting.
	Clean	Ea.	.37	45.50		45.87	
	Minimum Charge	Job		265		265	
36" x 36"							
	Demolish	Ea.		199		199	Includes material and labor to install a
	Install	Ea.	440	174		614	fiberglass shower stall with door,
	Demolish and Install	Ea.	440	373		813	mixing valve and shower head and
	Reinstall	Ea.		138.82		138.82	drain fitting.
	Clean	Ea.	.37	45.50		45.87	
	Minimum Charge	Job		265		265	
35" x 60"							
	Demolish	Ea.		199		199	Includes material and labor to install a
	Install	Ea.	725	239		964	handicap fiberglass shower stall with
	Demolish and Install	Ea.	725	438		1163	door & seat, mixing valve and shower
	Reinstall	Ea.		190.88		190.88	head and drain fitting.
	Clean	Ea.	.37	45.50		45.87	
	Minimum Charge	Job		265		265	
Accessories							
Single Door							
	Demolish	Ea.		25.50		25.50	Includes material and labor to install a
	Install	Ea.	149	57.50		206.50	shower door.
	Demolish and Install	Ea.	149	83		232	
	Reinstall	Ea.		46.01		46.01	
	Clean	Ea.	.63	15.15		15.78	
	Minimum Charge	Job		265		265	

For customer support on your Contractor's Pricing Guide: Residential Repair & Remodeling, call 800.448.8182.

Finish Mechanical

Shower

	Unit	Material	Labor	Equip.	Total	Specification
Shower Head						
Demolish	Ea.		6.65		6.65	Includes material and labor to install a
Install	Ea.	54.50	22		76.50	water saving shower head.
Demolish and Install	Ea.	54.50	28.65		83.15	
Reinstall	Ea.		17.68		17.68	
Clean	Ea.	.01	4.55		4.56	
Minimum Charge	Job		265		265	
Faucet Set						
Demolish	Ea.		33		33	Includes material and labor to install a
Install	Ea.	95	66.50		161.50	combination spout / diverter for a
Demolish and Install	Ea.	95	99.50		194.50	bathtub.
Reinstall	Ea.		53.04		53.04	
Clean	Ea.	.06	11.35		11.41	
Minimum Charge	Job		265		265	
Shower Pan						
Demolish	Ea.		29.50		29.50	Includes material and labor to install
Install	Ea.	118	66.50		184.50	the base portion of a fiberglass
Demolish and Install	Ea.	118	96		214	shower unit.
Reinstall	Ea.		53.04		53.04	
Clean	Ea.	.30	30.50		30.80	
Minimum Charge	Job		265		265	
Rough-in						
Install	Ea.	425	465		890	Includes labor and materials to install copper supply pipe with valves and drain, waste and vent pipe with all couplings, hangers and fasteners necessary.

Ductwork

	Unit	Material	Labor	Equip.	Total	Specification
Return Grill						
Under 10" Wide						
Demolish	Ea.		11.50		11.50	Includes material and labor to install a
Install	Ea.	29	29		58	10" x 10" air return register.
Demolish and Install	Ea.	29	40.50		69.50	
Reinstall	Ea.		23		23	
Clean	Ea.	.08	10.10		10.18	
Minimum Charge	Job		259		259	
12" to 20" Wide						
Demolish	Ea.		11.50		11.50	Includes material and labor to install a
Install	Ea.	72	30.50		102.50	16" x 16" air return register.
Demolish and Install	Ea.	72	42		114	
Reinstall	Ea.		24.36		24.36	
Clean	Ea.	.08	10.10		10.18	
Minimum Charge	Job		259		259	
Over 20" Wide						
Demolish	Ea.		17.45		17.45	Includes material and labor to install a
Install	Ea.	129	47		176	24" x 24" air return register.
Demolish and Install	Ea.	129	64.45		193.45	
Reinstall	Ea.		37.64		37.64	
Clean	Ea.	.08	11.35		11.43	
Minimum Charge	Job		259		259	

Finish Mechanical

Ductwork

	Unit	Material	Labor	Equip.	Total	Specification
Supply Grill						
Baseboard Type						
Demolish	Ea.		8.20		8.20	Includes material and labor to install a
Install	Ea.	10.50	26		36.50	14" x 6" baseboard air supply
Demolish and Install	Ea.	10.50	34.20		44.70	register.
Reinstall	Ea.		20.70		20.70	
Clean	Ea.	.08	10.10		10.18	
Minimum Charge	Job		259		259	
10" Wide						
Demolish	Ea.		9.60		9.60	Includes material and labor to install a
Install	Ea.	22	26		48	10" x 6" air supply register.
Demolish and Install	Ea.	22	35.60		57.60	
Reinstall	Ea.		20.70		20.70	
Clean	Ea.	.08	10.10		10.18	
Minimum Charge	Job		259		259	
12" to 15" Wide						
Demolish	Ea.		9.60		9.60	Includes material and labor to install a
Install	Ea.	22.50	30.50		53	14" x 8" air supply register.
Demolish and Install	Ea.	22.50	40.10		62.60	
Reinstall	Ea.		24.36		24.36	
Clean	Ea.	.08	10.10		10.18	
Minimum Charge	Job		259		259	
18" to 24" Wide						
Demolish	Ea.		11.50		11.50	Includes material and labor to install a
Install	Ea.	62	40		102	24" x 8" air supply register.
Demolish and Install	Ea.	62	51.50		113.50	
Reinstall	Ea.		31.85		31.85	
Clean	Ea.	.08	11.35		11.43	
Minimum Charge	Job		259		259	
30" to 36" Wide						
Demolish	Ea.		14.40		14.40	Includes material and labor to install a
Install	Ea.	80.50	37		117.50	30" x 8" air supply register.
Demolish and Install	Ea.	80.50	51.40		131.90	
Reinstall	Ea.		29.58		29.58	
Clean	Ea.	.08	22.50		22.58	
Minimum Charge	Job		259		259	
Diffuser						
12" x 12"						
Demolish	Ea.		11.50		11.50	Includes material and labor to install a
Install	Ea.	82	43		125	12" x 12" aluminum diffuser.
Demolish and Install	Ea.	82	54.50		136.50	
Reinstall	Ea.		34.51		34.51	
Clean	Ea.	.08	10.10		10.18	
Minimum Charge	Job		259		259	
14" x 14"						
Demolish	Ea.		11.50		11.50	Includes material and labor to install a
Install	Ea.	146	43		189	14" x 14" aluminum diffuser.
Demolish and Install	Ea.	146	54.50		200.50	
Reinstall	Ea.		34.51		34.51	
Clean	Ea.	.08	11.35		11.43	
Minimum Charge	Job		259		259	
18" x 18"						
Demolish	Ea.		17.45		17.45	Includes material and labor to install a
Install	Ea.	129	57.50		186.50	18" x 18" aluminum diffuser.
Demolish and Install	Ea.	129	74.95		203.95	
Reinstall	Ea.		46.01		46.01	
Clean	Ea.	.08	15.15		15.23	
Minimum Charge	Job		259		259	

For customer support on your Contractor's Pricing Guide: Residential Repair & Remodeling, call 800.448.8182.

Finish Mechanical

Ductwork

	Unit	Material	Labor	Equip.	Total	Specification
12" Round						
Demolish	Ea.		16		16	Includes material and labor to install a
Install	Ea.	25	43		68	12" diameter aluminum diffuser with a
Demolish and Install	Ea.	25	59		84	butterfly damper.
Reinstall	Ea.		34.51		34.51	
Clean	Ea.	.08	10.10		10.18	
Minimum Charge	Job		259		259	
14" to 20" Round						
Demolish	Ea.		21.50		21.50	Includes material and labor to install a
Install	Ea.	173	57.50		230.50	20" diameter aluminum diffuser with a
Demolish and Install	Ea.	173	79		252	butterfly damper.
Reinstall	Ea.		46.01		46.01	
Clean	Ea.	.08	11.35		11.43	
Minimum Charge	Job		259		259	
Flush-out / Sanitize						
Install	L.F.	.30	4.55		4.85	Includes labor and material costs to
Minimum Charge	Job		259		259	clean ductwork or flue.

Lighting Fixture

Duplex Receptacle

Electrical Per S.F.	Unit	Material	Labor	Equip.	Total	Specification
Residential						
Light Fixtures						
Install	S.F.	.26	.27		.53	Includes labor and material to replace house light fixtures. Apply S.F. cost to floor area including garages but not basements.

Circuits	Unit	Material	Labor	Equip.	Total	Specification
120V						
Outlet with Wiring						
Demolish	Ea.		30		30	Includes material and labor to install
Install	Ea.	9.55	37		46.55	general purpose outlet with cover, up
Demolish and Install	Ea.	9.55	67		76.55	to 20 feet of #12/2 wire.
Reinstall	Ea.		29.71		29.71	
Clean	Ea.		2.27		2.27	
Minimum Charge	Job		270		270	
GFCI Outlet w / Wiring						
Demolish	Ea.		30		30	Cost includes material and labor to
Install	Ea.	23	44		67	install 120V GFCI receptacle and wall
Demolish and Install	Ea.	23	74		97	plate, with up to 20 feet of #12/2
Reinstall	Ea.		35.12		35.12	wire.
Clean	Ea.		2.27		2.27	
Minimum Charge	Job		270		270	
Exterior Outlet w / Wiring						
Demolish	Ea.		30		30	Cost includes material and labor to
Install	Ea.	48	45		93	install 110V waterproof external
Demolish and Install	Ea.	48	75		123	receptacle and cover plate with up to
Reinstall	Ea.		36.03		36.03	20' of #12/2 wire.
Clean	Ea.		2.27		2.27	
Minimum Charge	Job		270		270	
240V Outlet w / Wiring						
Demolish	Ea.		30		30	Cost includes material and labor to
Install	Ea.	59.50	84.50		144	install 240V receptacle and cover
Demolish and Install	Ea.	59.50	114.50		174	plate with up to 20' of #12/2 wire.
Reinstall	Ea.		67.44		67.44	
Clean	Ea.		2.27		2.27	
Minimum Charge	Job		270		270	

261

For customer support on your Contractor's Pricing Guide: Residential Repair & Remodeling, call 800.448.8182.

Finish Electrical

Circuits

	Unit	Material	Labor	Equip.	Total	Specification
TV / Cable Jack w / Wiring						Includes material and labor to install
Demolish	Ea.		30		30	TV/cable outlet box and cover plate
Install	Ea.	18.80	34		52.80	including 20' of wiring.
Demolish and Install	Ea.	18.80	64		82.80	
Reinstall	Ea.		27.02		27.02	
Clean	Ea.		2.27		2.27	
Minimum Charge	Job		270		270	
Phone Jack with Wiring						Includes material and labor to install
Demolish	Ea.		30		30	phone jack and cover plate including
Install	Ea.	9.40	21		30.40	up to 20' of telephone cable.
Demolish and Install	Ea.	9.40	51		60.40	
Reinstall	Ea.		16.63		16.63	
Clean	Ea.		1.42		1.42	
Minimum Charge	Job		270		270	

Light Switch

	Unit	Material	Labor	Equip.	Total	Specification
Single						
Switch						Includes material and labor to install
Demolish	Ea.		6		6	single switch and wall plate. Wiring is
Install	Ea.	13	25.50		38.50	not included.
Demolish and Install	Ea.	13	31.50		44.50	
Reinstall	Ea.		20.59		20.59	
Clean	Ea.		2.27		2.27	
Minimum Charge	Job		270		270	
Plate						Includes material and labor to install a
Demolish	Ea.		3		3	one gang brown wall switch plate.
Install	Ea.	.42	6.75		7.17	
Demolish and Install	Ea.	.42	9.75		10.17	
Reinstall	Ea.		5.40		5.40	
Clean	Ea.	.03	.45		.48	
Minimum Charge	Job		270		270	
Double						
Switch						Includes material and labor to install
Demolish	Ea.		12		12	double switch and wall plate. Wiring
Install	Ea.	14.60	38.50		53.10	is not included.
Demolish and Install	Ea.	14.60	50.50		65.10	
Reinstall	Ea.		30.88		30.88	
Clean	Ea.	.03	7.25		7.28	
Minimum Charge	Job		270		270	
Plate						Includes material and labor to install a
Demolish	Ea.		3		3	two gang brown wall switch plate.
Install	Ea.	.83	10.20		11.03	
Demolish and Install	Ea.	.83	13.20		14.03	
Reinstall	Ea.		8.16		8.16	
Clean	Ea.	.03	.45		.48	
Minimum Charge	Job		270		270	
Triple						
Switch						Includes material and labor to install
Demolish	Ea.		18.20		18.20	triple switch and wall plate. Wiring is
Install	Ea.	8.30	45		53.30	not included.
Demolish and Install	Ea.	8.30	63.20		71.50	
Reinstall	Ea.		36.03		36.03	
Clean	Ea.	.03	10.90		10.93	
Minimum Charge	Job		270		270	

Finish Electrical

Light Switch

	Unit	Material	Labor	Equip.	Total	Specification
Plate						
Demolish	Ea.		3		3	Includes material and labor to install a
Install	Ea.	1.22	16.90		18.12	three gang brown wall switch plate.
Demolish and Install	Ea.	1.22	19.90		21.12	
Reinstall	Ea.		13.51		13.51	
Clean	Ea.	.03	.45		.48	
Minimum Charge	Job		270		270	
Dimmer						
Switch						
Demolish	Ea.		6		6	Includes material and labor to install
Install	Ea.	34.50	32		66.50	dimmer switch. Wiring is not included.
Demolish and Install	Ea.	34.50	38		72.50	
Reinstall	Ea.		25.43		25.43	
Clean	Ea.		2.27		2.27	
Minimum Charge	Job		270		270	
Plate						
Demolish	Ea.		3		3	Includes material and labor to install a
Install	Ea.	.42	6.75		7.17	one gang brown wall switch plate.
Demolish and Install	Ea.	.42	9.75		10.17	
Reinstall	Ea.		5.40		5.40	
Clean	Ea.	.03	.45		.48	
Minimum Charge	Job		270		270	
With Wiring						
Single						
Demolish	Ea.		30		30	Includes material and labor to install
Install	Ea.	17.05	31.50		48.55	single switch and plate including up to
Demolish and Install	Ea.	17.05	61.50		78.55	20' of romex cable.
Reinstall	Ea.		25.28		25.28	
Clean	Ea.		2.27		2.27	
Minimum Charge	Job		270		270	
Double						
Demolish	Ea.		30		30	Includes material and labor to install
Install	Ea.	27.50	54		81.50	double switch and plate including up
Demolish and Install	Ea.	27.50	84		111.50	to 20' of romex cable.
Reinstall	Ea.		43.23		43.23	
Clean	Ea.	.03	7.25		7.28	
Minimum Charge	Job		270		270	
Triple						
Demolish	Ea.		45		45	Includes material and labor to install
Install	Ea.	13.80	61		74.80	triple switch and plate including up to
Demolish and Install	Ea.	13.80	106		119.80	20' of romex cable.
Reinstall	Ea.		48.63		48.63	
Clean	Ea.	.03	10.90		10.93	
Minimum Charge	Job		270		270	
Dimmer						
Demolish	Ea.		30		30	Includes material and labor to install
Install	Ea.	39	37		76	dimmer switch and cover plate
Demolish and Install	Ea.	39	67		106	including up to 20' of romex cable.
Reinstall	Ea.		29.71		29.71	
Clean	Ea.		2.27		2.27	
Minimum Charge	Job		270		270	

Finish Electrical

Cover for Outlet / Switch

	Unit	Material	Labor	Equip.	Total	Specification
High Grade						
Demolish	Ea.		3		3	Includes material and labor to install
Install	Ea.	2.84	6.75		9.59	wall plate, stainless steel, 1 gang.
Demolish and Install	Ea.	2.84	9.75		12.59	
Reinstall	Ea.		5.40		5.40	
Clean	Ea.	.03	.45		.48	
Minimum Charge	Job		270		270	
Deluxe Grade						
Demolish	Ea.		3		3	Includes material and labor to install
Install	Ea.	6.15	6.75		12.90	wall plate, brushed brass, 1 gang.
Demolish and Install	Ea.	6.15	9.75		15.90	
Reinstall	Ea.		5.40		5.40	
Clean	Ea.	.03	.45		.48	
Minimum Charge	Job		270		270	

Residential Light Fixture

	Unit	Material	Labor	Equip.	Total	Specification
Ceiling						
Good Quality						
Demolish	Ea.		12.50		12.50	Includes material and labor to install
Install	Ea.	25	13.50		38.50	standard quality incandescent ceiling
Demolish and Install	Ea.	25	26		51	light fixture. Wiring is not included.
Reinstall	Ea.		10.81		10.81	
Clean	Ea.	.03	30.50		30.53	
Minimum Charge	Job		270		270	
Better Quality						
Demolish	Ea.		12.50		12.50	Includes material and labor to install
Install	Ea.	91.50	28.50		120	custom quality incandescent ceiling
Demolish and Install	Ea.	91.50	41		132.50	light fixture. Wiring is not included.
Reinstall	Ea.		22.75		22.75	
Clean	Ea.	.03	30.50		30.53	
Minimum Charge	Job		270		270	
Premium Quality						
Demolish	Ea.		12.50		12.50	Includes material and labor to install
Install	Ea.	370	28.50		398.50	deluxe quality incandescent ceiling
Demolish and Install	Ea.	370	41		411	light fixture. Wiring is not included.
Reinstall	Ea.		22.75		22.75	
Clean	Ea.	.03	30.50		30.53	
Minimum Charge	Job		270		270	
Recessed Ceiling						
Demolish	Ea.		25		25	Includes material and labor to install
Install	Ea.	72.50	18		90.50	standard quality recessed eyeball
Demolish and Install	Ea.	72.50	43		115.50	spotlight with housing. Wiring is not
Reinstall	Ea.		14.41		14.41	included.
Clean	Ea.	.01	3.79		3.80	
Minimum Charge	Job		270		270	
Recessed Eyeball						
Demolish	Ea.		25		25	Includes material and labor to install
Install	Ea.	84.50	19.30		103.80	custom quality recessed eyeball
Demolish and Install	Ea.	84.50	44.30		128.80	spotlight with housing. Wiring is not
Reinstall	Ea.		15.44		15.44	included.
Clean	Ea.	.01	3.79		3.80	
Minimum Charge	Job		270		270	

Finish Electrical

Residential Light Fixture	Unit	Material	Labor	Equip.	Total	Specification
Recessed Shower Type						
Demolish	Ea.		25		25	Includes material and labor to install
Install	Ea.	99	18		117	deluxe quality recessed eyeball
Demolish and Install	Ea.	99	43		142	spotlight with housing. Wiring is not
Reinstall	Ea.		14.41		14.41	included.
Clean	Ea.	.01	3.79		3.80	
Minimum Charge	Job		270		270	
Run Wiring						
Install	Ea.	16.35	21.50		37.85	Includes labor and material to install
Minimum Charge	Job		270		270	an electric fixture utility box with
						non-metallic cable.
Fluorescent						
Good Quality						
Demolish	Ea.		27.50		27.50	Includes material and labor to install
Install	Ea.	43.50	77		120.50	standard quality 12" wide by 48" long
Demolish and Install	Ea.	43.50	104.50		148	decorative surface mounted fluorescent
Reinstall	Ea.		61.76		61.76	fixture w/acrylic diffuser.
Clean	Ea.	.01	3.79		3.80	
Minimum Charge	Job		270		270	
Better Quality						
Demolish	Ea.		27.50		27.50	Includes material and labor to install
Install	Ea.	88.50	87		175.50	an interior fluorescent lighting fixture.
Demolish and Install	Ea.	88.50	114.50		203	
Reinstall	Ea.		69.73		69.73	
Clean	Ea.	.01	3.79		3.80	
Minimum Charge	Job		270		270	
Premium Quality						
Demolish	Ea.		27.50		27.50	Includes material and labor to install
Install	Ea.	93.50	102		195.50	deluxe quality 24" wide by 48" long
Demolish and Install	Ea.	93.50	129.50		223	decorative surface mounted fluorescent
Reinstall	Ea.		81.57		81.57	fixture w/acrylic diffuser.
Clean	Ea.	.01	3.79		3.80	
Minimum Charge	Job		270		270	
Run Wiring						
Install	Ea.	16.35	21.50		37.85	Includes labor and material to install
Minimum Charge	Job		270		270	an electric fixture utility box with
						non-metallic cable.
Exterior						
Good Quality						
Demolish	Ea.		24		24	Includes material and labor to install
Install	Ea.	67	34		101	standard quality incandescent outdoor
Demolish and Install	Ea.	67	58		125	light fixture. Wiring is not included.
Reinstall	Ea.		27.02		27.02	
Clean	Ea.	.08	15.15		15.23	
Minimum Charge	Job		270		270	
Better Quality						
Demolish	Ea.		24		24	Includes material and labor to install
Install	Ea.	118	34		152	custom quality incandescent outdoor
Demolish and Install	Ea.	118	58		176	light fixture. Wiring is not included.
Reinstall	Ea.		27.02		27.02	
Clean	Ea.	.08	15.15		15.23	
Minimum Charge	Job		270		270	
Premium Quality						
Demolish	Ea.		24		24	Includes material and labor to install
Install	Ea.	257	135		392	deluxe quality incandescent outdoor
Demolish and Install	Ea.	257	159		416	light fixture. Wiring is not included.
Reinstall	Ea.		108.08		108.08	
Clean	Ea.	.08	15.15		15.23	
Minimum Charge	Job		270		270	

Finish Electrical

Residential Light Fixture

Residential Light Fixture	Unit	Material	Labor	Equip.	Total	Specification
Outdoor w / Pole						Includes material and labor to install
Demolish	Ea.		24		24	post lantern type incandescent outdoor
Install	Ea.	295	135		430	light fixture. Wiring is not included.
Demolish and Install	Ea.	295	159		454	
Reinstall	Ea.		108.08		108.08	
Clean	Ea.	.08	15.15		15.23	
Minimum Charge	Job		270		270	
Run Wiring						Includes labor and material to install
Install	Ea.	16.35	21.50		37.85	an electric fixture utility box with
Minimum Charge	Job		270		270	non-metallic cable.
Track Lighting						Includes material and labor to install
Demolish	L.F.		8.35		8.35	track lighting.
Install	L.F.	22.50	22.50		45	
Demolish and Install	L.F.	22.50	30.85		53.35	
Reinstall	L.F.		18.01		18.01	
Clean	L.F.	.08	7.25		7.33	
Minimum Charge	Job		270		270	

Commercial Light Fixture

Commercial Light Fixture	Unit	Material	Labor	Equip.	Total	Specification
Fluorescent						
2 Bulb, 2'						Includes material and labor to install a
Demolish	Ea.		24		24	2' long surface mounted fluorescent
Install	Ea.	37	67.50		104.50	light fixture with two 40 watt bulbs.
Demolish and Install	Ea.	37	91.50		128.50	
Reinstall	Ea.		54.04		54.04	
Clean	Ea.	.37	26		26.37	
Minimum Charge	Job		270		270	
2 Bulb, 4'						Includes material and labor to install a
Demolish	Ea.		24		24	4' long surface mounted fluorescent
Install	Ea.	78	77		155	light fixture with two 40 watt bulbs.
Demolish and Install	Ea.	78	101		179	
Reinstall	Ea.		61.76		61.76	
Clean	Ea.	.37	26		26.37	
Minimum Charge	Job		270		270	
3 Bulb, 4'						Includes material and labor to install a
Demolish	Ea.		24		24	4' long pendant mounted fluorescent
Install	Ea.	107	77		184	light fixture with three 34 watt bulbs.
Demolish and Install	Ea.	107	101		208	
Reinstall	Ea.		61.76		61.76	
Clean	Ea.	.37	26		26.37	
Minimum Charge	Job		270		270	
4 Bulb, 4'						Includes material and labor to install a
Demolish	Ea.		24		24	4' long pendant mounted fluorescent
Install	Ea.	96.50	83		179.50	light fixture with four 34 watt bulbs.
Demolish and Install	Ea.	96.50	107		203.50	
Reinstall	Ea.		66.51		66.51	
Clean	Ea.	.37	26		26.37	
Minimum Charge	Job		270		270	
2 Bulb, 8'						Includes material and labor to install
Demolish	Ea.		30		30	an 8' long surface mounted
Install	Ea.	121	102		223	fluorescent light fixture with two 110
Demolish and Install	Ea.	121	132		253	watt bulbs.
Reinstall	Ea.		81.57		81.57	
Clean	Ea.	.37	26		26.37	
Minimum Charge	Job		270		270	

Finish Electrical

Commercial Light Fixture

	Unit	Material	Labor	Equip.	Total	Specification
Run Wiring						
Install	Ea.	16.35	21.50		37.85	Includes labor and material to install
Minimum Charge	Job		270		270	an electric fixture utility box with
						non-metallic cable.
Ceiling, Recessed						
Eyeball Type						
Demolish	Ea.		25		25	Includes material and labor to install
Install	Ea.	84.50	19.30		103.80	custom quality recessed eyeball
Demolish and Install	Ea.	84.50	44.30		128.80	spotlight with housing. Wiring is not
Reinstall	Ea.		15.44		15.44	included.
Clean	Ea.	.01	3.79		3.80	
Minimum Charge	Job		270		270	
Shower Type						
Demolish	Ea.		25		25	Includes material and labor to install
Install	Ea.	99	18		117	deluxe quality recessed eyeball
Demolish and Install	Ea.	99	43		142	spotlight with housing. Wiring is not
Reinstall	Ea.		14.41		14.41	included.
Clean	Ea.	.01	3.79		3.80	
Minimum Charge	Job		270		270	
Run Wiring						
Install	Ea.	16.35	21.50		37.85	Includes labor and material to install
Minimum Charge	Job		270		270	an electric fixture utility box with
						non-metallic cable.
Ceiling Grid Type						
2 Tube						
Demolish	Ea.		35.50		35.50	Cost includes material and labor to
Install	Ea.	59.50	102		161.50	install 2 lamp fluorescent light fixture
Demolish and Install	Ea.	59.50	137.50		197	in 2' x 4' ceiling grid system including
Reinstall	Ea.		81.57		81.57	lens and tubes.
Clean	Ea.	.37	26		26.37	
Minimum Charge	Job		270		270	
4 Tube						
Demolish	Ea.		40		40	Cost includes material and labor to
Install	Ea.	69	115		184	install 4 lamp fluorescent light fixture
Demolish and Install	Ea.	69	155		224	in 2' x 4' ceiling grid system including
Reinstall	Ea.		91.98		91.98	lens and tubes.
Clean	Ea.	.37	26		26.37	
Minimum Charge	Job		270		270	
Acrylic Diffuser						
Demolish	Ea.		6.25		6.25	Includes material and labor to install
Install	Ea.	102	16.40		118.40	fluorescent light fixture w/acrylic
Demolish and Install	Ea.	102	22.65		124.65	diffuser.
Reinstall	Ea.		13.10		13.10	
Clean	Ea.	.24	4.55		4.79	
Minimum Charge	Job		270		270	
Louver Diffuser						
Demolish	Ea.		7.50		7.50	Includes material and labor to install
Install	Ea.	114	16.40		130.40	fluorescent light fixture w/louver
Demolish and Install	Ea.	114	23.90		137.90	diffuser.
Reinstall	Ea.		13.10		13.10	
Clean	Ea.	.24	4.55		4.79	
Minimum Charge	Job		270		270	
Run Wiring						
Install	Ea.	16.35	21.50		37.85	Includes labor and material to install
Minimum Charge	Job		270		270	an electric fixture utility box with
						non-metallic cable.

Finish Electrical

Commercial Light Fixture		Unit	Material	Labor	Equip.	Total	Specification
Track Lighting							
	Demolish	L.F.		8.35		8.35	Includes material and labor to install
	Install	L.F.	22.50	22.50		45	track lighting.
	Demolish and Install	L.F.	22.50	30.85		53.35	
	Reinstall	L.F.		18.01		18.01	
	Clean	L.F.	.08	7.25		7.33	
	Minimum Charge	Job		270		270	
Emergency							
	Demolish	Ea.		43		43	Includes material and labor to install
	Install	Ea.	158	135		293	battery-powered emergency lighting
	Demolish and Install	Ea.	158	178		336	unit with two (2) lights, including
	Reinstall	Ea.		108.08		108.08	wiring.
	Clean	Ea.	.19	26		26.19	
	Minimum Charge	Job		270		270	
Run Wiring							
	Install	Ea.	16.35	21.50		37.85	Includes labor and material to install
	Minimum Charge	Job		270		270	an electric fixture utility box with non-metallic cable.

Fan		Unit	Material	Labor	Equip.	Total	Specification
Ceiling Paddle							
Good Quality							
	Demolish	Ea.		33.50		33.50	Includes material and labor to install
	Install	Ea.	108	54		162	standard ceiling fan.
	Demolish and Install	Ea.	108	87.50		195.50	
	Reinstall	Ea.		43.23		43.23	
	Clean	Ea.	.15	22.50		22.65	
	Minimum Charge	Job		270		270	
Better Quality							
	Demolish	Ea.		33.50		33.50	Includes material and labor to install
	Install	Ea.	269	54		323	better ceiling fan.
	Demolish and Install	Ea.	269	87.50		356.50	
	Reinstall	Ea.		43.23		43.23	
	Clean	Ea.	.15	22.50		22.65	
	Minimum Charge	Job		270		270	
Premium Quality							
	Demolish	Ea.		33.50		33.50	Includes material and labor to install
	Install	Ea.	310	67.50		377.50	deluxe ceiling fan.
	Demolish and Install	Ea.	310	101		411	
	Reinstall	Ea.		54.04		54.04	
	Clean	Ea.	.15	22.50		22.65	
	Minimum Charge	Job		270		270	
Light Kit							
	Install	Ea.	44	67.50		111.50	Includes labor and material to install a
	Minimum Charge	Job		270		270	ceiling fan light kit.
Bathroom Exhaust							
Ceiling Mounted							
	Demolish	Ea.		26		26	Cost includes material and labor to
	Install	Ea.	47.50	54		101.50	install 60 CFM ceiling mounted
	Demolish and Install	Ea.	47.50	80		127.50	bathroom exhaust fan, up to 10' of
	Reinstall	Ea.		43.10		43.10	wiring with one wall switch.
	Clean	Ea.	.08	9.10		9.18	
	Minimum Charge	Job		270		270	

Fan

	Unit	Material	Labor	Equip.	Total	Specification
Wall Mounted						
Demolish	Ea.		26		26	Cost includes material and labor to
Install	Ea.	69.50	54		123.50	install 110 CFM wall mounted
Demolish and Install	Ea.	69.50	80		149.50	bathroom exhaust fan, up to 10' of
Reinstall	Ea.		43.10		43.10	wiring with one wall switch.
Clean	Ea.	.08	9.10		9.18	
Minimum Charge	Job		270		270	
Lighted						
Demolish	Ea.		26		26	Cost includes material and labor to
Install	Ea.	56	54		110	install 50 CFM ceiling mounted lighted
Demolish and Install	Ea.	56	80		136	exhaust fan, up to 10' of wiring with
Reinstall	Ea.		43.10		43.10	one wall switch..
Clean	Ea.	.08	9.10		9.18	
Minimum Charge	Job		270		270	
Blower Heater w / Light						
Demolish	Ea.		26		26	Includes material and labor to install
Install	Ea.	86	90		176	ceiling blower type ventilator with
Demolish and Install	Ea.	86	116		202	switch, light and snap-on grill. Includes
Reinstall	Ea.		71.84		71.84	1300 watt/120 volt heater.
Clean	Ea.	.08	9.10		9.18	
Minimum Charge	Job		270		270	
Fan Heater w / Light						
Demolish	Ea.		26		26	Includes material and labor to install
Install	Ea.	116	90		206	circular ceiling fan heater unit, 1500
Demolish and Install	Ea.	116	116		232	watt, surface mounted.
Reinstall	Ea.		71.84		71.84	
Clean	Ea.	.08	9.10		9.18	
Minimum Charge	Job		270		270	
Whole House Exhaust						
36" Ceiling Mounted						
Install	Ea.	1100	135		1235	Includes material and labor to install a
Clean	Ea.	.02	6.80		6.82	whole house exhaust fan.
Minimum Charge	Job		270		270	
Kitchen Exhaust						
Standard Model						
Install	Ea.	73.50	36		109.50	Includes labor and materials to install
Clean	Ea.	.02	6.80		6.82	a bathroom or kitchen vent fan.
Minimum Charge	Job		270		270	
Low Noise Model						
Install	Ea.	103	36		139	Includes labor and materials to install
Clean	Ea.	.02	6.80		6.82	a bathroom or kitchen vent fan, low
Minimum Charge	Job		270		270	noise model.

Low Voltage Systems

	Unit	Material	Labor	Equip.	Total	Specification
Smoke Detector						
Pre-wired						
Demolish	Ea.		12.50		12.50	Includes material and labor to install
Install	Ea.	23	25.50		48.50	AC type smoke detector including
Demolish and Install	Ea.	23	38		61	wiring.
Reinstall	Ea.		20.27		20.27	
Clean	Ea.	.03	11.35		11.38	
Minimum Charge	Job		270		270	

269

Finish Electrical

Low Voltage Systems

	Unit	Material	Labor	Equip.	Total	Specification
Run Wiring						
Install	Ea.	16.35	21.50		37.85	Includes labor and material to install
Minimum Charge	Job		270		270	an electric fixture utility box with non-metallic cable.
Door Bell / Chimes						
Good Quality						
Demolish	Ea.		37.50		37.50	Includes material and labor to install
Install	Ea.	54	135		189	surface mounted 2 note door bell.
Demolish and Install	Ea.	54	172.50		226.50	Wiring is not included.
Reinstall	Ea.		108.08		108.08	
Clean	Ea.	.08	11.35		11.43	
Minimum Charge	Job		270		270	
Better Quality						
Demolish	Ea.		37.50		37.50	Includes material and labor to install
Install	Ea.	79	135		214	standard quality surface mounted 2
Demolish and Install	Ea.	79	172.50		251.50	note door bell. Wiring is not included.
Reinstall	Ea.		108.08		108.08	
Clean	Ea.	.08	11.35		11.43	
Minimum Charge	Job		270		270	
Premium Quality						
Demolish	Ea.		37.50		37.50	Includes material and labor to install
Install	Ea.	116	135		251	deluxe quality surface mounted 2 note
Demolish and Install	Ea.	116	172.50		288.50	door bell. Wiring is not included.
Reinstall	Ea.		108.08		108.08	
Clean	Ea.	.08	11.35		11.43	
Minimum Charge	Job		270		270	
Run Wiring						
Install	Ea.	16.35	21.50		37.85	Includes labor and material to install
Minimum Charge	Job		270		270	an electric fixture utility box with non-metallic cable.
Intercom System						
Intercom (master)						
Demolish	Ea.		100		100	Includes material and labor to install
Install	Ea.	780	270		1050	master intercom station with AM/FM
Demolish and Install	Ea.	780	370		1150	music and room monitoring. Wiring is
Reinstall	Ea.		216.16		216.16	not included.
Clean	Ea.	.08	30.50		30.58	
Minimum Charge	Job		270		270	
Intercom (remote)						
Demolish	Ea.		60		60	Includes material and labor to install
Install	Ea.	71.50	67.50		139	remote station for master intercom
Demolish and Install	Ea.	71.50	127.50		199	system. Wiring is not included.
Reinstall	Ea.		54.04		54.04	
Clean	Ea.	.08	11.35		11.43	
Minimum Charge	Job		270		270	
Security System						
Alarm Panel						
Demolish	Ea.		120		120	Includes material and labor to install a
Install	Ea.	172	270		442	6 zone burglar alarm panel.
Demolish and Install	Ea.	172	390		562	
Reinstall	Ea.		216.16		216.16	
Clean	Ea.	.08	11.35		11.43	
Minimum Charge	Job		270		270	

Finish Electrical

Low Voltage Systems	Unit	Material	Labor	Equip.	Total	Specification
Motion Detector						
Demolish	Ea.		86		86	Includes material and labor to install
Install	Ea.	220	235		455	an ultrasonic motion detector for a
Demolish and Install	Ea.	220	321		541	security system.
Reinstall	Ea.		187.97		187.97	
Clean	Ea.	.08	11.35		11.43	
Minimum Charge	Job		270		270	
Door Switch						
Demolish	Ea.		37.50		37.50	Includes material and labor to install a
Install	Ea.	67	102		169	door / window contact for a security
Demolish and Install	Ea.	67	139.50		206.50	system.
Reinstall	Ea.		81.57		81.57	
Clean	Ea.	.08	3.64		3.72	
Minimum Charge	Job		270		270	
Window Switch						
Demolish	Ea.		37.50		37.50	Includes material and labor to install a
Install	Ea.	67	102		169	door / window contact for a security
Demolish and Install	Ea.	67	139.50		206.50	system.
Reinstall	Ea.		81.57		81.57	
Clean	Ea.	.08	3.64		3.72	
Minimum Charge	Job		270		270	
Satellite System						
Dish						
Demolish	Ea.		167		167	Includes material, labor and
Install	Ea.	1900	450		2350	equipment to install 10' mesh dish.
Demolish and Install	Ea.	1900	617		2517	
Reinstall	Ea.		360.27		360.27	
Clean	Ea.	1.50	45.50		47	
Minimum Charge	Job		270		270	
Motor						
Demolish	Ea.		83.50		83.50	Includes material and labor to install a
Install	Ea.	289	225		514	television satellite dish motor unit.
Demolish and Install	Ea.	289	308.50		597.50	
Reinstall	Ea.		180.13		180.13	
Clean	Ea.	.08	11.35		11.43	
Minimum Charge	Job		270		270	
Television Antenna						
40' Pole						
Demolish	Ea.		94		94	Includes material and labor to install
Install	Ea.	21.50	169		190.50	stand-alone 1" - 2" metal pole, 10'
Demolish and Install	Ea.	21.50	263		284.50	long. Cost does not include antenna.
Reinstall	Ea.		135.10		135.10	
Clean	Ea.	.30	45.50		45.80	
Paint	Ea.	8.40	49		57.40	
Minimum Charge	Job		270		270	
Rotor Unit						
Demolish	Ea.		75		75	Includes material and labor to install a
Install	Ea.	70.50	67.50		138	television antenna rotor.
Demolish and Install	Ea.	70.50	142.50		213	
Reinstall	Ea.		54.04		54.04	
Clean	Ea.	.30	45.50		45.80	
Minimum Charge	Job		270		270	
Single Booster						
Demolish	Ea.		75		75	Includes material and labor to install
Install	Ea.	34	67.50		101.50	standard grade VHF television signal
Demolish and Install	Ea.	34	142.50		176.50	booster unit and all hardware required
Reinstall	Ea.		54.04		54.04	for connection to TV cable.
Clean	Ea.	.01	5.70		5.71	
Minimum Charge	Job		270		270	

For customer support on your Contractor's Pricing Guide: Residential Repair & Remodeling, call 800.448.8182.

| Gazebo | Gutter | Shutter |

Window Treatment	Unit	Material	Labor	Equip.	Total	Specification
Drapery						
Good Grade						
Install	L.F.	47	7.25		54.25	Includes material and labor to install
Reinstall	L.F.		5.82		5.82	good quality drapery.
Clean	L.F.	1.19	.18		1.37	
Minimum Charge	Job		182		182	
Premium Grade						
Install	L.F.	77	7.25		84.25	Includes material and labor to install
Reinstall	L.F.		5.82		5.82	premium quality drapery.
Clean	L.F.	1.19	.20		1.39	
Minimum Charge	Job		182		182	
Lining						
Install	L.F.	25.50	7.25		32.75	Includes material and labor to install
Reinstall	L.F.		5.82		5.82	drapery lining.
Clean	L.F.	1.20	.10		1.30	
Minimum Charge	Job		182		182	
Valance Board						
Demolish	L.F.		2.02		2.02	Includes material and labor to install a
Install	L.F.	7.70	14.55		22.25	drapery valance board.
Demolish and Install	L.F.	7.70	16.57		24.27	
Reinstall	L.F.		11.64		11.64	
Clean	L.F.	1.20	.36		1.56	
Minimum Charge	Job		182		182	
Rod						
Demolish	L.F.		5.05		5.05	Includes material and labor to install
Install	L.F.	7.05	8		15.05	standard traverse drapery hardware.
Demolish and Install	L.F.	7.05	13.05		20.10	
Reinstall	L.F.		6.39		6.39	
Clean	L.F.	.03	.34		.37	
Paint	L.F.	.08	.75		.83	
Minimum Charge	Job		182		182	
Blinds						
Mini-type						
Demolish	S.F.		1.02		1.02	Includes material and labor to install
Install	S.F.	6.50	.80		7.30	horizontal 1″ slat blinds.
Demolish and Install	S.F.	6.50	1.82		8.32	
Reinstall	S.F.		.64		.64	
Clean	S.F.	1.34	.04		1.38	
Minimum Charge	Job		182		182	

Window Treatment		Unit	Material	Labor	Equip.	Total	Specification
Vertical Type							
	Demolish	S.F.		1.02		1.02	Includes material and labor to install
	Install	S.F.	9.80	1.02		10.82	vertical 3" - 5" cloth or PVC blinds.
	Demolish and Install	S.F.	9.80	2.04		11.84	
	Reinstall	S.F.		.82		.82	
	Clean	S.F.	1.34	.04		1.38	
	Minimum Charge	Job		182		182	
Vertical Type per L.F.							
	Demolish	L.F.		1.68		1.68	Includes material and labor to install
	Install	L.F.	59	4.85		63.85	vertical 3" - 5" cloth or PVC blinds x
	Demolish and Install	L.F.	59	6.53		65.53	72" long.
	Reinstall	L.F.		3.88		3.88	
	Clean	L.F.	8.40	.25		8.65	
	Minimum Charge	Job		182		182	
Venetian Type							
	Demolish	S.F.		1.02		1.02	Includes material and labor to install
	Install	S.F.	5.40	.80		6.20	horizontal 2" slat blinds.
	Demolish and Install	S.F.	5.40	1.82		7.22	
	Reinstall	S.F.		.64		.64	
	Clean	S.F.	1.34	.04		1.38	
	Minimum Charge	Job		182		182	
Wood Shutter							
27" x 36", 4 Panel							
	Demolish	Ea.		32.50		32.50	Includes material and labor to install
	Install	Pr.	120	27.50		147.50	movable louver, up to 33" x 36", 4
	Demolish and Install	Pr.	120	60		180	panel shutter.
	Reinstall	Pr.		22.17		22.17	
	Clean	Ea.	.03	11.35		11.38	
	Paint	Ea.	1.35	28		29.35	
	Minimum Charge	Job		236		236	
33" x 36", 4 Panel							
	Demolish	Ea.		32.50		32.50	Includes material and labor to install
	Install	Pr.	120	27.50		147.50	movable louver, up to 33" x 36", 4
	Demolish and Install	Pr.	120	60		180	panel shutter.
	Reinstall	Pr.		22.17		22.17	
	Clean	Ea.	.03	11.35		11.38	
	Paint	Ea.	1.41	30		31.41	
	Minimum Charge	Job		236		236	
39" x 36", 4 Panel							
	Demolish	Ea.		32.50		32.50	Includes material and labor to install
	Install	Pr.	131	27.50		158.50	movable louver, up to 47" x 36", 4
	Demolish and Install	Pr.	131	60		191	panel shutter.
	Reinstall	Pr.		22.17		22.17	
	Clean	Ea.	.03	11.35		11.38	
	Paint	Ea.	1.41	30		31.41	
	Minimum Charge	Job		236		236	
47" x 36", 4 Panel							
	Demolish	Ea.		32.50		32.50	Includes material and labor to install
	Install	Pr.	131	27.50		158.50	movable louver, up to 47" x 36", 4
	Demolish and Install	Pr.	131	60		191	panel shutter.
	Reinstall	Pr.		22.17		22.17	
	Clean	Ea.	.03	11.35		11.38	
	Paint	Ea.	1.46	32.50		33.96	
	Minimum Charge	Job		236		236	

Stairs

Stairs		Unit	Material	Labor	Equip.	Total	Specification
Concrete Steps							
Precast							
	Demolish	Ea.		68.50	45.50	114	Includes material and labor to install
	Install	Flight	655	163	30.50	848.50	precast concrete front entrance stairs,
	Demolish and Install	Flight	655	231.50	76	962.50	4' wide, with 48" platform and 2
	Clean	Ea.		5.05		5.05	risers.
	Paint	Ea.	4.03	49		53.03	
	Minimum Charge	Job		430		430	
Metal							
Concrete Filled							
	Demolish	Ea.		20		20	Includes material and labor to install
	Install	Riser	670	74.50	3.51	748.01	steel, cement filled metal pan stairs
	Demolish and Install	Riser	670	94.50	3.51	768.01	with picket rail.
	Clean	Ea.	.04	3.03		3.07	
	Paint	Ea.	.72	2.90		3.62	
	Minimum Charge	Job		276		276	
Landing							
	Demolish	S.F.		5.05		5.05	Includes material and labor to install
	Install	S.F.	140	8.75	.41	149.16	pre-erected conventional steel pan
	Demolish and Install	S.F.	140	13.80	.41	154.21	landing.
	Clean	S.F.	.04	.65		.69	
	Paint	S.F.	.24	.83		1.07	
	Minimum Charge	Job		276		276	
Railing							
	Demolish	L.F.		2.79		2.79	Includes material and labor to install
	Install	L.F.	19.95	12.70	.60	33.25	primed steel pipe.
	Demolish and Install	L.F.	19.95	15.49	.60	36.04	
	Clean	L.F.	.30	2.27		2.57	
	Paint	L.F.	.81	6.50		7.31	
	Minimum Charge	Job		276		276	

Elevator

Elevator		Unit	Material	Labor	Equip.	Total	Specification
2 Stop Residential							
	Install	Ea.	11900	7225		19125	Includes labor, material and
	Clean	Ea.	.47	22.50		22.97	equipment to install 2 stop elevator for
	Minimum Charge	Job		720		720	residential use.
2 Stop Commercial							
	Install	Ea.	47000	14400		61400	Includes labor, material and
	Clean	Ea.	.47	22.50		22.97	equipment to install 2 stop elevator for
	Minimum Charge	Job		720		720	apartment or commercial building with
							2,000 lb/13 passenger capacity.
3 Stop Commercial							
	Install	Ea.	58000	22200		80200	Includes labor, material and
	Clean	Ea.	.47	22.50		22.97	equipment to install 3 stop elevator for
	Minimum Charge	Job		720		720	apartment or commercial building with
							2,000 lb/13 passenger capacity.
4 Stop Commercial							
	Install	Ea.	68000	30000		98000	Includes labor, material and
	Clean	Ea.	.47	22.50		22.97	equipment to install 4 stop elevator for
	Minimum Charge	Job		720		720	apartment or commercial building with
							2,000 lb/13 passenger capacity.

275

For customer support on your Contractor's Pricing Guide: Residential Repair & Remodeling, call 800.448.8182.

Elevator	Unit	Material	Labor	Equip.	Total	Specification
5 Stop Commercial						
Install	Ea.	78500	38000		116500	Includes labor, material and
Clean	Ea.	.47	22.50		22.97	equipment to install 5 stop elevator for
Minimum Charge	Job		720		720	apartment or commercial building with
						2,000 lb/13 passenger capacity.

Rangetop	Unit	Material	Labor	Equip.	Total	Specification
4-burner						
Demolish	Ea.		36.50		36.50	Cost includes material and labor to
Install	Ea.	360	90		450	install 4-burner electric surface unit.
Demolish and Install	Ea.	360	126.50		486.50	
Reinstall	Ea.		72.05		72.05	
Clean	Ea.	.02	34		34.02	
Minimum Charge	Job		91		91	
6-burner						
Demolish	Ea.		36.50		36.50	Includes material and labor to install a
Install	Ea.	1900	108		2008	6-burner cooktop unit.
Demolish and Install	Ea.	1900	144.50		2044.50	
Reinstall	Ea.		86.46		86.46	
Clean	Ea.	.02	34		34.02	
Minimum Charge	Job		91		91	
Premium Brand w / Grill						
Demolish	Ea.		36.50		36.50	Includes material and labor to install
Install	Ea.	1675	108		1783	electric downdraft cooktop with grill.
Demolish and Install	Ea.	1675	144.50		1819.50	
Reinstall	Ea.		86.46		86.46	
Clean	Ea.	.02	34		34.02	
Minimum Charge	Job		91		91	

Rangehood	Unit	Material	Labor	Equip.	Total	Specification
Vented						
30"						
Demolish	Ea.		33.50		33.50	Includes material and labor to install a
Install	Ea.	106	108		214	vented range hood.
Demolish and Install	Ea.	106	141.50		247.50	
Reinstall	Ea.		86.46		86.46	
Clean	Ea.	.02	9.75		9.77	
Minimum Charge	Job		91		91	
36"						
Demolish	Ea.		33.50		33.50	Includes material and labor to install a
Install	Ea.	140	108		248	vented range hood.
Demolish and Install	Ea.	140	141.50		281.50	
Reinstall	Ea.		86.46		86.46	
Clean	Ea.	.02	9.75		9.77	
Minimum Charge	Job		91		91	
42"						
Demolish	Ea.		33.50		33.50	Includes material and labor to install a
Install	Ea.	171	121		292	vented range hood.
Demolish and Install	Ea.	171	154.50		325.50	
Reinstall	Ea.		97.01		97.01	
Clean	Ea.	.02	9.75		9.77	
Minimum Charge	Job		91		91	

Improvements / Appliances / Treatments

Rangehood		Unit	Material	Labor	Equip.	Total	Specification
Ventless							
30″							
	Demolish	Ea.		33.50		33.50	Includes material and labor to install a
	Install	Ea.	122	67.50		189.50	ventless range hood.
	Demolish and Install	Ea.	122	101		223	
	Reinstall	Ea.		54.04		54.04	
	Clean	Ea.	.02	9.75		9.77	
	Minimum Charge	Job		91		91	
36″							
	Demolish	Ea.		33.50		33.50	Includes material and labor to install a
	Install	Ea.	198	108		306	ventless range hood.
	Demolish and Install	Ea.	198	141.50		339.50	
	Reinstall	Ea.		86.46		86.46	
	Clean	Ea.	.02	9.75		9.77	
	Minimum Charge	Job		91		91	
42″							
	Demolish	Ea.		33.50		33.50	Includes material and labor to install a
	Install	Ea.	188	90		278	ventless range hood.
	Demolish and Install	Ea.	188	123.50		311.50	
	Reinstall	Ea.		72.05		72.05	
	Clean	Ea.	.02	9.75		9.77	
	Minimum Charge	Job		91		91	

Oven		Unit	Material	Labor	Equip.	Total	Specification
Free Standing							
	Demolish	Ea.		36.50		36.50	Cost includes material and labor to
	Install	Ea.	695	72.50		767.50	install 30″ free standing electric range
	Demolish and Install	Ea.	695	109		804	with self-cleaning oven.
	Reinstall	Ea.		58.18		58.18	
	Clean	Ea.	.02	34		34.02	
	Minimum Charge	Job		91		91	
Single Wall Oven							
	Demolish	Ea.		36.50		36.50	Includes material and labor to install a
	Install	Ea.	1025	135		1160	built-in oven.
	Demolish and Install	Ea.	1025	171.50		1196.50	
	Reinstall	Ea.		108.08		108.08	
	Clean	Ea.	.02	34		34.02	
	Minimum Charge	Job		91		91	
Double Wall Oven							
	Demolish	Ea.		36.50		36.50	Includes material and labor to install a
	Install	Ea.	1375	236		1611	conventional double oven.
	Demolish and Install	Ea.	1375	272.50		1647.50	
	Reinstall	Ea.		188.48		188.48	
	Clean	Ea.	.02	34		34.02	
	Minimum Charge	Job		91		91	
Drop-in Type w / Range							
	Demolish	Ea.		36.50		36.50	Includes material and labor to install a
	Install	Ea.	940	90		1030	built-in cooking range with oven.
	Demolish and Install	Ea.	940	126.50		1066.50	
	Reinstall	Ea.		72.05		72.05	
	Clean	Ea.	.02	34		34.02	
	Minimum Charge	Job		91		91	

Improvements / Appliances / Treatments

Oven

	Unit	Material	Labor	Equip.	Total	Specification
Hi/Lo w / Range, Microwave						Includes material and labor to install a
Demolish	Ea.		36.50		36.50	double oven, one conventional, one
Install	Ea.	2550	470		3020	microwave.
Demolish and Install	Ea.	2550	506.50		3056.50	
Reinstall	Ea.		376.96		376.96	
Clean	Ea.	.02	34		34.02	
Minimum Charge	Job		91		91	
Hi/Lo w/ Range / Double Oven						Includes material and labor to install a
Demolish	Ea.		36.50		36.50	conventional double oven.
Install	Ea.	1375	236		1611	
Demolish and Install	Ea.	1375	272.50		1647.50	
Reinstall	Ea.		188.48		188.48	
Clean	Ea.	.02	34		34.02	
Minimum Charge	Job		91		91	
Countertop Microwave						Includes material and labor to install a
Demolish	Ea.		12.65		12.65	1.0 cubic foot microwave oven.
Install	Ea.	175	45.50		220.50	
Demolish and Install	Ea.	175	58.15		233.15	
Reinstall	Ea.		36.36		36.36	
Clean	Ea.	.02	17.05		17.07	
Minimum Charge	Job		91		91	
Cabinet / Wall Mounted Microwave						Includes material and labor to install a
Demolish	Ea.		25.50		25.50	space saver microwave oven.
Install	Ea.	860	135		995	
Demolish and Install	Ea.	860	160.50		1020.50	
Reinstall	Ea.		108.08		108.08	
Clean	Ea.	.02	17.05		17.07	
Minimum Charge	Job		91		91	

Dryer

	Unit	Material	Labor	Equip.	Total	Specification
Basic Grade						Includes material and labor to install
Demolish	Ea.		16.85		16.85	electric or gas dryer. Vent kit not
Install	Ea.	500	279		779	included.
Demolish and Install	Ea.	500	295.85		795.85	
Reinstall	Ea.		223.27		223.27	
Clean	Ea.	.02	34		34.02	
Minimum Charge	Job		91		91	
Good Grade						Includes material and labor to install
Demolish	Ea.		16.85		16.85	electric or gas dryer. Vent kit not
Install	Ea.	625	279		904	included.
Demolish and Install	Ea.	625	295.85		920.85	
Reinstall	Ea.		223.27		223.27	
Clean	Ea.	.02	34		34.02	
Minimum Charge	Job		91		91	
Premium Grade						Includes material and labor to install
Demolish	Ea.		16.85		16.85	electric or gas dryer. Vent kit not
Install	Ea.	1750	420		2170	included.
Demolish and Install	Ea.	1750	436.85		2186.85	
Reinstall	Ea.		334.91		334.91	
Clean	Ea.	.02	34		34.02	
Minimum Charge	Job		91		91	

Improvements / Appliances / Treatments

Dryer

	Unit	Material	Labor	Equip.	Total	Specification
Washer / Dryer Combination Unit						
Demolish	Ea.		16.85		16.85	Includes material and labor to install
Install	Ea.	1700	72.50		1772.50	electric washer/dryer combination
Demolish and Install	Ea.	1700	89.35		1789.35	unit, 27" wide, stackable or unitized.
Reinstall	Ea.		58.18		58.18	
Clean	Ea.	.02	34		34.02	
Minimum Charge	Job		91		91	

Washer

	Unit	Material	Labor	Equip.	Total	Specification
Basic Grade						
Demolish	Ea.		16.85		16.85	Includes material and labor to install
Install	Ea.	650	177		827	washing machine. New hoses not
Demolish and Install	Ea.	650	193.85		843.85	included.
Reinstall	Ea.		141.44		141.44	
Clean	Ea.	.02	34		34.02	
Minimum Charge	Job		91		91	
Good Grade						
Demolish	Ea.		16.85		16.85	Includes material and labor to install
Install	Ea.	1075	265		1340	washing machine. New hoses not
Demolish and Install	Ea.	1075	281.85		1356.85	included.
Reinstall	Ea.		212.16		212.16	
Clean	Ea.	.02	34		34.02	
Minimum Charge	Job		91		91	
Premium Grade						
Demolish	Ea.		16.85		16.85	Includes material and labor to install
Install	Ea.	1475	530		2005	washing machine with digital
Demolish and Install	Ea.	1475	546.85		2021.85	computer readouts and large capacity.
Reinstall	Ea.		424.32		424.32	New hoses not included.
Clean	Ea.	.02	34		34.02	
Minimum Charge	Job		91		91	
Washer / Dryer Combination Unit						
Demolish	Ea.		16.85		16.85	Includes material and labor to install
Install	Ea.	1700	72.50		1772.50	electric washer/dryer combination
Demolish and Install	Ea.	1700	89.35		1789.35	unit, 27" wide, stackable or unitized.
Reinstall	Ea.		58.18		58.18	
Clean	Ea.	.02	34		34.02	
Minimum Charge	Job		91		91	

Refrigerator

	Unit	Material	Labor	Equip.	Total	Specification
12 Cubic Foot						
Demolish	Ea.		50.50		50.50	Includes material and labor to install a
Install	Ea.	495	72.50		567.50	refrigerator.
Demolish and Install	Ea.	495	123		618	
Reinstall	Ea.		58.18		58.18	
Clean	Ea.	.02	34		34.02	
Minimum Charge	Job		91		91	
16 Cubic Foot						
Demolish	Ea.		50.50		50.50	Includes material and labor to install a
Install	Ea.	620	81		701	refrigerator.
Demolish and Install	Ea.	620	131.50		751.50	
Reinstall	Ea.		64.64		64.64	
Clean	Ea.	.02	34		34.02	
Minimum Charge	Job		91		91	

For customer support on your Contractor's Pricing Guide: Residential Repair & Remodeling, call 800.448.8182.

Refrigerator	Unit	Material	Labor	Equip.	Total	Specification
18 Cubic Foot						
Demolish	Ea.		50.50		50.50	Includes material and labor to install a
Install	Ea.	890	91		981	refrigerator.
Demolish and Install	Ea.	890	141.50		1031.50	
Reinstall	Ea.		72.72		72.72	
Clean	Ea.	.02	34		34.02	
Minimum Charge	Job		91		91	
21 Cubic Foot						
Demolish	Ea.		50.50		50.50	Includes material and labor to install a
Install	Ea.	1200	104		1304	refrigerator.
Demolish and Install	Ea.	1200	154.50		1354.50	
Reinstall	Ea.		83.11		83.11	
Clean	Ea.	.02	34		34.02	
Minimum Charge	Job		91		91	
24 Cubic Foot						
Demolish	Ea.		50.50		50.50	Includes material and labor to install a
Install	Ea.	1600	104		1704	refrigerator.
Demolish and Install	Ea.	1600	154.50		1754.50	
Reinstall	Ea.		83.11		83.11	
Clean	Ea.	.02	34		34.02	
Minimum Charge	Job		91		91	
27 Cubic Foot						
Demolish	Ea.		50.50		50.50	Includes material and labor to install a
Install	Ea.	2850	72.50		2922.50	refrigerator.
Demolish and Install	Ea.	2850	123		2973	
Reinstall	Ea.		58.18		58.18	
Clean	Ea.	.02	34		34.02	
Minimum Charge	Job		91		91	
Ice Maker						
Install	Ea.	87	66.50		153.50	Includes labor and material to install
Minimum Charge	Job		91		91	an automatic ice maker in a refrigerator.

Freezer	Unit	Material	Labor	Equip.	Total	Specification
15 Cubic Foot						
Demolish	Ea.		50.50		50.50	Cost includes material and labor to
Install	Ea.	705	72.50		777.50	install 15 cubic foot freezer.
Demolish and Install	Ea.	705	123		828	
Reinstall	Ea.		58.18		58.18	
Clean	Ea.	.02	34		34.02	
Minimum Charge	Job		91		91	
18 Cubic Foot						
Demolish	Ea.		50.50		50.50	Cost includes material and labor to
Install	Ea.	935	72.50		1007.50	install 18 cubic foot freezer.
Demolish and Install	Ea.	935	123		1058	
Reinstall	Ea.		58.18		58.18	
Clean	Ea.	.02	34		34.02	
Minimum Charge	Job		91		91	
21 Cubic Foot						
Demolish	Ea.		50.50		50.50	Includes material and labor to install a
Install	Ea.	790	72.50		862.50	15 to 23 cubic foot deep freezer.
Demolish and Install	Ea.	790	123		913	
Reinstall	Ea.		58.18		58.18	
Clean	Ea.	.02	34		34.02	
Minimum Charge	Job		91		91	

For customer support on your Contractor's Pricing Guide: Residential Repair & Remodeling, call 800.448.8182.

Improvements / Appliances / Treatments

Freezer

	Unit	Material	Labor	Equip.	Total	Specification
24 Cubic Foot						
Demolish	Ea.		50.50		50.50	Includes material and labor to install a
Install	Ea.	880	145		1025	24 cubic foot deep freezer.
Demolish and Install	Ea.	880	195.50		1075.50	
Reinstall	Ea.		116.35		116.35	
Clean	Ea.	.02	34		34.02	
Minimum Charge	Job		91		91	
Chest-type						
Demolish	Ea.		50.50		50.50	Cost includes material and labor to
Install	Ea.	705	72.50		777.50	install 15 cubic foot freezer.
Demolish and Install	Ea.	705	123		828	
Clean	Ea.	.02	34		34.02	
Minimum Charge	Job		91		91	

Dishwasher

	Unit	Material	Labor	Equip.	Total	Specification
4 Cycle						
Demolish	Ea.		40.50		40.50	Includes material and labor to install
Install	Ea.	460	166		626	good grade automatic dishwasher.
Demolish and Install	Ea.	460	206.50		666.50	
Reinstall	Ea.		133.10		133.10	
Clean	Ea.	.02	34		34.02	
Minimum Charge	Job		91		91	

Garbage Disposal

	Unit	Material	Labor	Equip.	Total	Specification
In Sink Unit						
Demolish	Ea.		37		37	Cost includes material and labor to
Install	Ea.	211	66.50		277.50	install 1/2 HP custom disposal unit.
Demolish and Install	Ea.	211	103.50		314.50	
Reinstall	Ea.		53.24		53.24	
Clean	Ea.	.01	9.75		9.76	
Minimum Charge	Job		91		91	

Trash Compactor

	Unit	Material	Labor	Equip.	Total	Specification
4 to 1 Compaction						
Demolish	Ea.		37.50		37.50	Includes material and labor to install a
Install	Ea.	1050	94		1144	trash compactor.
Demolish and Install	Ea.	1050	131.50		1181.50	
Reinstall	Ea.		75.39		75.39	
Clean	Ea.	.02	22.50		22.52	
Minimum Charge	Job		91		91	

Kitchenette Unit

	Unit	Material	Labor	Equip.	Total	Specification
Range, Refrigerator and Sink						
Demolish	Ea.		150		150	Includes material and labor to install
Install	Ea.	1975	555		2530	range/refrigerator/sink unit, 72"
Demolish and Install	Ea.	1975	705		2680	wide, base unit only.
Reinstall	Ea.		443.67		443.67	
Clean	Ea.	.37	17.05		17.42	
Minimum Charge	Job		91		91	

For customer support on your Contractor's Pricing Guide: Residential Repair & Remodeling, call 800.448.8182.

Improvements / Appliances / Treatments

Bath Accessories		Unit	Material	Labor	Equip.	Total	Specification
Towel Bar							
	Demolish	Ea.		8.40		8.40	Includes material and labor to install
	Install	Ea.	51.50	19.65		71.15	towel bar up to 24" long.
	Demolish and Install	Ea.	51.50	28.05		79.55	
	Reinstall	Ea.		15.71		15.71	
	Clean	Ea.		4.55		4.55	
	Minimum Charge	Job		236		236	
Toothbrush Holder							
	Demolish	Ea.		8.40		8.40	Includes material and labor to install a
	Install	Ea.	14.65	23.50		38.15	surface mounted tumbler and
	Demolish and Install	Ea.	14.65	31.90		46.55	toothbrush holder.
	Reinstall	Ea.		18.85		18.85	
	Clean	Ea.		4.55		4.55	
	Minimum Charge	Job		236		236	
Grab Rail							
	Demolish	Ea.		12.65		12.65	Includes material and labor to install a
	Install	Ea.	36	20.50		56.50	24" long grab bar. Blocking is not
	Demolish and Install	Ea.	36	33.15		69.15	included.
	Reinstall	Ea.		16.39		16.39	
	Clean	Ea.	.15	7.60		7.75	
	Minimum Charge	Job		236		236	
Soap Dish							
	Demolish	Ea.		8.40		8.40	Includes material and labor to install a
	Install	Ea.	5.80	23.50		29.30	surface mounted soap dish.
	Demolish and Install	Ea.	5.80	31.90		37.70	
	Reinstall	Ea.		18.85		18.85	
	Clean	Ea.		4.55		4.55	
	Minimum Charge	Job		236		236	
Toilet Paper Roller							
	Demolish	Ea.		8.40		8.40	Includes material and labor to install a
	Install	Ea.	19.40	15.70		35.10	surface mounted toilet tissue dispenser.
	Demolish and Install	Ea.	19.40	24.10		43.50	
	Reinstall	Ea.		12.57		12.57	
	Clean	Ea.	.15	11.35		11.50	
	Minimum Charge	Job		236		236	
Dispenser							
Soap							
	Demolish	Ea.		11.20		11.20	Includes material and labor to install a
	Install	Ea.	55.50	23.50		79	surface mounted soap dispenser.
	Demolish and Install	Ea.	55.50	34.70		90.20	
	Reinstall	Ea.		18.85		18.85	
	Clean	Ea.	.15	11.35		11.50	
	Minimum Charge	Job		236		236	
Towel							
	Demolish	Ea.		12.65		12.65	Includes material and labor to install a
	Install	Ea.	47.50	29.50		77	surface mounted towel dispenser.
	Demolish and Install	Ea.	47.50	42.15		89.65	
	Reinstall	Ea.		23.56		23.56	
	Clean	Ea.	.15	15.15		15.30	
	Minimum Charge	Job		236		236	
Seat Cover							
	Demolish	Ea.		8.40		8.40	Includes material and labor to install a
	Install	Ea.	32.50	31.50		64	surface mounted toilet seat cover
	Demolish and Install	Ea.	32.50	39.90		72.40	dispenser.
	Reinstall	Ea.		25.13		25.13	
	Clean	Ea.	.15	11.35		11.50	
	Minimum Charge	Job		236		236	

Bath Accessories

Bath Accessories		Unit	Material	Labor	Equip.	Total	Specification
Sanitary Napkin							
	Demolish	Ea.		21.50		21.50	Includes material and labor to install a
	Install	Ea.	43.50	49.50		93	surface mounted sanitary napkin
	Demolish and Install	Ea.	43.50	71		114.50	dispenser.
	Reinstall	Ea.		39.68		39.68	
	Clean	Ea.	.08	15.15		15.23	
	Minimum Charge	Job		236		236	
Partition							
Urinal							
	Demolish	Ea.		25.50		25.50	Includes material and labor to install a
	Install	Ea.	252	118		370	urinal screen.
	Demolish and Install	Ea.	252	143.50		395.50	
	Reinstall	Ea.		94.24		94.24	
	Clean	Ea.	.22	10.10		10.32	
	Minimum Charge	Job		236		236	
Toilet							
	Demolish	Ea.		75		75	Includes material and labor to install
	Install	Ea.	675	188		863	baked enamel standard size partition,
	Demolish and Install	Ea.	675	263		938	floor or ceiling mounted with one wall
	Reinstall	Ea.		150.78		150.78	and one door.
	Clean	Ea.	1.50	22.50		24	
	Minimum Charge	Job		236		236	
Handicap							
	Demolish	Ea.		75		75	Includes material and labor to install
	Install	Ea.	1075	188		1263	baked enamel handicap partition,
	Demolish and Install	Ea.	1075	263		1338	floor or ceiling mounted with one wall
	Reinstall	Ea.		150.78		150.78	and one door.
	Clean	Ea.	1.50	22.50		24	
	Minimum Charge	Job		236		236	

Fire Prevention / Protection

Fire Prevention / Protection		Unit	Material	Labor	Equip.	Total	Specification
Fire Extinguisher							
5# Carbon Dioxide							
	Install	Ea.	230			230	Includes labor and material to install one wall mounted 5 lb. carbon dioxide factory charged unit, complete with hose, horn and wall mounting bracket.
10# Carbon Dioxide							
	Install	Ea.	325			325	Includes labor and material to install one wall mounted 10 lb. carbon dioxide factory charged unit, complete with hose, horn and wall mounting bracket.
15# Carbon Dioxide							
	Install	Ea.	420			420	Includes labor and material to install one wall mounted 15 lb. carbon dioxide factory charged unit, complete with hose, horn and wall mounting bracket.

Improvements / Appliances / Treatments

Fire Prevention / Protection

Fire Prevention / Protection		Unit	Material	Labor	Equip.	Total	Specification
5# Dry Chemical							
	Install	Ea.	62.50			62.50	Includes labor and material to install one wall mounted 5 lb. dry chemical factory charged unit, complete with hose, horn and wall mounting bracket.
20# Dry Chemical							
	Install	Ea.	152			152	Includes labor and material to install one wall mounted 20 lb. dry chemical factory charged unit, complete with hose, horn and wall mounting bracket.

Gutters

Gutters		Unit	Material	Labor	Equip.	Total	Specification
Galvanized							
	Demolish	L.F.		1.68		1.68	Cost includes material and labor to install 5" galvanized steel gutter with enamel finish, fittings, hangers, and corners.
	Install	L.F.	2.45	4.14		6.59	
	Demolish and Install	L.F.	2.45	5.82		8.27	
	Reinstall	L.F.		3.31		3.31	
	Clean	L.F.		.49		.49	
	Paint	L.F.	.22	1.20		1.42	
	Minimum Charge	Job		259		259	
Aluminum							
	Demolish	L.F.		1.68		1.68	Cost includes material and labor to install 5" aluminum gutter with enamel finish, fittings, hangers, and corners.
	Install	L.F.	3.12	4.14		7.26	
	Demolish and Install	L.F.	3.12	5.82		8.94	
	Reinstall	L.F.		3.31		3.31	
	Clean	L.F.		.49		.49	
	Paint	L.F.	.22	1.20		1.42	
	Minimum Charge	Job		259		259	
Copper							
	Demolish	L.F.		1.68		1.68	Cost includes material and labor to install 5" half-round copper gutter with fittings, hangers, and corners.
	Install	L.F.	7.90	4.14		12.04	
	Demolish and Install	L.F.	7.90	5.82		13.72	
	Reinstall	L.F.		3.31		3.31	
	Clean	L.F.		.49		.49	
	Minimum Charge	Job		259		259	
Plastic							
	Demolish	L.F.		1.68		1.68	Cost includes material and labor to install 4" white vinyl gutter fittings, hangers, and corners.
	Install	L.F.	1.50	4.10		5.60	
	Demolish and Install	L.F.	1.50	5.78		7.28	
	Reinstall	L.F.		3.28		3.28	
	Clean	L.F.		.49		.49	
	Paint	L.F.	.22	1.20		1.42	
	Minimum Charge	Job		259		259	

Downspouts

Downspouts		Unit	Material	Labor	Equip.	Total	Specification
Galvanized							
	Demolish	L.F.		1.15		1.15	Cost includes material and labor to install 4" diameter downspout with enamel finish.
	Install	L.F.	2.33	3.57		5.90	
	Demolish and Install	L.F.	2.33	4.72		7.05	
	Reinstall	L.F.		2.86		2.86	
	Clean	L.F.		.49		.49	
	Paint	L.F.	.22	1.20		1.42	
	Minimum Charge	Job		259		259	

Improvements / Appliances / Treatments

Downspouts		Unit	Material	Labor	Equip.	Total	Specification
Aluminum							
	Demolish	L.F.		1.15		1.15	Includes material and labor to install
	Install	L.F.	2.35	3.70		6.05	aluminum enameled downspout.
	Demolish and Install	L.F.	2.35	4.85		7.20	
	Reinstall	L.F.		2.96		2.96	
	Clean	L.F.		.49		.49	
	Paint	L.F.	.22	1.20		1.42	
	Minimum Charge	Job		259		259	
Copper							
	Demolish	L.F.		1.15		1.15	Includes material and labor to install
	Install	L.F.	9.95	2.72		12.67	round copper downspout.
	Demolish and Install	L.F.	9.95	3.87		13.82	
	Reinstall	L.F.		2.18		2.18	
	Clean	L.F.		.49		.49	
	Paint	L.F.	.22	1.20		1.42	
	Minimum Charge	Job		259		259	
Plastic							
	Demolish	L.F.		1.15		1.15	Cost includes material and labor to
	Install	L.F.	2.13	2.47		4.60	install 2" x 3" vinyl rectangular
	Demolish and Install	L.F.	2.13	3.62		5.75	downspouts.
	Reinstall	L.F.		1.97		1.97	
	Clean	L.F.		.49		.49	
	Paint	L.F.	.22	1.20		1.42	
	Minimum Charge	Job		259		259	

Shutters		Unit	Material	Labor	Equip.	Total	Specification
Exterior							
25" High							
	Demolish	Ea.		12.65		12.65	Includes material and labor to install
	Install	Pr.	204	47		251	pair of exterior shutters 25" x 16",
	Demolish and Install	Pr.	204	59.65		263.65	unpainted, installed as fixed shutters
	Reinstall	Pr.		47.12		47.12	on concrete or wood frame.
	Clean	Ea.	.03	11.35		11.38	
	Paint	Ea.	1.78	21		22.78	
	Minimum Charge	Job		236		236	
39" High							
	Demolish	Ea.		12.65		12.65	Includes material and labor to install
	Install	Pr.	276	47		323	pair of shutters 39" x 16", unpainted,
	Demolish and Install	Pr.	276	59.65		335.65	installed as fixed shutters on concrete
	Reinstall	Pr.		47.12		47.12	or wood frame.
	Clean	Ea.	.03	11.35		11.38	
	Paint	Ea.	2.78	28.50		31.28	
	Minimum Charge	Job		236		236	
51" High							
	Demolish	Ea.		12.65		12.65	Includes material and labor to install
	Install	Pr.	350	47		397	pair of shutters 51" x 16", unpainted,
	Demolish and Install	Pr.	350	59.65		409.65	installed as fixed shutters on concrete
	Reinstall	Pr.		47.12		47.12	or wood frame.
	Clean	Ea.	.03	11.35		11.38	
	Paint	Ea.	3.61	29.50		33.11	
	Minimum Charge	Job		236		236	

For customer support on your Contractor's Pricing Guide: Residential Repair & Remodeling, call 800.448.8182.

Shutters	Unit	Material	Labor	Equip.	Total	Specification
59" High						
Demolish	Ea.		12.65		12.65	Includes material and labor to install
Install	Pr.	455	47		502	pair of shutters 59" x 16", unpainted,
Demolish and Install	Pr.	455	59.65		514.65	installed as fixed shutters on concrete
Reinstall	Pr.		47.12		47.12	or wood frame.
Clean	Ea.	.03	11.35		11.38	
Paint	Ea.	4.17	32.50		36.67	
Minimum Charge	Job		236		236	
67" High						
Demolish	Ea.		12.65		12.65	Includes material and labor to install
Install	Pr.	510	52.50		562.50	pair of shutters 67" x 16", unpainted,
Demolish and Install	Pr.	510	65.15		575.15	installed as fixed shutters on concrete
Reinstall	Pr.		52.36		52.36	or wood frame.
Clean	Ea.	.03	11.35		11.38	
Paint	Ea.	4.76	36		40.76	
Minimum Charge	Job		236		236	
Interior Wood						
23" x 24", 2 Panel						
Demolish	Ea.		12.65		12.65	Cost includes material and labor to
Install	Ea.	156	27.50		183.50	install 23" x 24", 4 panel interior
Demolish and Install	Ea.	156	40.15		196.15	movable pine shutter, 1-1/4" louver,
Reinstall	Ea.		22.17		22.17	knobs and hooks and hinges.
Clean	Ea.	.03	11.35		11.38	
Paint	Ea.	1.35	28		29.35	
Minimum Charge	Job		236		236	
27" x 24", 4 Panel						
Demolish	Ea.		12.65		12.65	Cost includes material and labor to
Install	Ea.	158	27.50		185.50	install 27" x 24", 4 panel interior
Demolish and Install	Ea.	158	40.15		198.15	movable pine shutter, 1-1/4" louver,
Reinstall	Ea.		22.17		22.17	knobs and hooks and hinges.
Clean	Ea.	.03	11.35		11.38	
Paint	Ea.	1.35	28		29.35	
Minimum Charge	Job		236		236	
31" x 24", 4 Panel						
Demolish	Ea.		12.65		12.65	Cost includes material and labor to
Install	Ea.	174	27.50		201.50	install 31" x 24", 4 panel interior
Demolish and Install	Ea.	174	40.15		214.15	movable pine shutter, 1-1/4" louver,
Reinstall	Ea.		22.17		22.17	knobs and hooks and hinges.
Clean	Ea.	.03	11.35		11.38	
Paint	Ea.	1.35	28		29.35	
Minimum Charge	Job		236		236	
35" x 24", 4 Panel						
Demolish	Ea.		12.65		12.65	Cost includes material and labor to
Install	Ea.	186	27.50		213.50	install 35" x 24", 4 panel interior
Demolish and Install	Ea.	186	40.15		226.15	movable pine shutter, 1-1/4" louver,
Reinstall	Ea.		22.17		22.17	knobs and hooks and hinges.
Clean	Ea.	.03	11.35		11.38	
Paint	Ea.	1.35	28		29.35	
Minimum Charge	Job		236		236	
39" x 24", 4 Panel						
Demolish	Ea.		12.65		12.65	Cost includes material and labor to
Install	Ea.	207	27.50		234.50	install 39" x 24", 4 panel interior
Demolish and Install	Ea.	207	40.15		247.15	movable pine shutter, 1-1/4" louver,
Reinstall	Ea.		22.17		22.17	knobs and hooks and hinges.
Clean	Ea.	.03	11.35		11.38	
Paint	Ea.	1.35	28		29.35	
Minimum Charge	Job		236		236	

Improvements / Appliances / Treatments

Shutters

	Unit	Material	Labor	Equip.	Total	Specification
47" x 24", 4 Panel						
Demolish	Ea.		12.65		12.65	Cost includes material and labor to
Install	Ea.	235	27.50		262.50	install 47" x 24", 4 panel interior
Demolish and Install	Ea.	235	40.15		275.15	movable pine shutter, 1-1/4" louver,
Reinstall	Ea.		22.17		22.17	knobs and hooks and hinges.
Clean	Ea.	.03	11.35		11.38	
Paint	Ea.	1.41	30		31.41	
Minimum Charge	Job		236		236	

Awning

	Unit	Material	Labor	Equip.	Total	Specification
Standard Aluminum						
36" Wide x 30" Deep						
Demolish	Ea.		12.65		12.65	Includes material and labor to install
Install	Ea.	196	39.50		235.50	one aluminum awning with
Demolish and Install	Ea.	196	52.15		248.15	weather-resistant finish.
Reinstall	Ea.		31.41		31.41	
Clean	Ea.		15.15		15.15	
Minimum Charge	Job		236		236	
48" Wide x 30" Deep						
Demolish	Ea.		12.65		12.65	Includes material and labor to install
Install	Ea.	242	47		289	one aluminum awning with
Demolish and Install	Ea.	242	59.65		301.65	weather-resistant finish.
Reinstall	Ea.		37.70		37.70	
Clean	Ea.		15.15		15.15	
Minimum Charge	Job		236		236	
60" Wide x 30" Deep						
Demolish	Ea.		25.50		25.50	Includes material and labor to install
Install	Ea.	276	52.50		328.50	one aluminum awning with
Demolish and Install	Ea.	276	78		354	weather-resistant finish.
Reinstall	Ea.		41.88		41.88	
Clean	Ea.		15.15		15.15	
Minimum Charge	Job		236		236	
72" Wide x 30" Deep						
Demolish	Ea.		25.50		25.50	Includes material and labor to install
Install	Ea.	305	59		364	one aluminum awning with
Demolish and Install	Ea.	305	84.50		389.50	weather-resistant finish.
Reinstall	Ea.		47.12		47.12	
Clean	Ea.		15.15		15.15	
Minimum Charge	Job		236		236	
Roll-up Type						
Demolish	S.F.		.91		.91	Includes material and labor to install
Install	S.F.	17.80	4.71		22.51	an aluminum roll-up awning.
Demolish and Install	S.F.	17.80	5.62		23.42	
Reinstall	S.F.		3.77		3.77	
Clean	S.F.	.01	.76		.77	
Minimum Charge	Job		236		236	
Canvas						
30" Wide						
Demolish	Ea.		7.60		7.60	Includes material and labor to install a
Install	Ea.	213	39.50		252.50	canvas window awning.
Demolish and Install	Ea.	213	47.10		260.10	
Reinstall	Ea.		31.41		31.41	
Clean	Ea.		4.55		4.55	
Minimum Charge	Job		236		236	

Improvements / Appliances / Treatments

Awning		Unit	Material	Labor	Equip.	Total	Specification
36" Wide							
	Demolish	Ea.		7.60		7.60	Includes material and labor to install a
	Install	Ea.	255	47		302	canvas window awning.
	Demolish and Install	Ea.	255	54.60		309.60	
	Reinstall	Ea.		37.70		37.70	
	Clean	Ea.		4.98		4.98	
	Minimum Charge	Job		236		236	
42" Wide							
	Demolish	Ea.		7.60		7.60	Includes material and labor to install a
	Install	Ea.	268	59		327	canvas window awning.
	Demolish and Install	Ea.	268	66.60		334.60	
	Reinstall	Ea.		47.12		47.12	
	Clean	Ea.		5.90		5.90	
	Minimum Charge	Job		236		236	
48" Wide							
	Demolish	Ea.		7.60		7.60	Includes material and labor to install a
	Install	Ea.	283	67.50		350.50	canvas window awning.
	Demolish and Install	Ea.	283	75.10		358.10	
	Reinstall	Ea.		53.85		53.85	
	Clean	Ea.		6.85		6.85	
	Minimum Charge	Job		236		236	
Cloth Canopy Cover							
8' x 10'							
	Demolish	Ea.		405		405	Includes material and labor to install
	Install	Ea.	340	236		576	one 8' x 10' cloth canopy patio cover
	Demolish and Install	Ea.	340	641		981	with front bar and tension support
	Reinstall	Ea.		188.48		188.48	rafters and 9" valance.
	Clean	Ea.	1.20	60.50		61.70	
	Minimum Charge	Job		470		470	
8' x 15'							
	Demolish	Ea.		405		405	Includes material and labor to install
	Install	Ea.	530	236		766	one 8' x 15' cloth canopy patio cover
	Demolish and Install	Ea.	530	641		1171	with front bar and tension support
	Reinstall	Ea.		188.48		188.48	rafters and 9" valance.
	Clean	Ea.	1.20	91		92.20	
	Minimum Charge	Job		470		470	

Vent		Unit	Material	Labor	Equip.	Total	Specification
Small Roof Turbine							
	Demolish	Ea.		12.65		12.65	Includes material and labor to install
	Install	Ea.	54.50	46.50		101	spinner ventilator, wind driven,
	Demolish and Install	Ea.	54.50	59.15		113.65	galvanized, 4" neck diam, 180 CFM.
	Reinstall	Ea.		37.27		37.27	
	Clean	Ea.		8.10		8.10	
	Minimum Charge	Job		265		265	
Large Roof Turbine							
	Demolish	Ea.		10.10		10.10	Includes material and labor to install
	Install	Ea.	64	66.50		130.50	spinner ventilator, wind driven,
	Demolish and Install	Ea.	64	76.60		140.60	galvanized, 8" neck diam, 360 CFM.
	Reinstall	Ea.		53.24		53.24	
	Clean	Ea.		9.10		9.10	
	Minimum Charge	Job		265		265	

Improvements / Appliances / Treatments

Vent

Vent		Unit	Material	Labor	Equip.	Total	Specification
Gable Louvered							
	Demolish	Ea.		25.50		25.50	Includes material and labor to install a
	Install	Ea.	66	15.70		81.70	vinyl gable end vent.
	Demolish and Install	Ea.	66	41.20		107.20	
	Reinstall	Ea.		12.57		12.57	
	Clean	Ea.		4.85		4.85	
	Paint	Ea.	1.12	6		7.12	
	Minimum Charge	Job		236		236	
Ridge Vent							
	Demolish	L.F.		1.30		1.30	Includes material and labor to install a
	Install	L.F.	4.33	3.34		7.67	mill finish aluminum ridge vent strip.
	Demolish and Install	L.F.	4.33	4.64		8.97	
	Minimum Charge	Job		236		236	
Foundation							
	Demolish	Ea.		9.10		9.10	Includes material and labor to install
	Install	Ea.	17.95	15.65		33.60	galvanized foundation block vent.
	Demolish and Install	Ea.	17.95	24.75		42.70	
	Reinstall	Ea.		12.53		12.53	
	Clean	Ea.	.08	9.10		9.18	
	Minimum Charge	Job		236		236	
Frieze							
	Demolish	L.F.		.90		.90	Includes material and labor to install
	Install	L.F.	1.52	1.88		3.40	pine frieze board.
	Demolish and Install	L.F.	1.52	2.78		4.30	
	Reinstall	L.F.		1.51		1.51	
	Clean	L.F.	.02	.30		.32	
	Minimum Charge	Job		236		236	
Water Heater Cap							
	Demolish	Ea.		18.40		18.40	Includes material and labor to install a
	Install	Ea.	41.50	64.50		106	galvanized steel water heater cap.
	Demolish and Install	Ea.	41.50	82.90		124.40	
	Reinstall	Ea.		51.76		51.76	
	Clean	Ea.	.53	22.50		23.03	

Cupolas

Cupolas		Unit	Material	Labor	Equip.	Total	Specification
22" x 22"							
	Demolish	Ea.		233		233	Includes material and labor to install a
	Install	Ea.	495	118		613	cedar cupola with aluminum roof
	Demolish and Install	Ea.	495	351		846	covering.
	Reinstall	Ea.		94.24		94.24	
	Minimum Charge	Job		236		236	
29" x 29"							
	Demolish	Ea.		233		233	Includes material and labor to install a
	Install	Ea.	690	118		808	cedar cupola with aluminum roof
	Demolish and Install	Ea.	690	351		1041	covering.
	Reinstall	Ea.		94.24		94.24	
	Minimum Charge	Job		236		236	
35" x 35"							
	Demolish	Ea.		233		233	Includes material and labor to install a
	Install	Ea.	970	157		1127	cedar cupola with aluminum roof
	Demolish and Install	Ea.	970	390		1360	covering.
	Reinstall	Ea.		125.65		125.65	
	Minimum Charge	Job		236		236	

For customer support on your Contractor's Pricing Guide: Residential Repair & Remodeling, call 800.448.8182.

Cupolas

	Unit	Material	Labor	Equip.	Total	Specification
47" x 47"						Includes material and labor to install a
Demolish	Ea.		233		233	cedar cupola with aluminum roof
Install	Ea.	1900	236		2136	covering.
Demolish and Install	Ea.	1900	469		2369	
Reinstall	Ea.		188.48		188.48	
Minimum Charge	Job		236		236	
Weather Vane						
Install	Ea.	150	59		209	Includes minimum labor and
Minimum Charge	Job		236		236	equipment to install residential type weathervane.

Aluminum Carport

	Unit	Material	Labor	Equip.	Total	Specification
20' x 20' Complete						Includes material, labor and
Demolish	Ea.		253		253	equipment to install an aluminum
Install	Ea.	9175	780	135	10090	carport. Foundations are not included.
Demolish and Install	Ea.	9175	1033	135	10343	
Reinstall	Ea.		622.56	108.19	730.75	
Clean	Ea.	.08	91		91.08	
Minimum Charge	Job		470		470	
24' x 24' Complete						Includes material, labor and
Demolish	Ea.		253		253	equipment to install an aluminum
Install	Ea.	12700	1050	180	13930	carport. Foundations are not included.
Demolish and Install	Ea.	12700	1303	180	14183	
Reinstall	Ea.		830.08	144.26	974.34	
Clean	Ea.	.08	182		182.08	
Minimum Charge	Job		470		470	
Metal Support Posts						Includes material and labor to install
Demolish	Ea.		3.37		3.37	metal support posts for a carport.
Install	Ea.	77	59		136	
Demolish and Install	Ea.	77	62.37		139.37	
Reinstall	Ea.		47.12		47.12	
Clean	Ea.	.08	3.79		3.87	
Paint	Ea.	.20	7.80		8	
Minimum Charge	Job		470		470	

Storage Shed

	Unit	Material	Labor	Equip.	Total	Specification
Aluminum						
Pre-fab (8' x 10')						
Demolish	Ea.		73.50		73.50	Includes material and labor to install
Install	Ea.	2150	700		2850	an aluminum storage shed.
Demolish and Install	Ea.	2150	773.50		2923.50	
Clean	Ea.	.50	45.50		46	
Minimum Charge	Ea.		550		550	
Pre-fab (10' x 12')						
Demolish	Ea.		135		135	Includes material and labor to install
Install	Ea.	2075	700		2775	an aluminum storage shed.
Demolish and Install	Ea.	2075	835		2910	
Clean	Ea.	.50	50.50		51	
Minimum Charge	Ea.		550		550	

Improvements / Appliances / Treatments

Storage Shed

Storage Shed	Unit	Material	Labor	Equip.	Total	Specification
Wood						
Pre-fab (8' x 10')						
Demolish	Ea.		73.50		73.50	Includes material and labor to install a
Install	Ea.	1925	700		2625	pre-fabricated wood storage shed.
Demolish and Install	Ea.	1925	773.50		2698.50	
Clean	Ea.	.50	45.50		46	
Minimum Charge	Ea.		550		550	
Pre-fab (10' x 12')						
Demolish	Ea.		135		135	Includes material and labor to install a
Install	Ea.	2400	700		3100	pre-fabricated wood storage shed.
Demolish and Install	Ea.	2400	835		3235	
Clean	Ea.	.50	50.50		51	
Minimum Charge	Ea.		550		550	

Gazebo

Gazebo	Unit	Material	Labor	Equip.	Total	Specification
Good Grade						
Demolish	S.F.		4.04		4.04	Includes material and labor to install a
Install	S.F.	78	35.50		113.50	gazebo.
Demolish and Install	S.F.	78	39.54		117.54	
Reinstall	S.F.		28.27		28.27	
Clean	S.F.	.06	2.42		2.48	
Minimum Charge	Job		236		236	
Custom Grade						
Demolish	S.F.		4.04		4.04	Includes material and labor to install a
Install	S.F.	109	35.50		144.50	gazebo.
Demolish and Install	S.F.	109	39.54		148.54	
Reinstall	S.F.		28.27		28.27	
Clean	S.F.	.06	2.42		2.48	
Minimum Charge	Job		236		236	
Bench Seating						
Demolish	L.F.		6.75		6.75	Includes material and labor to install
Install	L.F.	2.33	23.50		25.83	pressure treated bench seating.
Demolish and Install	L.F.	2.33	30.25		32.58	
Reinstall	L.F.		18.85		18.85	
Clean	L.F.	.06	1.82		1.88	
Paint	L.F.	.68	3.13		3.81	
Minimum Charge	Job		236		236	
Re-screen						
Install	S.F.	1.85	1.88		3.73	Includes labor and material to
Minimum Charge	Job		236		236	re-screen wood frame.

Pool

Pool	Unit	Material	Labor	Equip.	Total	Specification
Screen Enclosure						
Re-screen						
Install	S.F. Wall	.03	5.50		5.53	Includes material and labor to install a
Minimum Charge	Job		236		236	screen enclosure for a pool.
Screen Door						
Demolish	Ea.		9.35		9.35	Includes material and labor to install
Install	Ea.	340	78.50		418.50	wood screen door with aluminum cloth
Demolish and Install	Ea.	340	87.85		427.85	screen. Frame and hardware not
Reinstall	Ea.		62.83		62.83	included.
Clean	Ea.	.06	13.60		13.66	
Paint	Ea.	4.84	15.65		20.49	
Minimum Charge	Job		236		236	

291

Improvements / Appliances / Treatments

Pool		Unit	Material	Labor	Equip.	Total	Specification
Heater / Motor							
	Demolish	Ea.		47.50		47.50	Includes material and labor to install
	Install	Ea.	2675	610		3285	swimming pool heater, not incl. base
	Demolish and Install	Ea.	2675	657.50		3332.50	or pad, elec., 12 kW, 4,800 gal
	Reinstall	Ea.		487.07		487.07	pool, incl. pump.
	Clean	Ea.	1.50	45.50		47	
	Minimum Charge	Job		273		273	
Filter System							
	Demolish	Ea.		81		81	Includes material, labor and
	Install	Total	1450	425		1875	equipment to install sand filter system
	Demolish and Install	Total	1450	506		1956	tank including hook-up to existing
	Reinstall	Total		339.46		339.46	equipment. Pump and motor not
	Minimum Charge	Job		265		265	included.
Pump / Motor							
	Demolish	Ea.		47.50		47.50	Includes material and labor to install a
	Install	Ea.	1025	106		1131	22 GPM pump.
	Demolish and Install	Ea.	1025	153.50		1178.50	
	Reinstall	Ea.		84.86		84.86	
	Clean	Ea.	1.50	45.50		47	
	Minimum Charge	Job		273		273	
Pump Water							
	Install	Day		490	93	583	Includes labor and equipment to pump
	Minimum Charge	Job		231		231	water from a swimming pool.
Replaster							
	Install	S.F. Surf	1.03	12.15		13.18	Includes labor and material to
	Minimum Charge	Job		231		231	replaster gunite pool with a surface
							area of 500 to 600 S.F.
Solar Panels							
	Demolish	S.F.		1.96		1.96	Includes material and labor to install
	Install	S.F.	35	3.49		38.49	solar panel.
	Demolish and Install	S.F.	35	5.45		40.45	
	Reinstall	S.F.		2.79		2.79	
	Minimum Charge	Ea.		550		550	

Construction Clean-up	Unit	Material	Labor	Equip.	Total	Specification
Final						
Clean	S.F.		.32		.32	Includes labor for final site/job
Minimum Charge	Job		182		182	clean-up including detail cleaning.

293

Location Factors - Residential

Costs shown in *Residential Costs with RSMeans data* are based on national averages for materials and installation. To adjust these costs to a specific location, simply multiply the base cost by the factor for that city. The data is arranged alphabetically by state and postal zip code numbers. For a city not listed, use the factor for a nearby city with similar economic characteristics.

STATE	CITY	Residential
ALABAMA		
350-352	Birmingham	.84
354	Tuscaloosa	.84
355	Jasper	.83
356	Decatur	.83
357-358	Huntsville	.84
359	Gadsden	.81
360-361	Montgomery	.82
362	Anniston	.80
363	Dothan	.84
364	Evergreen	.79
365-366	Mobile	.82
367	Selma	.81
368	Phenix City	.82
369	Butler	.80
ALASKA		
995-996	Anchorage	1.21
997	Fairbanks	1.24
998	Juneau	1.22
999	Ketchikan	1.24
ARIZONA		
850,853	Phoenix	.87
851,852	Mesa/Tempe	.85
855	Globe	.84
856-857	Tucson	.85
859	Show Low	.86
860	Flagstaff	.88
863	Prescott	.87
864	Kingman	.85
865	Chambers	.87
ARKANSAS		
716	Pine Bluff	.78
717	Camden	.75
718	Texarkana	.78
719	Hot Springs	.74
720-722	Little Rock	.81
723	West Memphis	.78
724	Jonesboro	.77
725	Batesville	.75
726	Harrison	.76
727	Fayetteville	.74
728	Russellville	.75
729	Fort Smith	.82
CALIFORNIA		
900-902	Los Angeles	1.14
903-905	Inglewood	1.13
906-908	Long Beach	1.12
910-912	Pasadena	1.12
913-916	Van Nuys	1.15
917-918	Alhambra	1.16
919-921	San Diego	1.10
922	Palm Springs	1.12
923-924	San Bernardino	1.13
925	Riverside	1.14
926-927	Santa Ana	1.14
928	Anaheim	1.14
930	Oxnard	1.13
931	Santa Barbara	1.13
932-933	Bakersfield	1.12
934	San Luis Obispo	1.15
935	Mojave	1.13
936-938	Fresno	1.17
939	Salinas	1.20
940-941	San Francisco	1.32
942,956-958	Sacramento	1.19
943	Palo Alto	1.27
944	San Mateo	1.30
945	Vallejo	1.23
946	Oakland	1.29
947	Berkeley	1.32
948	Richmond	1.31
949	San Rafael	1.30
950	Santa Cruz	1.24
951	San Jose	1.30
952	Stockton	1.20
953	Modesto	1.18

STATE	CITY	Residential
CALIFORNIA (CONT'D)		
954	Santa Rosa	1.27
955	Eureka	1.24
959	Marysville	1.20
960	Redding	1.24
961	Susanville	1.24
COLORADO		
800-802	Denver	.89
803	Boulder	.91
804	Golden	.87
805	Fort Collins	.89
806	Greeley	.88
807	Fort Morgan	.88
808-809	Colorado Springs	.86
810	Pueblo	.86
811	Alamosa	.84
812	Salida	.83
813	Durango	.87
814	Montrose	.83
815	Grand Junction	.92
816	Glenwood Springs	.82
CONNECTICUT		
060	New Britain	1.09
061	Hartford	1.08
062	Willimantic	1.09
063	New London	1.09
064	Meriden	1.09
065	New Haven	1.09
066	Bridgeport	1.10
067	Waterbury	1.09
068	Norwalk	1.10
069	Stamford	1.11
D.C.		
200-205	Washington	.92
DELAWARE		
197	Newark	1.01
198	Wilmington	1.00
199	Dover	1.01
FLORIDA		
320,322	Jacksonville	.81
321	Daytona Beach	.83
323	Tallahassee	.81
324	Panama City	.81
325	Pensacola	.83
326,344	Gainesville	.80
327-328,347	Orlando	.82
329	Melbourne	.82
330-332,340	Miami	.81
333	Fort Lauderdale	.82
334,349	West Palm Beach	.82
335-336,346	Tampa	.81
337	St. Petersburg	.80
338	Lakeland	.78
339,341	Fort Myers	.80
342	Sarasota	.82
GEORGIA		
300-303,399	Atlanta	.88
304	Statesboro	.79
305	Gainesville	.82
306	Athens	.81
307	Dalton	.84
308-309	Augusta	.87
310-312	Macon	.83
313-314	Savannah	.84
315	Waycross	.80
316	Valdosta	.74
317,398	Albany	.83
318-319	Columbus	.82
HAWAII		
967	Hilo	1.18
968	Honolulu	1.22

STATE	CITY	Residential
STATES & POSS.		
969	Guam	.95
IDAHO		
832	Pocatello	.89
833	Twin Falls	.89
834	Idaho Falls	.87
835	Lewiston	.98
836-837	Boise	.90
838	Coeur d'Alene	.99
ILLINOIS		
600-603	North Suburban	1.23
604	Joliet	1.23
605	South Suburban	1.23
606-608	Chicago	1.25
609	Kankakee	1.16
610-611	Rockford	1.12
612	Rock Island	.97
613	La Salle	1.13
614	Galesburg	1.03
615-616	Peoria	1.06
617	Bloomington	1.03
618-619	Champaign	1.04
620-622	East St. Louis	1.01
623	Quincy	1.03
624	Effingham	1.04
625	Decatur	1.03
626-627	Springfield	1.04
628	Centralia	1.03
629	Carbondale	1.00
INDIANA		
460	Anderson	.90
461-462	Indianapolis	.93
463-464	Gary	1.03
465-466	South Bend	.90
467-468	Fort Wayne	.89
469	Kokomo	.91
470	Lawrenceburg	.86
471	New Albany	.87
472	Columbus	.91
473	Muncie	.90
474	Bloomington	.93
475	Washington	.91
476-477	Evansville	.90
478	Terre Haute	.90
479	Lafayette	.91
IOWA		
500-503,509	Des Moines	.92
504	Mason City	.85
505	Fort Dodge	.84
506-507	Waterloo	.87
508	Creston	.88
510-511	Sioux City	.89
512	Sibley	.76
513	Spencer	.78
514	Carroll	.88
515	Council Bluffs	.87
516	Shenandoah	.88
520	Dubuque	.88
521	Decorah	.87
522-524	Cedar Rapids	.94
525	Ottumwa	.87
526	Burlington	.94
527-528	Davenport	.99
KANSAS		
660-662	Kansas City	.99
664-666	Topeka	.87
667	Fort Scott	.91
668	Emporia	.84
669	Belleville	.82
670-672	Wichita	.81
673	Independence	.90
674	Salina	.80
675	Hutchinson	.80
676	Hays	.83
677	Colby	.86
678	Dodge City	.83
679	Liberal	.82
KENTUCKY		
400-402	Louisville	.86
403-405	Lexington	.87

STATE	CITY	Residential
KENTUCKY (CONT'D)		
406	Frankfort	.90
407-409	Corbin	.83
410	Covington	.85
411-412	Ashland	.88
413-414	Campton	.87
415-416	Pikeville	.90
417-418	Hazard	.86
420	Paducah	.88
421-422	Bowling Green	.87
423	Owensboro	.87
424	Henderson	.88
425-426	Somerset	.83
427	Elizabethtown	.83
LOUISIANA		
700-701	New Orleans	.86
703	Thibodaux	.82
704	Hammond	.78
705	Lafayette	.86
706	Lake Charles	.86
707-708	Baton Rouge	.85
710-711	Shreveport	.82
712	Monroe	.80
713-714	Alexandria	.81
MAINE		
039	Kittery	.87
040-041	Portland	.92
042	Lewiston	.91
043	Augusta	.90
044	Bangor	.90
045	Bath	.88
046	Machias	.87
047	Houlton	.88
048	Rockland	.89
049	Waterville	.88
MARYLAND		
206	Waldorf	.92
207-208	College Park	.89
209	Silver Spring	.90
210-212	Baltimore	.93
214	Annapolis	.91
215	Cumberland	.91
216	Easton	.87
217	Hagerstown	.90
218	Salisbury	.85
219	Elkton	.94
MASSACHUSETTS		
010-011	Springfield	1.07
012	Pittsfield	1.03
013	Greenfield	1.05
014	Fitchburg	1.08
015-016	Worcester	1.10
017	Framingham	1.12
018	Lowell	1.12
019	Lawrence	1.13
020-022, 024	Boston	1.18
023	Brockton	1.10
025	Buzzards Bay	1.08
026	Hyannis	1.07
027	New Bedford	1.10
MICHIGAN		
480,483	Royal Oak	.98
481	Ann Arbor	.99
482	Detroit	1.02
484-485	Flint	.93
486	Saginaw	.89
487	Bay City	.90
488-489	Lansing	.90
490	Battle Creek	.88
491	Kalamazoo	.87
492	Jackson	.90
493,495	Grand Rapids	.88
494	Muskegon	.86
496	Traverse City	.82
497	Gaylord	.86
498-499	Iron Mountain	.86
MINNESOTA		
550-551	Saint Paul	1.07
553-555	Minneapolis	1.08
556-558	Duluth	.99

Location Factors - Residential

STATE	CITY	Residential
MINNESOTA (CONT'D)		
559	Rochester	.99
560	Mankato	.96
561	Windom	.91
562	Willmar	.94
563	St. Cloud	1.02
564	Brainerd	.93
565	Detroit Lakes	.90
566	Bemidji	.91
567	Thief River Falls	.90
MISSISSIPPI		
386	Clarksdale	.73
387	Greenville	.81
388	Tupelo	.74
389	Greenwood	.77
390-392	Jackson	.83
393	Meridian	.79
394	Laurel	.78
395	Biloxi	.83
396	McComb	.74
397	Columbus	.75
MISSOURI		
630-631	St. Louis	1.03
633	Bowling Green	.98
634	Hannibal	.95
635	Kirksville	.91
636	Flat River	.98
637	Cape Girardeau	.91
638	Sikeston	.89
639	Poplar Bluff	.89
640-641	Kansas City	1.02
644-645	St. Joseph	.97
646	Chillicothe	.96
647	Harrisonville	.99
648	Joplin	.91
650-651	Jefferson City	.92
652	Columbia	.89
653	Sedalia	.90
654-655	Rolla	.95
656-658	Springfield	.88
MONTANA		
590-591	Billings	.88
592	Wolf Point	.87
593	Miles City	.89
594	Great Falls	.88
595	Havre	.84
596	Helena	.88
597	Butte	.86
598	Missoula	.85
599	Kalispell	.86
NEBRASKA		
680-681	Omaha	.90
683-685	Lincoln	.91
686	Columbus	.89
687	Norfolk	.89
688	Grand Island	.87
689	Hastings	.89
690	McCook	.83
691	North Platte	.88
692	Valentine	.83
693	Alliance	.83
NEVADA		
889-891	Las Vegas	1.02
893	Ely	1.02
894-895	Reno	.90
897	Carson City	.90
898	Elko	.99
NEW HAMPSHIRE		
030	Nashua	.98
031	Manchester	.97
032-033	Concord	.97
034	Keene	.93
035	Littleton	.93
036	Charleston	.92
037	Claremont	.92
038	Portsmouth	.95

STATE	CITY	Residential
NEW JERSEY		
070-071	Newark	1.21
072	Elizabeth	1.22
073	Jersey City	1.19
074-075	Paterson	1.20
076	Hackensack	1.19
077	Long Branch	1.16
078	Dover	1.19
079	Summit	1.20
080,083	Vineland	1.16
081	Camden	1.17
082,084	Atlantic City	1.20
085-086	Trenton	1.18
087	Point Pleasant	1.17
088-089	New Brunswick	1.23
NEW MEXICO		
870-872	Albuquerque	.85
873	Gallup	.85
874	Farmington	.85
875	Santa Fe	.85
877	Las Vegas	.85
878	Socorro	.85
879	Truth/Consequences	.84
880	Las Cruces	.83
881	Clovis	.84
882	Roswell	.84
883	Carrizozo	.84
884	Tucumcari	.85
NEW YORK		
100-102	New York	1.39
103	Staten Island	1.35
104	Bronx	1.36
105	Mount Vernon	1.18
106	White Plains	1.24
107	Yonkers	1.24
108	New Rochelle	1.21
109	Suffern	1.18
110	Queens	1.36
111	Long Island City	1.39
112	Brooklyn	1.40
113	Flushing	1.38
114	Jamaica	1.37
115,117,118	Hicksville	1.25
116	Far Rockaway	1.37
119	Riverhead	1.26
120-122	Albany	1.02
123	Schenectady	1.03
124	Kingston	1.14
125-126	Poughkeepsie	1.22
127	Monticello	1.15
128	Glens Falls	.97
129	Plattsburgh	.99
130-132	Syracuse	.98
133-135	Utica	.96
136	Watertown	.94
137-139	Binghamton	.99
140-142	Buffalo	1.09
143	Niagara Falls	1.04
144-146	Rochester	1.02
147	Jamestown	.95
148-149	Elmira	.97
NORTH CAROLINA		
270,272-274	Greensboro	.93
271	Winston-Salem	.93
275-276	Raleigh	.93
277	Durham	.93
278	Rocky Mount	.91
279	Elizabeth City	.90
280	Gastonia	.95
281-282	Charlotte	.94
283	Fayetteville	.93
284	Wilmington	.92
285	Kinston	.91
286	Hickory	.92
287-288	Asheville	.93
289	Murphy	.94
NORTH DAKOTA		
580-581	Fargo	.89
582	Grand Forks	.88
583	Devils Lake	.91
584	Jamestown	.89
585	Bismarck	.89

297

STATE	CITY	Residential
NORTH DAKOTA (CONT'D)		
586	Dickinson	.88
587	Minot	.87
588	Williston	.88
OHIO		
430-432	Columbus	.91
433	Marion	.92
434-436	Toledo	.96
437-438	Zanesville	.90
439	Steubenville	.93
440	Lorain	.95
441	Cleveland	.97
442-443	Akron	.97
444-445	Youngstown	.94
446-447	Canton	.93
448-449	Mansfield	.91
450	Hamilton	.92
451-452	Cincinnati	.91
453-454	Dayton	.92
455	Springfield	.92
456	Chillicothe	.94
457	Athens	.90
458	Lima	.90
OKLAHOMA		
730-731	Oklahoma City	.83
734	Ardmore	.81
735	Lawton	.82
736	Clinton	.81
737	Enid	.82
738	Woodward	.77
739	Guymon	.81
740-741	Tulsa	.82
743	Miami	.80
744	Muskogee	.80
745	McAlester	.74
746	Ponca City	.79
747	Durant	.78
748	Shawnee	.79
749	Poteau	.79
OREGON		
970-972	Portland	1.00
973	Salem	.99
974	Eugene	.99
975	Medford	.98
976	Klamath Falls	.98
977	Bend	1.01
978	Pendleton	.99
979	Vale	.96
PENNSYLVANIA		
150-152	Pittsburgh	1.00
153	Washington	.96
154	Uniontown	.94
155	Bedford	.91
156	Greensburg	.94
157	Indiana	.95
158	Dubois	.92
159	Johnstown	.90
160	Butler	.93
161	New Castle	.93
162	Kittanning	.92
163	Oil City	.93
164-165	Erie	.94
166	Altoona	.89
167	Bradford	.95
168	State College	.92
169	Wellsboro	.92
170-171	Harrisburg	.95
172	Chambersburg	.89
173-174	York	.91
175-176	Lancaster	.93
177	Williamsport	.90
178	Sunbury	.92
179	Pottsville	.92
180	Lehigh Valley	1.02
181	Allentown	1.03
182	Hazleton	.92
183	Stroudsburg	.95
184-185	Scranton	.96
186-187	Wilkes-Barre	.92
188	Montrose	.92
189	Doylestown	1.08

STATE	CITY	Residential
PENNSYLVANIA (CONT'D)		
190-191	Philadelphia	1.17
193	Westchester	1.09
194	Norristown	1.08
195-196	Reading	.96
PUERTO RICO		
009	San Juan	.77
RHODE ISLAND		
028	Newport	1.07
029	Providence	1.07
SOUTH CAROLINA		
290-292	Columbia	.95
293	Spartanburg	.94
294	Charleston	.95
295	Florence	.91
296	Greenville	.93
297	Rock Hill	.93
298	Aiken	.94
299	Beaufort	.84
SOUTH DAKOTA		
570-571	Sioux Falls	.90
572	Watertown	.85
573	Mitchell	.79
574	Aberdeen	.88
575	Pierre	.84
576	Mobridge	.78
577	Rapid City	.87
TENNESSEE		
370-372	Nashville	.84
373-374	Chattanooga	.84
375,380-381	Memphis	.85
376	Johnson City	.77
377-379	Knoxville	.81
382	McKenzie	.72
383	Jackson	.75
384	Columbia	.78
385	Cookeville	.70
TEXAS		
750	McKinney	.80
751	Waxahachie	.80
752-753	Dallas	.83
754	Greenville	.80
755	Texarkana	.81
756	Longview	.79
757	Tyler	.81
758	Palestine	.77
759	Lufkin	.79
760-761	Fort Worth	.82
762	Denton	.85
763	Wichita Falls	.84
764	Eastland	.83
765	Temple	.81
766-767	Waco	.84
768	Brownwood	.81
769	San Angelo	.79
770-772	Houston	.83
773	Huntsville	.80
774	Wharton	.80
775	Galveston	.82
776-777	Beaumont	.85
778	Bryan	.78
779	Victoria	.81
780	Laredo	.80
781-782	San Antonio	.81
783-784	Corpus Christi	.83
785	McAllen	.83
786-787	Austin	.79
788	Del Rio	.80
789	Giddings	.80
790-791	Amarillo	.79
792	Childress	.81
793-794	Lubbock	.81
795-796	Abilene	.82
797	Midland	.84
798-799,885	El Paso	.80
UTAH		
840-841	Salt Lake City	.85
842,844	Ogden	.83
843	Logan	.83

Location Factors - Residential

STATE	CITY	Residential
UTAH (CONT'D)		
845	Price	.84
846-847	Provo	.84
VERMONT		
050	White River Jct.	.94
051	Bellows Falls	.99
052	Bennington	.98
053	Brattleboro	.99
054	Burlington	.93
056	Montpelier	.97
057	Rutland	.92
058	St. Johnsbury	.95
059	Guildhall	.95
VIRGINIA		
220-221	Fairfax	1.01
222	Arlington	1.02
223	Alexandria	1.03
224-225	Fredericksburg	.99
226	Winchester	1.00
227	Culpeper	1.02
228	Harrisonburg	.85
229	Charlottesville	.88
230-232	Richmond	1.02
233-235	Norfolk	.96
236	Newport News	.95
237	Portsmouth	.86
238	Petersburg	.95
239	Farmville	.86
240-241	Roanoke	1.00
242	Bristol	.90
243	Pulaski	.85
244	Staunton	.86
245	Lynchburg	.98
246	Grundy	.82
WASHINGTON		
980-981,987	Seattle	1.05
982	Everett	1.05
983-984	Tacoma	1.02
985	Olympia	1.00
986	Vancouver	.97
988	Wenatchee	.95
989	Yakima	.99
990-992	Spokane	.97
993	Richland	.97
994	Clarkston	.93
WEST VIRGINIA		
247-248	Bluefield	.92
249	Lewisburg	.91
250-253	Charleston	.95
254	Martinsburg	.89
255-257	Huntington	.96
258-259	Beckley	.92
260	Wheeling	.92
261	Parkersburg	.91
262	Buckhannon	.91
263-264	Clarksburg	.91
265	Morgantown	.92
266	Gassaway	.92
267	Romney	.91
268	Petersburg	.92
WISCONSIN		
530,532	Milwaukee	1.05
531	Kenosha	1.06
534	Racine	1.04
535	Beloit	1.00
537	Madison	1.00
538	Lancaster	.99
539	Portage	.97
540	New Richmond	.97
541-543	Green Bay	1.02
544	Wausau	.97
545	Rhinelander	.95
546	La Crosse	.95
547	Eau Claire	.98
548	Superior	.96
549	Oshkosh	.96
WYOMING		
820	Cheyenne	.85
821	Yellowstone Nat. Pk.	.83
822	Wheatland	.79

STATE	CITY	Residential
WYOMING (CONT'D)		
823	Rawlins	.86
824	Worland	.83
825	Riverton	.82
826	Casper	.82
827	Newcastle	.84
828	Sheridan	.85
829-831	Rock Springs	.88
CANADIAN FACTORS (reflect Canadian currency)		
ALBERTA		
	Calgary	1.07
	Edmonton	1.07
	Fort McMurray	1.09
	Lethbridge	1.07
	Lloydminster	1.02
	Medicine Hat	1.02
	Red Deer	1.02
BRITISH COLUMBIA		
	Kamloops	.99
	Prince George	.98
	Vancouver	1.01
	Victoria	1.00
MANITOBA		
	Brandon	1.06
	Portage la Prairie	.97
	Winnipeg	.95
NEW BRUNSWICK		
	Bathurst	.89
	Dalhousie	.90
	Fredericton	.94
	Moncton	.90
	Newcastle	.89
	Saint John	.99
NEWFOUNDLAND		
	Corner Brook	1.05
	St. John's	1.04
NORTHWEST TERRITORIES		
	Yellowknife	1.11
NOVA SCOTIA		
	Bridgewater	.92
	Dartmouth	1.01
	Halifax	1.00
	New Glasgow	1.00
	Sydney	.99
	Truro	.92
	Yarmouth	1.00
ONTARIO		
	Barrie	1.10
	Brantford	1.09
	Cornwall	1.08
	Hamilton	1.05
	Kingston	1.08
	Kitchener	1.03
	London	1.06
	North Bay	1.15
	Oshawa	1.05
	Ottawa	1.05
	Owen Sound	1.09
	Peterborough	1.06
	Sarnia	1.09
	Sault Ste. Marie	1.03
	St. Catharines	1.04
	Sudbury	1.01
	Thunder Bay	1.06
	Timmins	1.05
	Toronto	1.07
	Windsor	1.05
PRINCE EDWARD ISLAND		
	Charlottetown	.90
	Summerside	.94
QUEBEC		
	Cap-de-la-Madeleine	1.06
	Charlesbourg	1.06
	Chicoutimi	1.09
	Gatineau	1.06
	Granby	1.06

STATE	CITY	Residential
QUEBEC (CONT'D)		
	Hull	1.06
	Joliette	1.06
	Laval	1.06
	Montreal	1.05
	Quebec City	1.05
	Rimouski	1.10
	Rouyn-Noranda	1.06
	Saint-Hyacinthe	1.06
	Sherbrooke	1.06
	Sorel	1.06
	Saint-Jerome	1.06
	Trois-Rivieres	1.17
SASKATCHEWAN		
	Moose Jaw	.89
	Prince Albert	.88
	Regina	1.07
	Saskatoon	1.03
YUKON		
	Whitehorse	1.03

For customer support on your Contractor's Pricing Guide: Residential Repair & Remodeling, call 800.448.8182.

A	Area
ASTM	American Society for Testing and Materials
B.F.	Board feet
Carp.	Carpenter
C.F.	Cubic feet
CWJ	Composite wood joist
C.Y.	Cubic yard
Ea.	Each
Equip.	Equipment
Exp.	Exposure
Ext.	Exterior
F	Fahrenheit
Ft.	Foot, feet
Gal.	Gallon
Hr.	Hour
in.	Inch, inches
Inst.	Installation
Int.	Interior
Lb.	Pound
L.F.	Linear feet
LVL	Laminated veneer lumber
Mat.	Material
Max.	Maximum
MBF	Thousand board feet
MBM	Thousand feet board measure
MSF	Thousand square feet
Min.	Minimum
O.C.	On center
O&P	Overhead and profit
OWJ	Open web wood joist
Oz.	Ounce
Pr.	Pair
Quan.	Quantity
S.F.	Square foot
Sq.	Square, 100 square feet
S.Y.	Square yard
V.L.F.	Vertical linear feet
'	Foot, feet
"	Inch, inches
o	Degrees

303

For customer support on your Contractor's Pricing Guide: Residential Repair & Remodeling, call 800.448.8182.

Cloud-based access to North America's leading construction database

ans data Online

Cloud-based access to North America's leading construction database

RSMeans Data Online

Cloud-based access to North America's
leading construction cost database.

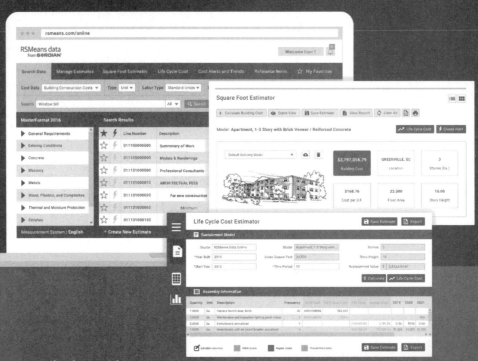

Plan your next budget with
accurate prices and insight.

Estimate in great detail or
at a conceptual level.

Validate unfamiliar costs
or scopes of work.

Maintain existing buildings
with schedules and costs.

RSMeans data
from G⦾RDIAN®

Learn more at rsmeans.com/online

rsmeans.com/core

RSMeans Data Online Core

The Core tier of RSMeans Data Online provides reliable construction cost data along with the tools necessary to quickly access costs at the material or task level. Users can create unit line estimates with RSMeans data from Gordian and share them with ease from the web-based application.

Key Features:

- Search
- Estimate
- Share

Data Available:

16 datasets and packages curated by project type

RSMeans Data Online Complete

The Complete tier of RSMeans Data Online is designed to provide the latest in construction costs with comprehensive tools for projects of varying scopes. Unit, assembly or square foot price data is available 24/7 from the web-based application. Harness the power of RSMeans data from Gordian with expanded features and powerful tools like the square foot model estimator and cost trends analysis.

rsmeans.com/complete

Key Features:

- Alerts
- Square Foot Estimator
- Trends

Data Available:

20 datasets and packages curated by project type

rsmeans.com/completeplus

RSMeans Data Online Complete Plus

The Complete Plus tier of RSMeans Data Online was developed to take project planning and estimating to the next level with predictive cost data. Users gain access to an exclusive set of tools and features and the most comprehensive database in the industry. Leverage everything RSMeans Data Online has to offer with the all-inclusive Complete Plus tier.

Key Features:

- Predictive Cost Data
- Life Cycle Costing
- Full History

Data Available:

Full library and renovation models

RSMeans data Online

Cloud-based access to North America's leading construction database

2019 Seminar Schedule   📞 877-620-6245

Note: call for exact dates, locations, and details as some cities are subject to change.

Location	Dates	Location	Dates
Seattle, WA	January and August	San Francisco, CA	June
Dallas/Ft. Worth, TX	January	Bethesda, MD	June
Austin, TX	February	Dallas, TX	September
Jacksonville, FL	February	Raleigh, NC	October
Anchorage, AK	March and September	Baltimore, MD	November
Las Vegas, NV	March	Orlando, FL	November
Washington, DC	April and September	San Diego, CA	December
Charleston, SC	April	San Antonio, TX	December
Toronto	May		
Denver, CO	May		

Gordian also offers a suite of online RSMeans data self-paced offerings. Check our website RSMeans.com/products/training.aspx for more information.

Facilities Construction Estimating

In this two-day course, professionals working in facilities management can get help with their daily challenges to establish budgets for all phases of a project.

Some of what you'll learn:
• Determining the full scope of a project
• Identifying the scope of risks and opportunities
• Creative solutions to estimating issues
• Organizing estimates for presentation and discussion
• Special techniques for repair/remodel and maintenance projects
• Negotiating project change orders

Who should attend: facility managers, engineers, contractors, facility tradespeople, planners, and project managers.

Mechanical & Electrical Estimating

This two-day course teaches attendees how to prepare more accurate and complete mechanical/electrical estimates, avoid the pitfalls of omission and double-counting, and understand the composition and rationale within the RSMeans mechanical/electrical database.

Some of what you'll learn:
• The unique way mechanical and electrical systems are interrelated
• M&E estimates—conceptual, planning, budgeting, and bidding stages
• Order of magnitude, square foot, assemblies, and unit price estimating
• Comparative cost analysis of equipment and design alternatives

Who should attend: architects, engineers, facilities managers, mechanical and electrical contractors, and others who need a highly reliable method for developing, understanding, and evaluating mechanical and electrical contracts.

Construction Cost Estimating: Concepts and Practice

This one or two day introductory course to improve estimating skills and effectiveness starts with the details of interpreting bid documents and ends with the summary of the estimate and bid submission.

Some of what you'll learn:
• Using the plans and specifications to create estimates
• The takeoff process—deriving all tasks with correct quantities
• Developing pricing using various sources; how subcontractor pricing fits in
• Summarizing the estimate to arrive at the final number
• Formulas for area and cubic measure, adding waste and adjusting productivity to specific projects
• Evaluating subcontractors' proposals and prices
• Adding insurance and bonds
• Understanding how labor costs are calculated
• Submitting bids and proposals

Who should attend: project managers, architects, engineers, owners' representatives, contractors, and anyone who's responsible for budgeting or estimating construction projects.

317

Assessing Scope of Work for Facilities Construction Estimating

This two-day practical training program addresses the vital importance of understanding the scope of projects in order to produce accurate cost estimates for facility repair and remodeling.

Some of what you'll learn:
- Discussions of site visits, plans/specs, record drawings of facilities, and site-specific lists
- Review of CSI divisions, including means, methods, materials, and the challenges of scoping each topic
- Exercises in scope identification and scope writing for accurate estimating of projects
- Hands-on exercises that require scope, take-off, and pricing

Who should attend: corporate and government estimators, planners, facility managers, and others who need to produce accurate project estimates.

Maintenance & Repair Estimating for Facilities

This two-day course teaches attendees how to plan, budget, and estimate the cost of ongoing and preventive maintenance and repair for existing buildings and grounds.

Some of what you'll learn:
- The most financially favorable maintenance, repair, and replacement scheduling and estimating
- Auditing and value engineering facilities
- Preventive planning and facilities upgrading
- Determining both in-house and contract-out service costs
- Annual, asset-protecting M&R plan

Who should attend: facility managers, maintenance supervisors, buildings and grounds superintendents, plant managers, planners, estimators, and others involved in facilities planning and budgeting.

Practical Project Management for Construction Professionals

In this two-day course, acquire the essential knowledge and develop the skills to effectively and efficiently execute the day-to-day responsibilities of the construction project manager.

Some of what you'll learn:
- General conditions of the construction contract
- Contract modifications: change orders and construction change directives
- Negotiations with subcontractors and vendors
- Effective writing: notification and communications
- Dispute resolution: claims and liens

Who should attend: architects, engineers, owners' representatives, and project managers.

Life Cycle Cost Estimating for Facility Asset Managers

Life Cycle Cost Estimating will take the attendee through choosing the correct RSMeans database to use and then correctly applying RSMeans data to their specific life cycle application. Conceptual estimating through RSMeans new building models, conceptual estimating of major existing building projects through RSMeans renovation models, pricing specific renovation elements, estimating repair, replacement and preventive maintenance costs today and forward up to 30 years will be covered.

Some of what you'll learn:
- Cost implications of managing assets
- Planning projects and initial & life cycle costs
- How to use RSMeans data online

Who should attend: facilities owners and managers and anyone involved in the financial side of the decision making process in the planning, design, procurement, and operation of facility real assets.

Please bring a laptop with ability to access the internet.

Building Systems and the Construction Process

This one-day course was written to assist novices and those outside the industry in obtaining a solid understanding of the construction process - from both a building systems and construction administration approach.

Some of what you'll learn:
- Various systems used and how components come together to create a building
- Start with foundation and end with the physical systems of the structure such as HVAC and Electrical
- Focus on the process from start of design through project closeout

This training session requires you to bring a laptop computer to class.

Who should attend: building professionals or novices to help make the crossover to the construction industry; suited for anyone responsible for providing high level oversight on construction projects.

Training for our Online Estimating Solution

Construction estimating is vital to the decision-making process at each state of every project. Our online solution works the way you do. It's systematic, flexible and intuitive. In this one-day class you will see how you can estimate any phase of any project faster and better.

Some of what you'll learn:
- Customizing our online estimating solution
- Making the most of RSMeans "Circle Reference" numbers
- How to integrate your cost data
- Generating reports, exporting estimates to MS Excel, sharing, collaborating and more

Also offered as a self-paced or on-site training program!

Training for our CD Estimating Solution

This one-day course helps users become more familiar with the functionality of the CD. Each menu, icon, screen, and function found in the program is explained in depth. Time is devoted to hands-on estimating exercises.

Some of what you'll learn:
- Searching the database using all navigation methods
- Exporting RSMeans data to your preferred spreadsheet format
- Viewing crews, assembly components, and much more
- Automatically regionalizing the database

This training session requires you to bring a laptop computer to class.

When you register for this course you will receive an outline for your laptop requirements.

Also offered as a self-paced or on-site training program!

Site Work Estimating with RSMeans data

This one-day program focuses directly on site work costs. Accurately scoping, quantifying, and pricing site preparation, underground utility work, and improvements to exterior site elements are often the most difficult estimating tasks on any project. Some of what you'll learn:
- Evaluation of site work and understanding site scope including: site clearing, grading, excavation, disposal and trucking of materials, backfill and compaction, underground utilities, paving, sidewalks, and seeding & planting.
- Unit price site work estimates—Correct use of RSMeans site work cost data to develop a cost estimate.
- Using and modifying assemblies—Save valuable time when estimating site work activities using custom assemblies.

Who should attend: Engineers, contractors, estimators, project managers, owner's representatives, and others who are concerned with the proper preparation and/or evaluation of site work estimates.

Please bring a laptop with ability to access the internet.

Facilities Estimating Using the CD

This two-day class combines hands-on skill-building with best estimating practices and real-life problems. You will learn key concepts, tips, pointers, and guidelines to save time and avoid cost oversights and errors.

Some of what you'll learn:
- Estimating process concepts
- Customizing and adapting RSMeans cost data
- Establishing scope of work to account for all known variables
- Budget estimating: when, why, and how
- Site visits: what to look for and what you can't afford to overlook
- How to estimate repair and remodeling variables

This training session requires you to bring a laptop computer to class.

Who should attend: facility managers, architects, engineers, contractors, facility tradespeople, planners, project managers, and anyone involved with JOC, SABRE, or IDIQ.

Registration Information

Register early to save up to $100!!!

Register 45+ days before date of a class and save $50 off each class. This savings cannot be combined with any other promotional or discounting of the regular price of classes!

How to register

By Phone
Register by phone at 877-620-6245

Online
Register online at
RSMeans.com/products/seminars.aspx

Note: Purchase Orders or Credits Cards are required to register.

Two-day seminar registration fee - $1,200.

One-Day Construction Cost Estimating or Building Systems and the Construction Process - $765.

Government pricing

All federal government employees save off the regular seminar price. Other promotional discounts cannot be combined with the government discount. Call 781-422-5115 for government pricing.

CANCELLATION POLICY:

If you are unable to attend a seminar, substitutions may be made at any time before the session starts by notifying the seminar registrar at 1-781-422-5115 or your sales representative.

If you cancel twenty-one (21) days or more prior to the seminar, there will be no penalty and your registration fees will be refunded. These cancellations must be received by the seminar registrar or your sales representative and will be confirmed to be eligible for cancellation.

If you cancel fewer than twenty-one (21) days prior to the seminar, you will forfeit the registration fee.

In the unfortunate event of an RSMeans cancellation, RSMeans will work with you to reschedule your attendance in the same seminar at a later date or will fully refund your registration fee. RSMeans cannot be responsible for any non-refundable travel expenses incurred by you or another as a result of your registration, attendance at, or cancellation of an RSMeans seminar.

Any on-demand training modules are not eligible for cancellation, substitution, transfer, return or refund.

AACE approved courses

Many seminars described and offered here have been approved for 14 hours (1.4 recertification credits) of credit by the AACE International Certification Board toward meeting the continuing education requirements for recertification as a Certified Cost Engineer/Certified Cost Consultant.

AIA Continuing Education

We are registered with the AIA Continuing Education System (AIA/CES) and are committed to developing quality learning activities in accordance with the CES criteria. Many seminars meet the AIA/CES criteria for Quality Level 2. AIA members may receive 14 learning units (LUs) for each two-day RSMeans course.

Daily course schedule

The first day of each seminar session begins at 8:30 a.m. and ends at 4:30 p.m. The second day begins at 8:00 a.m. and ends at 4:00 p.m. Participants are urged to bring a hand-held calculator since many actual problems will be worked out in each session.

Continental breakfast

Your registration includes the cost of a continental breakfast and a morning and afternoon refreshment break. These informal segments allow you to discuss topics of mutual interest with other seminar attendees. (You are free to make your own lunch and dinner arrangements.)

Hotel/transportation arrangements

We arrange to hold a block of rooms at most host hotels. To take advantage of special group rates when making your reservation, be sure to mention that you are attending the RSMeans Institute data seminar. You are, of course, free to stay at the lodging place of your choice. (Hotel reservations and transportation arrangements should be made directly by seminar attendees.)

Important

Class sizes are limited, so please register as soon as possible.

Note: Pricing subject to change.